Nonlinear Least Squares for Inverse Problems

Scientific Computation

For other titles published in this series, go to
http:/www.springer.com/series/718

G. Chavent

Nonlinear Least Squares for Inverse Problems

Theoretical Foundations and Step-by-Step Guide for Applications

With 25 Figures

Springer

Guy Chavent
Ceremade, Université Paris-Dauphine
75775 Paris Cedex 16
France

and

Inria-Rocquencourt
BP 105, 78153 Le Chesnay Cedex
France
Guy.Chavent@inria.fr

ISSN 1434-8322
ISBN 978-94-007-3060-1 e-ISBN 978-90-481-2785-6
DOI 10.1007/978-90-481-2785-6
Springer Dordrecht Heidelberg London New York

Cover design: eStudio Calamar S.L.

Printed on acid-free paper

Springer is part of Springer Science+Business Media (www.springer.com)

To my wife Annette

Preface

The domain of inverse problems has experienced a rapid expansion, driven by the increase in computing power and the progress in numerical modeling. When I started working on this domain years ago, I became somehow frustrated to see that my friends working on modeling where producing existence, uniqueness, and stability results for the solution of their equations, but that I was most of the time limited, because of the nonlinearity of the problem, to prove that my least squares objective function was differentiable.... But with my experience growing, I became convinced that, after the inverse problem has been properly trimmed, the *final least squares problem*, the one solved on the computer, should be *Quadratically (Q)-wellposed*, that is, both *wellposed* and *optimizable*: optimizability ensures that a global minimizer of the least squares function can actually be found using efficient local optimization algorithms, and wellposedness that this minimizer is stable with respect to perturbation of the data.

But the vast majority of inverse problems are nonlinear, and the classical mathematical tools available for their analysis fail to bring answers to these crucial questions: for example, compactness will ensure existence, but provides no uniqueness results, and brings no information on the presence or absence of parasitic local minima or stationary points ....

This book is partly a consequence of this early frustration: a first objective is to present a *geometrical theory* for the analysis of NLS problems from the point of view of *Q-wellposedness*: for an attainable set with finite curvature, this theory provides an estimation of the size of the admissible parameter set and of the error level on the data for which Q-wellposedness holds. The various regularization techniques used to trim the inverse problem can then be checked against their ability to produce the desirable Q-wellposed problems.

The second objective of the book is to give a detailed presentation of important practical issues for the resolution of NLS problems: sensitivity functions and adjoint state methods for the computations of derivatives, choice of optimization parameters (calibration, sensitivity analysis, multiscale and/or adaptive parameterization), organization of the inversion code, and choice of the descent step for the minimization algorithm. Most of this material is seldom presented in detail, because it is quite elementary from the mathematical point of view, and has usually to be rediscovered by trial-and-error!

As one can see from these objectives, this book does not pretend to give an exhaustive panorama of nonlinear inverse problems, but merely to present the author's view and experience on the subject. Alternative approaches, when known, are mentioned and referenced, but not developed. The book is organized in two parts, which can be read independently:

Part I (Chaps. 1–5) is devoted to the step-by-step resolution and analysis of NLS inverse problems. It should be of interest to scientists of various application fields interested in the practical resolution of inverse problems, as well as to applied mathematicians interested also in their analysis. The required background is a good knowledge of Hilbert spaces, and some notions of functional analysis if one is interested in the infinite dimensional examples. The elements of the geometrical theory of Part II, which are necessary for the Q-wellposedness analysis, are presented without demonstration, but in an as-intuitive-as-possible way, at the beginning of Chap. 4, so that it is not necessary to read Part II, which is quite technical.

Part II (Chaps. 6–8) presents the geometric theory of quasi-convex and strictly quasi-convex sets, which are the basis for the results of Chaps. 4 and 5. These sets possess a neighborhood where the projection is well-behaved, and can be recognized by their finite curvature and limited deflection. This part should be of interest to those more interested in the theory of projection on nonconvex sets. It requires familiarity with Hilbert spaces and functional analysis. The material of Part II was scattered in various papers with different notations. It is presented for the first time in this book in a progressive and coherent approach, which benefits from substantial enhancements and simplifications in the definition of strictly quasi-convex sets.

To facilitate a top-to-bottom approach of the subject, each chapter starts with an overview of the concepts and results developed herein – at the price of some repetition between the overview and the main corpus of the chapter.... Also, we have tried to make the index more user-friendly, all indexed words or expressions are emphasized in the text (but not all emphasized words are indexed!).

I express my thanks to my colleagues, and in particular to François Clement, Karl Kunisch, and Hend Benameur for the stimulating discussions we had over all these years, and for the pleasure I found interacting with them.

March 2009
Lyon

GUY CHAVENT

Contents

Part I

Nonlinear Least Squares

Chapter 1 provides an overview of the book: it shows how various finite and infinite dimensional inverse problems can be cast in the same NLS mould, reviews the difficulties to be expected, and hints at the places in the book where they are addressed.

Chapter 2 details the sensitivity function and adjoint state approaches for the computation of derivatives (gradient of the objective function, Jacobian of the forward map). Various examples of discrete adjoint state calculations are given.

Chapter 3 is devoted to the issues of parameterization: choice of dimensionless parameters and data, use of singular value decomposition of the linearized problem to assess the number of parameters that can be retrieved for a given error level on the data, choice of a parameterization to reduce the number of unknown parameters, organization of the inversion code for an easy experimentation with various parameterization, choice of an adapted descent step to enhance robustness and performance of descent minimization algorithms. Special care is given to the discretization of distributed parameters: multiscale parameterization is shown to restore optimizability for certain class of "nicely nonlinear" inverse problems, and its adaptive variant based on refinement indicators is presented, which in addition allows to explain the data with a small number of degrees of freedom, thus avoiding overparameterization.

The Q-wellposedness of NLS problems is the subject of Chap. 4. It begins with a summary of the properties of the linear case, which are to be generalized. Two classes of NLS problems are then defined, based on properties of the first and second directional derivatives of the direct map: the class of finite curvature (FC) least squares problems, whose attainable set has a finite curvature, and the subclass of finite curvature/limited deflection (FC/LD) problems, whose attainable set is a strictly quasi-convex set introduced in Chap. 7. Linearly stable FC/LD problems are shown to be Q-wellposed. Application are given to the estimation of finite dimensional parameters, and to the estimation of the diffusion parameter in 1D and 2D elliptic equation

Chapter 5 is devoted to the practically important problem of restoring Q-wellposedness by regularization. The usual Levenberg–Marquardt–Tychonov (LMT) regularization is considered first. After recalling the convergence results to the minimum norm solution for the linear case, when both data error and regularization parameter go to zero, we show that these results generalize completely to the nonlinear case for FC/LD problems. For general nonlinear least square, where small ϵ does not guarantee

Q-wellposedness any more, we give an estimation of the minimum amount of regularization that does so. Then it is shown how *state-space regularization* can handle some cases where LMT regularization fails, when the attainable set has an infinite curvature. Roughly speaking, state-space regularization amounts to smooth the data before solving the inverse problem. Finally, an example of desirable *adapted regularization* is given, where the a-priori information brought by the regularization term can be chosen such that it does not conflict with the information conveyed from the data by the model.

Chapter 1

Nonlinear Inverse Problems: Examples and Difficulties

We present in this chapter the nonlinear least-squares (NLS) approach to parameter estimation and inverse problems, and analyze the difficulties associated with their theoretical and numerical resolution.

We begin in Sect. 1.1 with a simple finite dimensional example of nonlinear parameter estimation problem: the estimation of four parameters in the Knott–Zoeppritz equations. This example will reveal the structure of inverse problem, and will be used to set up the terminology.

Then we define in Sect. 1.2 an abstract framework for NLS problems, which contains the structure underlying the example of Sect. 1.1. Next we review in Sect. 1.3 the difficulties associated with the resolution of NLS problems, and hint at possible remedies and their location in the book.

Finally, Sects. 1.4–1.6 describe infinite dimensional parameter estimation problems of increasing difficulty, where the unknown is the source or diffusion coefficient function of an elliptic equation, to which the analysis developed in Chaps. 2–5 will be applied.

Examples of time marching problems are given in Sects. 2.8 and 2.9 of Chap. 2

G. Chavent, *Nonlinear Least Squares for Inverse Problems: Theoretical Foundations and Step-by-Step Guide for Applications*, Scientific Computation,
DOI 10.1007/978-90-481-2785-6_1,

1.1 Example 1: Inversion of Knott–Zoeppritz Equations

We begin with an example that is intrinsically finite dimensional, where the model consists in a sequence of simple algebraic calculations. The problem occurs in the amplitude versus angle (AVA) processing of seismic data, where the densities ρ_j, compressional velocity $V_{\mathrm{P},j}$, and shear velocities $V_{\mathrm{S},j}$ on each side $j = 1, 2$ of an interface are to be retrieved from the measurement of the compressional (P–P) reflection coefficient R_i at a collection $\theta_i, i = 1, \ldots, q$ of given incidence angles.

The P-P reflection coefficient R at incidence angle θ is given by the Knott–Zoeppritz equations ([1], pp. 148–151). They are made of quite complicated algebraic formulas involving many trigonometric functions. An in-depth analysis of the formula shows that R depends in fact only on the following four dimensionless combinations of the material parameters [50]:

$$\left\{\begin{array}{lll} e_\rho = \dfrac{\rho_1 - \rho_2}{\rho_1 + \rho_2} & & \text{(density contrast)}, \\[2ex] e_{\mathrm{P}} = \dfrac{V_{\mathrm{P},1}^2 - V_{\mathrm{P},2}^2}{V_{\mathrm{P},1}^2 + V_{\mathrm{P},2}^2} & & \text{(P-velocity contrast)}, \\[2ex] e_{\mathrm{S}} = \dfrac{V_{\mathrm{S},1}^2 - V_{\mathrm{S},2}^2}{V_{\mathrm{S},1}^2 + V_{\mathrm{S},2}^2} & & \text{(S-velocity contrast)}, \\[2ex] \chi = \dfrac{V_{\mathrm{S},1}^2 + V_{\mathrm{S},2}^2}{2}\Big(\dfrac{1}{V_{\mathrm{P},1}^2} + \dfrac{1}{V_{\mathrm{P},2}^2}\Big) & & \text{(background parameter)}, \end{array}\right. \tag{1.1}$$

so that R is given by the relatively simple sequence of calculations:

$$\left\{\begin{array}{l} e = e_S + e_\rho \\ f = 1 - e_\rho^2 \\ S_1 = \chi(1 + e_P) \\ S_2 = \chi(1 - e_P) \\ T_1 = 2/(1 - e_S) \\ T_2 = 2/(1 + e_S) \\ q^2 = S_1 \sin^2\theta \\ M_1 = \sqrt{S_1 - q^2} \\ M_2 = \sqrt{S_2 - q^2} \\ N_1 = \sqrt{T_1 - q^2} \\ N_2 = \sqrt{T_2 - q^2} \\ D = eq^2 \\ A = e_\rho - D \\ K = D - A \\ B = 1 - K \\ C = 1 + K \\ P = M_1(B^2 N_1 + f N_2) + 4eDM_1M_2N_1N_2 \\ Q = M_2(C^2 N_2 + f N_1) + 4q^2A^2 \\ R = (P - Q)/(P + Q). \end{array}\right. \tag{1.2}$$

We call *parameter vector* the vector

$$x = (e_\rho, e_P, e_S, \chi) \in I\!R^4 \tag{1.3}$$

of all quantities that are input to the calculation, and *state vector* the vector

$$y = (e, f, S_1, S_2, \ \dots \ , P, Q, R) \in I\!R^{19} \tag{1.4}$$

made of all quantities one has to compute to solve the state equations (here at a given incidence angle θ).

We have supposed in the above formulas that the incidence angle θ is smaller than the critical angle, so that the reflection coefficient R computed by formula (1.2) is real, but the least squares formulation that follows can be extended without difficulty to postcritical incidence angles with complex reflection coefficients.

If now we are given a sequence $R_i^m, i = 1, \dots, q$ of "measured" reflection coefficients corresponding to a sequence $\theta_1, \dots, \theta_q$ of known (precritical) incidence angles, we can set up a *data vector*

$$z = (R_1^m, \dots, R_q^m) \in I\!R^q, \tag{1.5}$$

which is to be compared to the *output vector*

$$v = (R_1, \dots, R_q) \in I\!R^q \tag{1.6}$$

of reflection coefficients computed by formula (1.2). We can write

$$v = M \begin{bmatrix} y_1 \\ \cdot \\ \cdot \\ \cdot \\ y_q \end{bmatrix}, \tag{1.7}$$

with y_i given by (1.4) for $\theta = \theta_i$, $i = 1, \dots, q$. The operator M that selects the last component R_i of each vector y_i in the *state vector* $(y_1, \dots, y_q) \in I\!R^{19q}$ is called the *measurement* or *observation operator*. It is here a simple matrix with q rows, $p = 19q$ columns and 0 or 1 entries.

We can now set up the problem of estimating $x \in I\!R^4$ from the knowledge of $z \in I\!R^q$. To compare v and z, we have to choose first for each $i = 1, \dots, q$ a unit Δz_i to measure how much the model output $v_i = R_i$ deviates from the data $z_i = R_i^m$. The misfit of the model parameter x to the data z can then be defined by

$$J(x) = \frac{1}{2} \sum_{i=1}^{q} \frac{|v_i - z_i|^2}{\Delta z_i^2}, \tag{1.8}$$

with v given by (1.7). It is expected that minimization of J with respect to x will allow to carry over to x some of the information we have on z. In practice, one has also some a-priori information on the parameters $e_\rho, e_\mathrm{P}, e_\mathrm{S}$, and χ, at least some bounds. Hence, one restrains the minimization of J to the *set of admissible parameters*

$$C = \{x = (e_\rho, e_\mathrm{P}, e_\mathrm{S}, \chi) \mid x_{i,\min} \le x_i \le x_{i,\max},\ i = 1 \dots 4\}, \tag{1.9}$$

where $x_{i,\min}$ and $x_{i,\max}$ are the given lower and upper bounds.

In the case where the measurement errors are independent Gaussian variables with known standard deviations σ_i, one can choose $\Delta z_i = \sigma_i$, in which case the minimization of J over C produces a maximum likelihood estimator for x.

Remark

Other choices are possible for the data misfit: one could, for example, replace the Euclidean norm $\|v\| = (\sum_1^q v_i^2)^{1/2}$ on the data space by the r-norm $\|v\| = (\sum_1^q v_i^r)^{1/r}$ for some $r > 1$, as such a norm is known to be less sensitive to outliers in the data for $1 < r < 2$. The results of Chaps. 2 and 3 on derivation and parameterization generalize easily to these norms. But the results on Q-wellposedness and regularization of Chap. 4 and 5 hold true only for the Euclidean norm.

1.2 An Abstract NLS Inverse Problem

The inversion of the Knott–Zoeppritz equations presented in Sect. 1.1 is a special case of the following abstract setting, which will be used throughout this book to handle NLS inverse problems:

- E will be the *parameter space*
- C the *set of admissible parameters*
- $\varphi : C \rightsquigarrow F$ the *forward* or *direct* or *input–output map*
- F the *data space*
- $z \in F$ the *data* to be inverted

The *inverse problem* consists then in searching a parameter $x \in C$ whose image $\varphi(x)$ by the forward map φ is the data z. But in applications z will never belong to the attainable set $\varphi(C)$, because φ never represents absolutely accurately the process under study (model errors), and because of the measurement errors on z. The inverse problems has then to be solved in the least squares sense as an NLS problem:

$$\hat{x} \quad \text{minimizes} \quad J(x) = \frac{1}{2}\|\varphi(x) - z\|_F^2 \quad \text{over} \quad C. \tag{1.10}$$

This formulation contains the inversion of the Knott–Zoeppritz equations (Sect. 1.1). It suffices for that to choose

$$\begin{cases} E = I\!R^4 \text{ with the norm } \|x\|_E = (\sum_1^4 x_i^2)^{1/2}, \\ C \quad \text{given by (1.9)}, \\ F = I\!R^q \text{ with the norm } \|v\|_F = (\sum_1^q v_i^2/\Delta z_i^2)^{1/2}, \\ \varphi \quad : \quad x \rightsquigarrow v \quad \text{with } v \text{ given by (1.7)}. \end{cases} \tag{1.11}$$

But it is much more general, and allows also to handle infinite dimensional problems, as we shall see in Sects. 1.4–1.6. So we shall suppose throughout this book that the following minimum set of hypothesis holds:

$$\begin{cases} E & = & \text{Banach space, with norm} \quad \| \, \|_E, \\ C & \subset & E \quad \text{with } C \text{ convex and closed,} \\ F & = & \text{Hilbert space, with norm} \quad \| \, \|_F, \\ z & \in & F \\ \varphi & : & C \rightsquigarrow F \text{ is differentiable along segments of C,} \\ \text{and} & : & \exists \alpha_M \geq 0 \ \text{ s.t. } \ \forall x_0, x_1 \in C, \quad \forall t \in [0,1] \\ & & \|D_t\varphi((1-t)x_0 + tx_1)\| \leq \ \alpha_M \|x_1 - x_0\|. \end{cases} \tag{1.12}$$

These hypothesis are satisfied in most inverse problems – and in all those we address in this book, including of course the inversion of the Knott–Zoeppritz equations of Sect. 1.1 – but they are far from being sufficient to ensure good properties of the NLS problem (1.10).

1.3 Analysis of NLS Problems

We review in this section the theoretical and practical difficulties associated with the theoretical and numerical resolution of (1.10), and present the tools developed in this book for their solution.

1.3.1 Wellposedness

The first question is wellposedness: does (1.10) has a unique solution $\hat{x}$ that depends continuously on the data z? It is the first question, but it is also a difficult one because of the nonlinearity of φ. It is most likely to be answered negatively, as the word *ill-posed* is almost automatically associated with *inverse problem*. To see where the difficulties arise from, we can conceptually split the resolution of (1.10) into two consecutive steps:

Projection step: given $z \in F$, find a projection $\hat{X}$ of z on $\varphi(C)$
Preimage step: given $\hat{X} \in \varphi(C)$, find one preimage $\hat{x}$ of $\hat{X}$ by φ

Difficulties can – and usually will – arise in both steps:

1. There can be more than one preimage $\hat{x}$ of $\hat{X}$ if φ is not injective

2. Even if φ is injective, its inverse $\varphi^{-1}: \ \hat{X} \rightarrow \hat{x}$ may not be continuous over $\varphi(C)$

3. The projection $\hat{X}$ does not necessarily exist if $\varphi(C)$ is not closed

4. The projection on $\varphi(C)$ can be nonunique and not continuous as $\varphi(C)$ is possibly nonconvex because of the nonlinearity of φ

We consider first the unrealistic case of attainable data as where there are no measurement or model error:

$$z = \hat{X}, \tag{1.13}$$

and recall the corresponding identifiability and stability properties, which concern only the preimage step:

Definition 1.3.1 *Let φ and C be given. The parameter x is*

- Identifiable *if φ is injective on C, that is,*

$$x_0, x_1 \in C \text{ and } \varphi(x_0) = \varphi(x_1) \implies x_0 = x_1 \tag{1.14}$$

- Stable *if there exists a constant $k \geq 0$ such that*

$$\|x_0 - x_1\| \leq k \, \|\varphi(x_0) - \varphi(x_1)\| \quad \forall x_0, x_1 \in C \tag{1.15}$$

The vast majority of available results concern the case of attainable data only: most of them are identifiability results (see, e.g., [46] for the estimation of coefficients in partial differential equations from distributed or boundary observation), and a much smaller number are stability results (see, e.g., [47, 70]).

But these results fall short to imply the wellposedness of NLS problem (1.10), as they do not say anything when *z is outside the attainable set $\varphi(C)$*.

In the more realistic case where one takes into account the model and measurement errors, the data z can be outside the output set $\varphi(C)$, and the problem of the projection on $\varphi(C)$ arises.

In the *linear case*, difficulty 3 (possible lack of existence of the projection) already exists. But it can be cured simultaneously with difficulties 1 and 2 by regularization (see *LMT-Regularization* in Sect. 1.3.4), because the projection on the convex set $\varphi(C)$, when it exists, is necessarily unique and Lipschitz continuous!

But for *nonlinear* inverse problems, because of the possible nonuniqueness and noncontinuity of the projection (difficulty 4), one cannot expect wellposedness to hold for all z of F. So we have to limit the existence, uniqueness, and stability requirement to some neighborhood ϑ of $\varphi(C)$ in F.

To handle this situation, we introduce in Chap. 6 a first generalization of convex sets to *quasi-convex* sets $D \subset F$, which precisely possess a neighborhood ϑ on which the projection, when it exists, is unique and Lipschitz stable [20].

1.3.2 Optimizability

The next question that arises as soon as one considers solving (1.10) on a computer is the possibility to use performant local optimization techniques to find the global minimum of the objective function J. It is known that such algorithms converge to stationary points of J, and so it would be highly desirable to be able to recognize the case where J has no parasitic stationary points on C, that is, stationary points where J is strictly larger than its global minimum over C. It will be convenient to call *unimodal* functions with this property, and *optimizable* the associated least squares problem. Convex functions, for example, are unimodal, and linear least squares problems are optimizable.

Stationary points of J are closely related to stationary points of the *distance to z* function over $\varphi(C)$, whose minimum gives the projection of z on $\varphi(C)$. Hence we introduce in Chap. 7 a generalization of convex sets to *strictly quasi-convex* (s.q.c.) sets $D \subset F$, which possess a neighborhood ϑ on which the *distance to z* function is unimodal over D [19].

Sufficient conditions for a set to be s.q.c. are developed in Chap. 8, in particular, in the case of interest for NLS problems where D is the attainable set $\varphi(C)$.

As it will turn out (Theorem 7.2.10), s.q.c. sets are indeed quasi-convex sets, and so requiring that the output set is s.q.c. will solve both the wellposedness and optimizability problems at once.

1.3.3 Output Least Squares Identifiability and Quadratically Wellposed Problems

Wellposedness and optimizability are combined in the following definition [16, 17], which tries to carry over the good properties of the linear case to the nonlinear case:

Definition 1.3.2 *Let φ and C be given. The parameter x is* OLS-identifiable *in C from a measurement z of $\varphi(x)$ – or equivalently the NLS problem (1.10) is* Quadratically (Q-)wellposed *– if and only if $\varphi(C)$ possesses an open neighborhood ϑ such that*

(i) *Existence and uniqueness: for every $z \in \vartheta$, problem (1.10) has a unique solution $\hat{x}$*

(ii) *Unimodality: for every $z \in \vartheta$, the objective function $x \rightsquigarrow J(x)$ has no parasitic stationary point*

(iii) *Local stability: the mapping $z \rightsquigarrow \hat{x}$ is locally Lipschitz continuous from $(\vartheta, \|\cdot\|_F)$ to $(C, \|\cdot\|_E)$*

The class of FC problems (Definition 4.2.1) represents a first step toward Q-wellposedness. The subclass of finite curvature/limited deflection (FC/LD) problems (Definition 4.2.2) represents a further step in this direction: in this class, the images by φ of the segments of C turn less than $\pi/2$ (Corollary 4.2.5) – which can always be achieved by reducing the size of the admissible set C – and the attainable set is s.q.c. (Corollary 4.2.5). The projection on the attainable set is then, when it exists, unique, unimodal, and stable (Proposition 4.2.7) for the arc length "distance" (4.27) on $\varphi(C)$, and the solution set is closed and convex.

Combining this with the *linear identifiability* and/or *stability* properties of Sect. 4.3, one obtains Theorem 4.4.1, our main result, which gives a sufficient conditions for Q-wellposedness or OLS-identifiability – and hence for both wellposedness *and* optimizability. These sufficient conditions are written in terms of the first and second directional derivatives of φ, and give an estimation of the localization and stability constants, and of the size of the neighborhood ϑ.

The case of finite dimensional parameters is considered in Sect. 4.5, where it is shown in Theorem 4.5.1 that *linear identifiability* (i.e., identifiability of the linearized problem) implies *local* (i.e., for small enough C) OLS-identifiability or Q-wellposedness. This shows the importance of (linear) identifiability results for the analysis of finite dimensional NLS problems.

The techniques available for the determination of the curvature, deflection, and stability constants required for the application of Theorem 4.4.1 are reviewed in Sect. 4.7. These quantities can sometimes be determined by calculus, but can also be approximated numerically when the number of parameters is small.

The sufficient conditions are then applied in Sects. 4.8 and 4.9 to analyze the Q-wellposedness of the estimation of the diffusion coefficient in one- and two-dimensional elliptic equations, as described in Sects. 1.4 and 1.6, when an H^1 measurement of the solution is available.

1.3.4 Regularization

OLS-identifiability is a very strong property, so one cannot expect that it will apply to the original NLS problem, when this one is known to be ill-posed.

On the other hand, the only NLS problems that should reasonably be solved on the computer at the very end are those where OLS-identifiability holds, as this ensures the following:

– Any local optimization algorithm will converge to the global minimum

– The identified parameter is stable with respect to data perturbations

which are the only conditions under which one can trust the results of the computation.

Hence the art of inverse problems consists in adding information of various nature to the original problem, until a Q-wellposed NLS problem is achieved, or equivalently OLS-identifiability holds: this is the *regularization* process. The sufficient conditions for OLS-identifiability of Chap. 4 will then allow to check whether the regularized problem has these desirable properties, and help to make decisions concerning the choice of the regularization method.

Adapted Regularization

If the added a-priori information is (at least partly) incompatible with the one conveyed from the data by the model, the optimal regularized parameter has to move away from its true value to honor both information. So one should ideally try to add information that tends to conflict as little as possible with the one coming from the data. Such an *adapted regularization* is recommended whenever it is possible. But it requires a refined analysis of the forward map φ, which is not always possible, and has to be considered case by case. We give in Sect. 5.4 an example of adapted regularization for the estimation of the diffusion coefficient in a two-dimensional elliptic equation.

But, in general, this hand-taylored approach is impossible, and the information has to be added by brute force:

Regularization by Parameterization

This is one of the most commonly used approach: instead of searching for the parameter x, one searches for a parameter x_{opt} related to x by

$$x = \psi(x_{\text{opt}}), \qquad x_{\text{opt}} \in C_{\text{opt}} \quad \text{with} \quad \psi(C_{\text{opt}}) \subset C, \tag{1.16}$$

where ψ is the *parameterization map*, and C_{opt} is the *admissible parameter set* for the *optimization variables* x_{opt} (Sect. 3.3). Satisfying the inclusion $\psi(C_{\text{opt}}) \subset C$ is not always easy, and it has to be kept in mind when choosing the parameterization mapping ψ.

Parameterization is primarily performed to reduce the number of unknown parameters, but it has also the effect, when x is a function, to impose regularity constraints, which have usually a stabilizing effect on the inverse problem. The regularized problem is now

$$\widehat{x_{\text{opt}}} \quad \text{minimizes} \quad J \circ \psi(x_{\text{opt}}) = \frac{1}{2}\|\varphi(\psi(x_{\text{opt}})) - z\|_F^2 \quad \text{over} \quad C_{\text{opt}}. \tag{1.17}$$

We address in Chap. 3 some practical aspects of parameterization:

– How to calibrate the parameters and the data

– How to use the singular value decomposition of the Jacobian $D = \varphi(x)'$ to estimate the number of independent parameters that can be estimated for a given level of uncertainty on the data

– How to use multiscale approximations and/or refinement indicators in order to determine an *adaptive* parameterization, which makes the problem optimizable, explains the data up to the noise level, *and* does not lead to over-parameterization

– How to organize the inversion code to make experimentation with various parameterizations more easy

– How to use in minimization algorithms a descent step adapted to the nonlinear least square structure of the problem

Regularization by Size Reduction of C

Another natural way to add a-priori information is to search for the parameter x in a smaller set:

$$\boldsymbol{C} \subset C, \tag{1.18}$$

and to *define* the regularized problem simply by

$$\hat{x} \quad \text{minimizes} \quad J(x) = \frac{1}{2}\|\varphi(x) - z\|_F^2 \quad \text{over} \quad \boldsymbol{C}. \tag{1.19}$$

The set $\boldsymbol{C}$ can be defined as follows:

- *Either by adding new constraint (s).* when C is made of functions defined over a domain $\Omega \in I\!R^n$, on can want to add the a priori information that the parameter is bounded in a space of smooth functions. This can be implemented by choosing a subspace $\mathcal{E} \subset E$ with the desired smoothness, equipped with the stronger norm $\|\cdot\|_{\mathcal{E}}$, and by restricting the search for x to the set

$$\boldsymbol{C} = \{x \in \mathcal{E} \mid x \in C \text{ and } \|x\|_{\mathcal{E}} \leq M\}. \tag{1.20}$$

  Such smoothness constraints are often used to ensure that the regularized problem (1.19) is a FC least squares problem. FC problems were introduced in [28] under the less descriptive name of "weakly nonlinear" problems, and are defined in Sect. 4.2 (Definition 4.2.1). They represent a first necessary step towards Q-wellposedness.

  For example, in the estimation of the diffusion coefficient a in a two-dimensional elliptic equation (Sect. 1.6), the natural admissible set C is defined by (1.66):

$$C = \{a : \Omega \rightsquigarrow I\!R \mid 0 < a_m \leq a(\xi) \leq a_M \text{ a.e. on } \Omega\}.$$

  The first regularization step we shall take in Sect. 4.9 will be to replace C by the set defined in (4.106), which we rewrite here as:

$$\begin{aligned} \boldsymbol{C} = \{\, a \in \mathcal{E} \mid\; & a_m \leq a(\xi) \leq a_M \quad \forall \xi \in \overline{\Omega}, \\ & |a(\xi_1) - a(\xi_0)| \leq b_M \|\xi_1 - \xi_0\| \quad \forall \xi_0, \xi_1 \in \overline{\Omega} \,\}, \end{aligned} \tag{1.21}$$

  where $\mathcal{E}$ is a (sub)space of the space of uniformly Lipschitz-continuous functions over $\overline{\Omega}$. These constraints will ensure (Theorem 4.9.4, point two) that (1.19) is a finite curvature problem.

- *Or by tightening existing constraints.* This is particularly useful in the case where (1.10) is already an FC problem. In this case, Corollary 4.2.5 shows that tightening the constraints – that is, reducing the size of $\boldsymbol{C}$ – will produce a FC/LD problem (Definition 4.2.2), whose attainable set $\varphi(\boldsymbol{C})$ is s.q.c., so that the projection on $\varphi(\boldsymbol{C})$ is wellposed on some of its neighborhoods (Proposition 4.2.7). This will make the problem amenable to LMT-regularization (see *LMT regularization* below and Sect. 5.1.2).

For example, the second regularization step we shall take in Sect. 4.9 for the same diffusion coefficient problem as earlier will be to reduce, in the admissible set (1.21), the interval between the lower and upper bounds on a until the condition

$$a_M - a_m \leq \frac{\pi}{4} a_m \tag{1.22}$$

is satisfied. Proposition 4.9.3 will then ensure that the deflection condition is satisfied, and make the problem ready for LMT regularization.

Levenberg–Marquardt–Tykhonov (LMT) Regularization

This approach is the oldest one, as the work by Levenberg [54] goes back to the forties, and that of Marquardt [62] to the sixties. It was popularized in the seventies by Tikhonov and the Russian school [75, 63], who developed the theory for infinite dimensional systems. It consists in stabilizing the original problem (1.10) by providing additional information in the form of a *regularizing functional* to be added to the least-squares functional $J(x)$. We shall limit ourselves to the case of quadratic regularizing functionals, which are by far the most widely used, and suppose that

$$\text{the parameter space } E \text{ is an Hilbert space,} \tag{1.23}$$

and that we have chosen

$$x_0 \in C \qquad \text{and} \qquad \epsilon > 0, \tag{1.24}$$

where x_0 is an *a priori guess* of the parameter, and $\epsilon > 0$ is the *regularization parameter.* The a priori guess x_0 represents the additional information provided, and ϵ measures the confidence one has in this a priori guess. The LMT-regularized version of (1.10) is

$$\hat{x}_\epsilon \text{ minimizes } J_\epsilon(x) = J(x) + \frac{\epsilon^2}{2}\|x - x_0\|_E^2 \text{ over } C. \tag{1.25}$$

Its properties have been studied extensively when φ is linear, and we refer, for example, to the monographs [8, 43, 59, 63, 38, 55, 12]. The main result for linear problems is that the regularized problem (1.25) is wellposed as soon as $\epsilon > 0$, and that the solution $\hat{x}_{\epsilon,\delta}$ of (1.25) corresponding to noise corrupted data z_δ converges to the x_0-*minimum norm solution* of (1.10) – that is, the solution of (1.10) closest to x_0, provided that such a solution exists, and that ϵ goes to zero more slowly than the noise level $\delta = \|z_\delta - z\|$. We give at the beginning of Chaps. 4 and 5 a short summary of properties of linear least-squares problems.

We shall see that all the nice properties of LMT regularization for the linear case extend to the class of FC/LC problems, where the attainable set has a FC, and the size of the admissible set has been chosen small enough for the deflection condition to hold (see *Regularization by size reduction* above). LMT regularization of FC/LC problems is studied in Sect. 5.1 and is applied in Sect. 5.2 to the the source identification problem of Sect. 1.5.

But for general nonlinear problems, the situation is much less brilliant. In particular, there is no reason for the regularized problem (1.25) to be Q-wellposed for small ϵ's! Hence when $\epsilon \to 0$, problem (1.25) may have more than one solution or stationary point: convergence results can still be obtained for sequences of adequately chosen solutions [37, 67], but optimizability by local minimization algorithms is lost. We give in Sect. 5.1.3 an estimation of the minimum amount of regularization to be added to restore Q-wellposedness.

State-Space Regularization

When the original problem (1.10) has no finite curvature, which is alas almost a generic situation for infinite dimensional parameters, it is difficult to find a classical regularization process that ensures Q-wellposednes of the regularized problem for small ϵ's. One approach is then to decompose (see Sect. 1.3.5) the forward map φ into the resolution of a state equation (1.33) followed by the application of an observation operator (1.34)

$$\varphi = M \circ \phi, \tag{1.26}$$

where $\phi : x \rightsquigarrow y$ is the solution mapping of the state equation, and $M : y \rightsquigarrow v$ is the observation operator.

The *state-space regularized* version of problem (1.10) is then [26, 29]

$$\hat{x}_\epsilon \text{ minimizes } J_\epsilon(x) = J(x) + \frac{\epsilon^2}{2}\|\phi(x) - \hat{y}_\epsilon\|^2 \text{ over } C \cap B_\epsilon, \tag{1.27}$$

where B_ϵ is a localization constraint, and $\hat{y}_\epsilon$ is the solution of the auxiliary unconstrained problem

$$\hat{y}_\epsilon \quad \text{minimizes} \quad \frac{1}{2}\|M(y) - z\|_F^2 + \frac{\epsilon^2}{2}\|y - y_0\|_Y^2 \quad \text{over } Y, \tag{1.28}$$

where y_0 is an a priori guess for the state vector.

For example, in the parameter estimation problems of Sects. 1.4 and 1.6, one can think of x as being the diffusion coefficient a, of y as being the solution of the elliptic equation in the Sobolev space $H^1(\Omega)$, and of v as being a measure of y in $L^2(\Omega)$. State-space regularization would then consist in using first LMT regularization to compute a smoothed version $y_\epsilon \in H^1(\Omega)$ of the data $z \in L^2(\Omega)$ by solving the auxiliary problem (1.28), and then to use this smoothed data as extra information in the regularized NLS problem (1.27).

Smoothing data before inverting them has been a long-time favorite in the engineering community, and so state-space regularization is after all not so unnatural. It is studied in Sect. 5.3. The main result is demonstrated in Sect. 5.3.1: problems (1.27) and (1.28) remain Q-wellposed when $\epsilon \to 0$, provided one can choose the state-space Y as follows:

– The problem of estimating $x \in C$ from a measure of $\phi(x)$ in Y is Q-wellposed (this requires usually that the the norm on Y is strong enough)

– The observation mapping M is *linear* and *injective*

This covers, for example, the case of the parameter estimation problems of Sects. 1.4 and 1.6 when only L^2 observations are available.

A partial result for the case of point and/or boundary measurements, which do not satisfy the injectivity condition, is given in Sect. 5.3.2.

Common Features

All the regularized problems can be cast into the same form as the original problem (1.10). It suffices to perform the following substitutions in (1.10):

$$\text{regularization by parameterization} \quad \left\{ \begin{array}{rcl} C & \leftarrow & C_{\text{opt}} \\ \varphi(x) & \leftarrow & \varphi(\psi(x_{\text{opt}})) \\ F & \leftarrow & F \\ \| \ \|_F & \leftarrow & \| \ \|_F \\ z & \leftarrow & z \end{array} \right. \tag{1.29}$$

$$\text{reduction of } C \quad \left\{ \begin{array}{rcl} C & \leftarrow & \mathbf{C} \\ \varphi(x) & \leftarrow & \varphi(x) \\ F & \leftarrow & F \\ \| \ \|_F & \leftarrow & \| \ \|_F \\ z & \leftarrow & z \end{array} \right. \tag{1.30}$$

$$\text{LMT regularization} \quad \left\{ \begin{array}{rcl} C & \leftarrow & C \\ \varphi(x) & \leftarrow & (\varphi(x), x) \\ F & \leftarrow & F_\epsilon = F \times E \\ \| \ \|_F & \leftarrow & \| \ \|_{F_\epsilon} = (\| \ \|_F^2 + \frac{\epsilon^2}{2} \| \ \|_E^2)^{1/2} \\ z & \leftarrow & (z, x_0) \end{array} \right. \tag{1.31}$$

$$\text{state-space regularization} \quad \left\{ \begin{array}{rcl} C & \leftarrow & C \\ \varphi(x) & \leftarrow & (M\,\phi(x), \phi(x)) \\ F & \leftarrow & F_\epsilon = F \times Y \\ \| \ \|_F & \leftarrow & \| \ \|_{F_\epsilon} = (\| \ \|_F^2 + \frac{\epsilon^2}{2} \| \ \|_Y^2)^{1/2} \\ z & \leftarrow & (z, y_\epsilon) \end{array} \right. \tag{1.32}$$

If more than one regularization technique are applied simultaneously, the property still holds with obvious adaptions.

Hence the same abstract formulation (1.10) can be used to represent the original NLS problem as well as any of its regularized versions (1.19), (1.17), (1.25), or (1.27).

1.3.5 Derivation

The last difficulty we shall address is a very practical one, and appears as soon as one wants to implement a local optimization algorithm to solve a NLS problem of the form (1.10): one needs to provide the optimization code with the gradient ∇J of the objective function, or the derivative or Jacobian $D = \varphi'(x)$ of the forward map.

To analyze this difficulty, it is convenient to describe φ at a finer level by introducing a *state-space decomposition*. So we shall suppose, when needed, that $v = \varphi(x)$ is evaluated in two steps:

$$\left\{ \begin{array}{l} \text{given } x \in C \text{ solve:} \\ e(x, y) = 0 \\ \text{with respect to } y \text{ in } Y, \end{array} \right. \tag{1.33}$$

followed by

$$\text{set:} \quad v = M(y), \tag{1.34}$$

where now

- y is the *state*, Y is an affine *state-space* with tangent vector space δY

- $e(x,y) = 0$ is the *state equation*, where e maps $C \times Y$ into the *right-hand side space* Z. For finite dimensional problems, $\delta Y = Z = I\!R^p$. But it will be convenient to use distinct spaces δY and Z for infinite dimensional problems
- $M : Y \rightsquigarrow F$ is the (possibly nonlinear) *measurement* or *observation operator*

We shall always suppose that the following minimum set of hypothesis holds:

$$\left\{\begin{array}{l} Y \text{ is an affine space,} \\ \text{its tangent space } \delta Y \text{ is equipped with the norm } \|\ \|_Y, \\ \forall x \in C, \text{ (1.33) has a unique solution } y \in Y, \\ \forall x \in C, \text{ the operator } \partial e/\partial y(x,y) \in \mathcal{L}(\delta Y, Z) \text{ is invertible,} \\ M \text{ is a continuously derivable application from } Y \text{ to } F. \end{array}\right. \tag{1.35}$$

For linear state equations, the invertibility of $\partial e/\partial y(x,y)$ is equivalent to wellposedness of the state equation. For nonlinear problems, the condition is required if one wants to solve the state equation by an iterative Newton scheme. Hence it will be satisfied in practice as soon as the forward model under consideration is stable and numerically solvable.

For example, if one wants that (1.33) and (1.34) define the forward map φ associated with the Knott–Zoeppriz equations of Sect. 1.1, one has to choose

$$\left\{\begin{array}{l} e(x,y) = 0 \quad \text{to be (1.2) repeated } q \text{ times ,} \\ y = (e_i, f_i \ \ldots \ Q_i, R_i \ , \ i = 1 \ \ldots \ q), \\ Y = \delta Y = Z = I\!R^{19q}, \\ M \text{ defined by (1.7).} \end{array}\right. \tag{1.36}$$

We describe and compare in Chap. 2 the sensitivity functions and adjoint state approaches for the computation of D and ∇J. Though calculating a derivative is a simple mathematical operation, it can become quite intricate when the function to derivate is defined through a set of equations, and so we give a step-by-step illustration of these calculations on various examples.

1.4 Example 2: 1D Elliptic Parameter Estimation Problem

Our second example is infinite dimensional: it consists in estimating the coefficient a as a function of the space variable ξ in the one-dimensional elliptic equation:

$$-(au_\xi)_\xi = \sum_{j\in J} g_j\, \delta(\xi - \xi_j), \qquad \xi \in \Omega, \tag{1.37}$$

over the domain $\Omega = [0, 1]$, with a right-hand side made of Dirac sources:

$$\begin{cases} \xi_j \text{ denotes the location of the } j\text{th source,} \\ g_j \in I\!R \text{ denotes the amplitude of the } j\text{th source,} \\ J \text{ is a finite set of source indexes.} \end{cases} \tag{1.38}$$

We complement (1.37) with Dirichlet boundary conditions:

$$u(0) = 0, \qquad u(1) = 0. \tag{1.39}$$

Equations (1.37) and (1.38) models, for example, the temperature u in a one-dimensional slab at thermal equilibrium, heated by point sources of amplitude g_i at locations x_i, and whose temperature is maintained equal to zero at each end, in which case a is the thermal diffusion coefficient.

A physically and mathematically useful quantity is

$$q(\xi) = -a(\xi)\, u_\xi(\xi), \qquad \xi \in \Omega, \tag{1.40}$$

which represents the (heat) flux distribution over Ω, counted positively in the direction of increasing ξ's. Equations (1.37) and (1.38) are simple in the sense that the flux q can be computed by a closed form formula

$$q(\xi) = \widetilde{H} - H(\xi), \tag{1.41}$$

where

$$H(\xi) = \int_0^\xi \sum_{j\in J} g_j\, \delta(\xi - \xi_j)\, d\xi \tag{1.42}$$

is the primitive of the right-hand side, and where

$$\widetilde{H} = \frac{\int_0^1 a^{-1}(\xi)\, H(\xi)\, d\xi}{\int_0^1 a^{-1}(\xi)\, d\xi} \in I\!R \tag{1.43}$$

is an a^{-1}-weighted mean of H. The function H is piecewise constant on Ω, and so is also q. The solution u is then given by

$$u_\xi(\xi) = a^{-1}(\xi) q(\xi), \tag{1.44}$$

$$u(\xi) = \int_0^{\xi} a^{-1}(\zeta) q(\zeta)\, d\zeta. \tag{1.45}$$

We see on (1.41) through (1.45) that u_ξ and u depend almost linearly on a^{-1}, which suggests that the search for a^{-1} could be better behaved than the search for a. So we decide to choose $b = a^{-1}$ for parameter in this example, and define the *parameter space* and *admissible parameter set* as

$$E = L^2(\Omega), \tag{1.46}$$
$$C = \{b : \Omega \rightsquigarrow I\!R \mid 0 < b_m \leq b(\xi) \leq b_M \text{ a.e. on } \Omega\}, \tag{1.47}$$

where b_m and b_M are given lower and upper bounds on $b = a^{-1}$.

Then for any $b \in C$, the *state equations* (1.37) and (1.39) have a unique solution u in the *state-space*:

$$Y = \{w \in L^\infty(\Omega) \mid w' \in L^\infty(\Omega)\,,\ w(0) = 0\,,\ w(1) = 0\}, \tag{1.48}$$

and because of the boundary conditions $w(0) = 0\,, w(1) = 0$, the Poincaré inequality $|v|_{L^2} \leq |v'|_{L^2}$ holds in Y. Hence we can equip Y with the norm

$$\|v\|_Y = |v'|_{L^2}. \tag{1.49}$$

To estimate b, we suppose we are given an element z of the *data space* $F = L^2(\Omega)$, which provides us with some information on the temperature u.

We consider first the (quite unrealistic) case of a "distributed H^1-observation," where one is able to measure the derivative of u all over Ω, so that z is a measure of u'. This corresponds to the *observation operator* and *forward map*:

$$M \ :\ u \in Y \rightsquigarrow v = u' \in F, \tag{1.50}$$
$$\varphi \ :\ b \in C \rightsquigarrow v = u' \in F. \tag{1.51}$$

Here M is an isometry from Y onto F, and so the observation does not incur any loss of information: it is the most favorable case. We shall prove in Sect. 4.8 that the associated NLS problem (1.10) is a linearly stable FC/FD problem, and hence Q-wellposed, on the subset D of C defined in (4.90), provided the size of D is small enough (Sect. 4.8.4). This is an example of regularization by **size reduction** of the admissible set.

The case of a weaker "distributed L^2-observation," where one measures only u over Ω:

$$M \ :\ u \in Y \rightsquigarrow v = u \in F, \tag{1.52}$$
$$\varphi \ :\ b \in C \rightsquigarrow v = u \in F \tag{1.53}$$

is considered in Sect. 5.3, where it is proven that the combination of the previous Q-wellposedness results for H^1 observation with the *state-space regularization* approach (1.27) and (1.28) can still produces a sequence of Q-wellposed problems.

For more realistic boundary or point measurements, the problem of the Q-wellposedness of the NLS problems (1.10) or its regularized versions is completely open.

1.5 Example 3: 2D Elliptic Nonlinear Source Estimation Problem

We consider the nonlinear elliptic equation:

$$\begin{cases} -\Delta u + k(u) \;=\; f \quad \text{in } \Omega, \\ u \;=\; 0 \quad \text{on a part } \partial\Omega_{\mathrm{D}} \quad \text{of } \partial\Omega, \\ \dfrac{\partial u}{\partial \nu} \;=\; g \quad \text{on } \partial\Omega_{\mathrm{N}} \;=\; \partial\Omega \setminus \partial\Omega_{\mathrm{D}}, \end{cases} \tag{1.54}$$

where

- Δ is the Laplacian with respect to the space variables $\xi_1 \cdots \xi_m$
- $\Omega \subset I\!R^m$ is a domain with boundary $\partial\Omega$
- $\partial\Omega_{\mathrm{D}}$ and $\partial\Omega_{\mathrm{N}}$ form a partition of $\partial\Omega$
- and ν is the outer normal to Ω

We make the following hypothesis on the nonlinearity k, the right-hand sides f, g and the space domain Ω:

$$\begin{cases} k \in W^{2,\infty}(I\!R), \text{ is nondecreasing,} \\ \text{with } k(0) \;=\; 0,\; k'(\zeta) \geq 0 \;\; \forall \zeta \in I\!R, \end{cases} \tag{1.55}$$

$$f \in L^2(\Omega), \qquad g \in L^2(\partial\Omega_{\mathrm{N}}) \tag{1.56}$$

$$\begin{cases} \Omega \subset I\!R^m,\; m = 1, 2, \text{ or } 3, \text{ with } C^{1,1} \text{ boundary } \partial\Omega, \\ \partial\Omega_{\mathrm{D}} \text{ nonempty and both open and closed in } \partial\Omega. \end{cases} \tag{1.57}$$

The last condition in (1.57) means that there are no point on $\partial\Omega$, where the Dirichlet boundary $\partial\Omega_{\mathrm{D}}$ and the Neumann boundary $\partial\Omega_{\mathrm{N}}$ meet (this excludes, e.g., the case where $\partial\Omega_{\rm D}$ and $\partial\Omega_{\rm N}$ are the two halves of the same circle). In the two-dimensional case, it is satisfied, for example, if $\partial\Omega_{\rm N}$ is the external boundary of Ω, and $\partial\Omega_{\rm D}$ is the boundary of a hole in the domain, as the boundary of a source or sink, for example.

The elliptic equation (1.54) is nonlinear, but with a nonlinearity in the lowest order term only. It admits a unique solution $u \in Y$ [56], where

$$Y = \{w \in L^2(\Omega) \mid \frac{\partial w}{\partial \xi_i} \in L^2(\Omega)\ ,\ i = 1 \dots m\ ,\ w = 0 \text{ on } \partial\Omega_{\rm D}\}. \tag{1.58}$$

The *source problem* consists here in estimating the source terms f and/or g of the elliptic problem (1.54) from a measurement z in the data space $F = L^2(\Omega)$ of its solution u (the observation operator M is then simply the canonical injection from the subspace Y of H^1 into L^2). We shall denote by

$$C \subset E \stackrel{\text{def}}{=} L^2(\Omega) \times L^2(\partial\Omega_{\rm N}), \quad \text{closed, convex, bounded} \tag{1.59}$$

the set of admissible f and g, to be chosen later, and by

$$z \in F \stackrel{\text{def}}{=} L^2(\Omega), \tag{1.60}$$

a (noisefree) measure of the solution u of (1.54).

Estimation of f and g in C from the measure z of u amounts to solve the nonlinear least-squares problem

$$(\hat{f},\ \hat{g}) \quad \text{minimizes} \quad \frac{1}{2}\ |\varphi(f,g) - z|^2_{L^2(\Omega)} \quad \text{over} \ \ C, \tag{1.61}$$

where

$$\varphi\ :\ (f,g) \in C \rightsquigarrow u \in F = L^2(\Omega) \quad \text{solution of (1.54)} \tag{1.62}$$

is the forward map, which is obviously injective. Hence one is here in the case of identifiable parameters.

Given a sequence $z_n \in L^2(\Omega), n = 1, 2 \dots,$ of noise corrupted measurements of z and another sequence $\epsilon_n > 0$ of regularization coefficients such that

$$|z_n - z|_{L^2(\Omega)} \to 0, \qquad \epsilon_n \to 0,$$

and the LMT-regularization of (1.61) is

$$\begin{cases} (\hat{f}_n,\ \hat{g}_n) \quad \text{minimizes} \\ \dfrac{1}{2}\ |\varphi(f,g) - z_n\Big|^2_{L^2(\Omega)} + \dfrac{\epsilon_n^2}{2}|f - f_0|^2_{L^2(\Omega)} + \dfrac{\epsilon_n^2}{2}|g - g_0|^2_{L^2(\partial\Omega_N)} \\ \text{over}\ \ C, \end{cases} \tag{1.63}$$

We show in Sect. 5.2 that (1.61) is a FC problem. This will imply, under adequate hypothesis on the size of C, that it is also a FC/LD problem. Hence its LMT-regularization (1.63) is Q-wellposed for large n, and $\left(\hat{f}_n, \hat{g}_n\right)$ will converge to the solution $\left(\hat{f}, \hat{g}\right)$ of (1.61).

1.6 Example 4: 2D Elliptic Parameter Estimation Problem

In this last example, we consider a two-dimensional linear elliptic equation:

$$\begin{cases} -\nabla\cdot(a\nabla u) = 0 \quad \text{in } \Omega, \\ u = 0 \quad \text{on } \partial\Omega_{\mathrm{D}}, \qquad a\,\dfrac{\partial u}{\partial \nu} = 0 \quad \text{on } \partial\Omega_{\mathrm{N}} \\ \int_{\partial\Omega_j} a\,\dfrac{\partial u}{\partial \nu} = Q_j,\; u|_{\partial\Omega_j} = c_j = \text{constant}, \quad j \in J, \end{cases} \tag{1.64}$$

where

- a is the diffusion coefficient function to be estimated
- $\nabla\cdot$ and ∇ are the divergence and gradient operators with respect to the space variables ξ_1, ξ_2
- Ω is a two-dimensional domain with boundary $\partial\Omega$ partitioned into $\partial\Omega_{\mathrm{D}}$, $\partial\Omega_{\mathrm{N}}$ and $\partial\Omega_j, j \in J$
- J is a finite set of indices
- $Q_j \in I\!R,\;\; j \in J$ is the total injection ($Q_j \geq 0$) or production ($Q_j \leq 0$) rate through the boundary $\partial\Omega_j$
- the condition $u|_{\partial\Omega_j} = c_j = \text{constant}$ means that the solution u take a constant, but unknown value c_j on each boundary $\partial\Omega_j$: the values of c_j will change if the intensities Q_j of the sources or the diffusion coefficient a change
- ν is the exterior normal to Ω

We make the hypothesis that the Dirichlet boundary satisfies

$$\partial\Omega_{\mathrm{D}} \text{ is non empty.} \tag{1.65}$$

In the context of fluid flow through porous media, the solution u of (1.64) is the fluid pressure, $\partial\Omega_j$ represents the boundary of the jth well, and Q_j its injection or production rate. In two-dimensional problems as the one we are considering here, a is the *transmissivity*, that is, the product of the permeability of the porous medium with its thickness. It is not accessible to direct measurements, and so it is usually estimated indirectly from the available pressure measurements via the resolution of an inverse problem, which we describe now.

The natural *admissible parameter* set is

$$C \stackrel{\text{def}}{=} \{a : \Omega \leadsto I\!R \mid 0 < a_m \leq a(\xi) \leq a_M \text{ a.e. on } \Omega\}, \tag{1.66}$$

where a_m and a_M are given lower and upper bounds on a, Then for any $a \in C$, (1.64) admits a unique solution u in the *state-space*:

$$Y = \{w \in L^2(\Omega) \mid \frac{\partial w}{\partial \xi_i} \in L^2(\Omega)\,,\, i = 1, 2\ ,\ w|_{\partial\Omega_D} = 0\}. \tag{1.67}$$

Because of the hypothesis (1.65), we can equip Y with the norm

$$\|w\|_Y = |\nabla w|_{I\!L^2(\Omega)}, \tag{1.68}$$

where

$$I\!L^2(\Omega) \stackrel{\text{def}}{=} L^2(\Omega) \times L^2(\Omega),\ \|v\|_{I\!L^2(\Omega)} = \left(|v_1|^2_{L^2(\Omega)} + |v_2|^2_{L^2(\Omega)}\right)^{1/2}. \tag{1.69}$$

As in the one-dimensional parameter estimation problem of Sect. 1.4, we shall consider both H^1 and L^2 observations.

The H^1 observation is obtained by choosing as observation space:

$$M\ :\ w \in Y \leadsto \nabla w\ \in F. \tag{1.70}$$

Here M is an isometry from Y to F, which means that no information is lost in the measurement process. The NLS problem for the estimation of the diffusion coefficient $a \in C$ from a data $z \in I\!L^2(\Omega)$ is then

$$\hat{a}\quad \text{minimizes}\quad \frac{1}{2}\,|\nabla u_a - z|^2_{I\!L^2(\Omega)}\quad \text{over}\ \ C, \tag{1.71}$$

where

$$u_a \text{ is the solution of the elliptic equation (1.64)}\ . \tag{1.72}$$

The possibility of applying the sufficient condition for Q-wellposedness of Chap. 4 to the case of H^1 observation is studied in Sect. 4.9: despite the availability of this rich (and highly unrealistic) observation, we shall see that there is not much hope to satisfy these sufficient conditions when a is infinite dimensional. But when C is replaced by the smaller set made of uniformly Lipschitz functions defined in (4.106), still denoted by C, one can prove the identifiability of the linearized problem, but stability and Q-wellposedness are out of reach.

A first approach is then to use brute force, and reduce the search to finite dimensional subsets $\boldsymbol{C}$ to restore Q-wellposedness for (1.71): when the dimension n of $\boldsymbol{C}$ goes to infinity, the size of $\boldsymbol{C}$ has to go to zero, as well as the thickness of the neighborhood ϑ of the attainable set, and the stability constant of the $z \rightsquigarrow \hat{a}$ inverse mapping over ϑ blows up to infinity.

A second approach is to determine which information on the parameter cannot be retrieved from the data, and to supply this information by an **adapted regularization** term. This is shown in Sect. 5.4 to restore stability of a on C, and to provide Q-wellposedness on finite dimensional subsets $\boldsymbol{C}$ of C. The size constraint on $\boldsymbol{C}$ and the stability constant of the $z \rightsquigarrow \hat{a}$ inverse mapping over ϑ are now independent of the dimension of $\boldsymbol{C}$, the only effect of increasing this dimension being to decrease the thickness of the neighborhood ϑ of the attainable set.

The case of an L^2 observation is obtained by choosing as observation space

$$F = L^2(\Omega) \text{ equipped with the norm } \|v\|_F = \|v\|_{L^2(\Omega)}, \tag{1.73}$$

and as measurement operator

$$M \ : \ w \in Y \rightsquigarrow w \ \in F. \tag{1.74}$$

Here M is the canonical injection from H^1 into L^2, which means that we have no information on the derivatives of the solution u. The NLS problem for the estimation of a in C from a data $z \in L^2(\Omega)$ is then

$$\hat{a} \quad \text{minimizes} \quad \frac{1}{2}\,|u_a - z|^2_{L^2(\Omega)} \quad \text{over} \ \ C, \tag{1.75}$$

where u_a is still given by (1.72). This case is studied in Sect. 5.3 by combining the above mentioned results for an H^1 observation with the state-space regularization technique introduced in **State-space Regularization** of Sect. 1.3.4.

Chapter 2

Computing Derivatives

We address in this chapter a practical aspect of the numerical resolution of NLS problems, namely the computation of the gradient of the objective function or the Jacobian of the forward map, after discretization has occurred. This calculation has to be computed both accurately, so that the optimization algorithm has a chance to work properly, and efficiently, in order to keep computation time as low as possible.

When the problem at hand is infinite dimensional, we suppose that it has been reduced first to finite dimension by choosing the following:

- A discretization of the state equation $e(x, y) = 0$
- A discretization of the measurement operator M
- A discretization of the objective function J
- A finite dimensional parameterization of the unknown parameter x

We refer to Chap. 3 for the last point, but we do not discuss here the the other discretizations, and we simply suppose that they are made according to the state of the art, so that the resulting discrete objective function is a reasonable approximation to the continuous one. We also do not discuss the convergence of the parameter estimated with the discrete model to the one estimated using the infinite dimensional model, and we refer for this to the book by Banks, Kunisch, and Ito [7].

So our starting point is the finite dimensional NLS problem, which is to be solved on the computer: the unknown parameter x is a vector of $I\!R^n$, the

G. Chavent, *Nonlinear Least Squares for Inverse Problems: Theoretical Foundations and Step-by-Step Guide for Applications*, Scientific Computation,
DOI 10.1007/978-90-481-2785-6_2,

state y a vector of $I\!R^P$, and the output $v = \varphi(x)$ a vector of $I\!R^q$. We shall discuss different techniques, such as the sensitivity function and adjoint state approaches, for the computation of the derivatives required by local optimization methods, and give on various examples a step-by-step presentation of their implementation.

The derivatives computed in this way are called *discrete*: they are the *exact derivatives* of the *discrete objective function*. For infinite dimensional problems, they are obtained by following the **first discretize then differentiate** rule. When the discrete equations that describe φ are too complicated, one can be tempted to break the rule, and to first calculate the derivatives on the continuous model, which is usually simpler, at least formally, and then only to discretize the resulting equations and formula to obtain the derivatives. The derivatives computed in this way are called *discretized*, and are only *approximate derivatives* of the *discrete objective function*. This is discussed in Sect. 2.8 on an example.

The material of this chapter is not new, but it is seldom presented in a detailed way, and we believe it can be useful to practitioners.

2.1 Setting the Scene

The evaluation of the objective function $J(x)$ for a given parameter vector x requires the computation of the *forward map* $\varphi(x)$. In most applications, as we have seen in Chap. 1, this involves the resolution of one or more possibly nonlinear equations:

$$\begin{cases} \text{given } x \in C \text{ solve} \\ e(x, y) = 0 \\ \text{with respect to } y \text{ in } Y, \end{cases} \tag{2.1}$$

followed by

$$\begin{cases} \text{set } \varphi(x) = M(y), \\ \text{where } M \text{ is the observation operator.} \end{cases} \tag{2.2}$$

The subset C of $I\!R^n$ is the convex set of admissible parameters, the variable $y \in Y$ describes completely the state of the system under consideration, the state-space Y is an affine space of dimension p with tangent space $\delta Y = I\!R^p$, and the observation operator M describes which (usually small) part of the state variable can actually be measured.

The use of an affine state-space will allow to incorporate into Y some of the linear conditions that define y, as, for example, the Dirichlet boundary condition in a partial differential equations (Sect. 2.6), or the initial condition in an evolutionary problem (Sect. 2.9), and simplify the determination of the adjoint state equation.

In the most frequent case where Y is a vector space, one has simply $Y = \delta Y = I\!R^p$.

The above decomposition of the forward map φ as a state equation followed by an observation operator is not uniquely defined. For example, one can always enlarge the state-space to vectors of the form $(y, v) \in Y \times I\!R^q$ and consider that φ is obtained by the new state equation

$$\begin{cases} \text{given } x \in C \text{ solve} \\ e(x,y) = 0, \qquad v = M(y) \\ \text{with respect to } (y,v) \text{ in } Y \times I\!R^q, \end{cases} \tag{2.3}$$

followed by

$$\text{set } \varphi(x) = v. \tag{2.4}$$

Equation (2.4) corresponds to the observation operator $(y, v) \to v$, which is now simply a (linear) selection operator.

So we will always suppose in the sequel that the decomposition (2.1) and (2.2) has been chosen in such a way that it corresponds to some computational reality, with the state equation being the "hard part" of the model, where most of the computational effort rests, and the observation operator being the "soft part," given by simple and explicit formulas. We shall also suppose that the decomposition satisfies the minimum set of hypothesis (1.35).

Once the decomposition (2.1) and (2.2) of the forward map and the norm $\|\cdot\|_F$ on the data space have been chosen, one is finally faced with the numerical resolution of the *inverse problem*:

$$\hat{x} \quad \text{minimizes} \quad J(x) = \frac{1}{2}\|\varphi(x) - z\|_F^2 \quad \text{over } C. \tag{2.5}$$

Among the possible methods of solution for problem (2.5), optimization algorithms are in good place, although other possible approaches exist, as, for example, the resolution of the associated optimality condition (see, e.g., [14, 52, 53]). We can pick from two large classes of optimization algorithms:

- *Global algorithms [77]*, as simulated annealing or genetic algorithms, perform a (clever) sampling of the parameter space. They converge to the global minimum of J under quite general conditions, and are very user-friendly as the only input they require is a code that computes $J(x)$. The price is the large number of function evaluation required (easily over one hundred thousand iterations for as few as ten parameters), and so the use of these algorithms tends to be limited to problems where the product *size of* x by *computation time of* J is not too large.

- *Local algorithms [13, 68]* use additional information on the derivatives of J to move from the current estimate to the next according to the local shape of the graph of J. They converge only to the nearest *stationary point* (Definition 4.0.9), but need much less iterations, which makes them applicable to large size problems. They are also much less user-friendly, as they require the user to provide the derivatives of J or φ, which requires delicate calculations and coding.

As we have seen in Sect. 1.3.4, the art of regularization consists in replacing the original ill-posed inverse problem by a Q-wellposed regularized problem, which can be solved by local optimization techniques (we refer to Chap. 5 for examples of this process). Hence the final optimization problem (2.5) is likely to be Q-wellposed, in which case local algorithms are guaranteed to converge to the global minimum. This is why we shall concentrate on local algorithms, and try to make them a little more user-friendly by addressing the problem of the computation of the gradient ∇J or the Jacobian $D = \varphi'(x)$ they require as input.

To begin with, notice that the calculation of the gradient by finite difference

$$\frac{\partial J}{\partial x_j}(x) \simeq \frac{J(x + h\, e_j) - J(x)}{h}, \tag{2.6}$$

where e_j is the jth basis vector of $I\!R^n$, cumulates all disadvantages: it is computationally expensive (the number of evaluation of J required is proportional to the size n of the parameter vector x), and it is not precise (making it precise would require to adjust the size of the step h by trial and error, for each component x_j, until a compromise between truncation errors – h too large – and rounding errors – h too small – is found, but this would be still more computionally intensive, and is never done in practice).

We present now the two methods of choice for the computation of ∇J or D: the sensitivity functions and the adjoint approaches.

2.2 The Sensitivity Functions Approach

In this approach, the *Jacobian* or *sensitivity matrix* D of the forward map φ at x is computed *column by column*. The jth column s_j of D gives the sensitivity of the model output to the parameter x_j, and it is called the jth *output sensitivity function*. Derivation of the state equation (2.1) with respect to x_j gives

$$\frac{\partial e}{\partial x_j}(x,y) + \frac{\partial e}{\partial y}(x,y).\frac{\partial y}{\partial x_j} = 0, \qquad j = 1 \ldots n. \tag{2.7}$$

The vectors $\partial y/\partial x_j \in \delta Y = \mathbb{R}^p$ represent here the sensitivity of the state y to each parameter x_j, and they are called the *state sensitivity functions*. They are obtained by solving the n linearized state equation (2.7). They can then be combined with the derivative of the observation operator $M'(x)$ to compute the n *output sensitivity functions* s_j:

$$s_j = M'(y)\,\frac{\partial y}{\partial x_j}, \qquad j = 1 \ldots n. \tag{2.8}$$

With the Jacobian $D = [s_1\, s_2\, \ldots\, s_n]$ computed, the gradient of J is immediately given by:

$$\nabla J = D^{\mathrm{T}}(\varphi(x) - z). \tag{2.9}$$

This approach is the most widely used, as it involves only the natural task of derivating the chain of equations and formula that define the forward map φ. With the chosen "state-space" decomposition of φ, the computational cost resides in the resolution of the n linear systems (2.7). Hence the sensitivity function approach allows to compute both ∇J and $D = \varphi'(x)$ at an additional cost proportional to the number n of parameters.

2.3 The Adjoint Approach

The adjoint approach provides an efficient way to compute the *gradient with respect to* x *of* $G(x,\varphi(x))$:

– For any *scalar valued* differentiable function $G(x,v)$, which is an *explicit function* of its arguments x and v

– For any mapping $\varphi(x)$ given by a *state-space decomposition* of the form (2.1) and (2.2)

We shall use the shorthand notation ∇G for this gradient. Depending on the derivative one wants to compute, different choices are possible for $G(x, v)$:

- If one chooses

$$G(x, v) = \frac{1}{2}\|v - z\|_F^2 \quad \text{(independant of } v \text{ !)}, \tag{2.10}$$

  then

$$G(x, \varphi(x)) = J(x) \quad \forall x \in C, \tag{2.11}$$

  so that

$$\nabla G = \nabla J,$$

  and the adjoint approach computes the *gradient* of J.

- If one chooses

$$G(x, v) = \langle v, e_i \rangle_F, \tag{2.12}$$

  where e_i is the ith basis vector of $I\!R^q$, and $E = I\!R^n$ and $F = I\!R^q$ are equipped with usual Euclidian scalar products, then

$$G(x, \varphi(x)) = \langle \varphi(x), e_i \rangle_F \quad \forall x \in C,$$

  so that

$$\nabla G = \varphi'(x)^{\mathrm{T}} e_i = D^{\mathrm{T}} e_i = r_i^{\mathrm{T}},$$

  where r_i is the ith row of $D = \varphi'(x)$. In that case, the adjoint approach computes the *Jacobian* D of φ *row by row.*

- If, given a vector $g_v \in I\!R^q$, one chooses

$$G(x, v) = \langle v, g_v \rangle_F, \tag{2.13}$$

  where now $E = I\!R^n$ and $F = I\!R^q$ are equipped with scalar products $\langle\, ,\rangle_E$ and $\langle\, ,\rangle_F$, then similarly $G(x, \varphi(x)) = \langle \varphi(x), g_v \rangle_F$, and

$$\nabla G = g_x \in I\!R^n, \text{ where } g_x = \varphi'(x)^{\mathrm{T}} g_v = D^{\mathrm{T}}, \tag{2.14}$$

  where gradient and transposition are relative to the chosen scalar products on E and F. Hence the adjoint approach will compute the result g_x of the action of the transposed Jacobian D^{T} on any vector g_v without having to assemble the whole matrix D, transpose it, and perform the matrix $\times$ vector product $D^{\mathrm{T}} g_v$.

Remark 2.3.1 *Because of formula (2.9), the choice (2.13) for G with $g_v = \varphi(x) - z$ leads to the computation of ∇J, as did the choice (2.10). But both choices will produce the same final formula for ∇J.* ■

Remark 2.3.2 *The adjoint approach with the choice (2.13) for G is used when it comes to change parameters in an optimization problem: the gradient of J with respect to the optimization parameter vector x_{opt} is given by*

$$\nabla_{x_{\mathrm{opt}}} J = \psi'(x_{\mathrm{opt}})^T \nabla_{x_{\mathrm{sim}}} J, \tag{2.15}$$

where x_{sim} is the simulation parameter vector, and ψ is the $x_{\mathrm{opt}} \rightsquigarrow x_{\mathrm{sim}}$ mapping (see Sect. 3.3). An example of such a calculation is given in Sect. 3.8.2. ■

We explain now how the adjoint approach computes ∇G once a state-space decomposition (1.33) and (1.34) of φ has been chosen: knowing that G is obtained by (1) solving for y the (possibly nonlinear) *direct* equation $e(x, y) = 0$ and (2) evaluating the *explicit* real-valued function $G(x, M(y))$, we want to show that the vector ∇G can always be obtained by (1) solving for λ a (linear) *adjoint* equation and (2) evaluating ∇G from x, y, and λ by simple *explicit* gradient formulas.

We remark for that purpose that minimizing $G(x, M(y_x))$ with respect to x amounts to minimize $G(x, M(y))$ with respect to (x, y) under the constraint that $e(x, y) = 0$. This suggest to consider the Lagrangian function associated to this constrained optimization problem:

$$\mathcal{L}(x, y, \lambda) = G(x, M(y)) + \langle e(x, y), \lambda \rangle_Z, \tag{2.16}$$

where the Lagrange multiplier $\lambda \in Z = I\!R^p$ is called the adjoint variable of y.

The interest of this Lagrangian function is that it provides a convenient way to calculate ∇G; we state the corresponding theorem in an abstract framework, with $E, \delta Y, Z, F$ instead of $I\!R^n, I\!R^p, I\!R^p, I\!R^q$, as this will occasionally allow us to use it for infinite dimensional problems:

Proposition 2.3.3 *Let (1.33) and (1.34) be a state-space decomposition for φ satisfying (1.12) and (1.35), let $(x, v) \in E \times F \rightsquigarrow G(x, v) \in I\!R$ be a given real-valued differentiable function, and $\mathcal{L}(x, y, \lambda)$ be the associated Lagrangian defined by (2.16).*

Then $x \in C \rightsquigarrow G(x, \varphi(x)) \in I\!R$ *is differentiable, and its gradient* ∇G *is given by the* gradient equation*:*

$$\langle \nabla G, \delta x \rangle_E = \frac{\partial \mathcal{L}}{\partial x}(x, y, \lambda)\delta x \quad \forall \delta x \in E, \tag{2.17}$$

where

* $y \in Y$ *is the solution of the* (direct) state equation

$$e(x, y) = 0, \tag{2.18}$$

* $\lambda \in F$ *is the solution of the* adjoint state equation

$$\frac{\partial \mathcal{L}}{\partial y}(x, y, \lambda)\delta y = 0 \quad \forall \delta y \in \delta Y. \tag{2.19}$$

In this context, y *is called the* (direct) state, *and* λ *the* adjoint state.

Proof. Because of the third hypothesis in (1.35), the solution y_x of the state equation (1.33) is uniquely defined for any $x \in C$. For this value $y = y_x$ of the state, one has $M(y_x) = \varphi(x)$ and $e(x, y_x) = 0$, so that the definition (2.16) of the Lagrangian reduces to

$$\mathcal{L}(x, y_x, \lambda) = G(x, \varphi(x)) \quad \forall x \in C, \ \forall \lambda \in Z. \tag{2.20}$$

But G and M are by hypothesis differentiable functions, and the implicit function theorem implies, using hypothesis (1.35), the differentiability of $x \rightsquigarrow y_x$. This proves the differentiability of $x \in C \rightsquigarrow G(x, M(y_x)) = G(x, \varphi(x)) \in I\!R$. So we can differentiate (2.20) with respect to x for a fixed λ:

$$\frac{\partial \mathcal{L}}{\partial x}(x, y_x, \lambda)\, \delta x + \frac{\partial \mathcal{L}}{\partial y}(x, y_x, \lambda)\, \delta y_x = \langle \nabla G, \delta x \rangle_E,$$

which reduces to the gradient equation (2.17) as soon as λ is a solution of the adjoint equation (2.19). So the theorem will be proved if we check that the adjoint equation (2.19) defines uniquely λ. Using the Definition (2.16) of the Lagrangian, the adjoint equation becomes ($v = M(y)$):

$$\begin{aligned}
\langle \nabla_v G(x, v), \delta v \rangle_F + & \left\langle \frac{\partial e}{\partial y}(x, y)\delta y, \lambda \right\rangle_Z && = 0, \\
\langle \nabla_v G(x, v), M'(y)\delta y \rangle_F + & \left\langle \frac{\partial e}{\partial y}(x, y)\delta y, \lambda \right\rangle_Z && = 0, \\
\langle M'(y)^T \nabla_v G(x, v), \delta y \rangle_Y + & \left\langle \delta y, \frac{\partial e}{\partial y}(x, y)^T \lambda \right\rangle_Y && = 0, \\
M'(y)^T \nabla_v G(x, v) \quad + & \quad \frac{\partial e}{\partial y}(x, y)^T \lambda && = 0.
\end{aligned} \tag{2.21}$$

Hypothesis (1.35) ensures that $\partial e/\partial y(x,y)$ is an isomorphism from δY to Z, so that λ is uniquely defined by (2.21). ■

We return now to the finite dimensional case where $E = I\!R^n$, $\delta Y = Z = I\!R^p$, and $F = I\!R^q$. The computational cost in the adjoint approach is in the resolution of the adjoint equation, the gradient being then obtained by simple explicit formulas. Hence we see that the adjoint approach allows to compute ∇J (or any row r_i of $D = \varphi'(x)$) at the sole additional cost of the resolution of the adjoint equation, independently of the number n of parameters. This makes the resolution of problems with a very large number of parameters possible by using gradient-based optimization algorithms. As for the computation of the Jacobian $D = \varphi'(x)$, it is made in the adjoint approach row-by-row, and so its cost is proportional to the number q of observations, but *independent of the number n of parameters.* This feature is extremely useful when it comes to determine the number of independant parameters that can be retrieved for a given combination of model and data (Sect. 3.2).

The above costs have been estimated under the assumption that storing the direct state y is possible, and that the corresponding cost is negligible. This is not always true, especially for large-size time-dependent problems, in which case the adjoint approach may require to compromise between storing and recomputing the direct state (see Grievank [40] for an optimal strategy).

Remark 2.3.4 *We have made explicit in (2.21) the variational formulation (2.19) for the adjoint equation. One could as well explicit the variational formula (2.17) for ∇G using the Definition (2.16) of the Lagrangian:*

$$\nabla G = \nabla_x G(x, M(y)) + \frac{\partial e}{\partial x}(x,y)^T \lambda. \tag{2.22}$$

It is, however, not advisable to use formulas (2.21) and (2.22) in practice, as they require to write down the matrices $\frac{\partial e}{\partial y}(x,y)$ and $\frac{\partial e}{\partial x}(x,y)$, which can be very large. Moreover, matrix transposition requires some thinking when it is made with respect to weighted scalar products.

Despite their abstract appearance, the variational formulations (2.17) (2.19) of the adjoint and gradient equations, which are based on the explicit Lagrangian function (2.16), are the most convenient to use in the applications. ■

2.4 Implementation of the Adjoint Approach

We give in this section a step-by-step presentation of the calculations required for the determination of the adjoint equations and the gradient formula prior to coding. These calculations are delicate, tedious, and error prone, and lot of effort is currently developing in the automatic differentiation community to develop the applicability of the "reverse mode" of automatic differentiation codes, which generates automatically the code for the adjoint equation when given the code for the forward map [41, 40]. A less ambitious approach toward automatization is that of [32], where the state equation (not the code!) is the input to a formal calculus code, whose output is made of the formula for the adjoint and gradient equations, which have then to be coded. As of today, these automated approaches work only on relatively simple problems, and most of the adjoint equations have to be established by hand. In any case, it is recommended to check that the gradient computed by the adjoint approach coincides, up to a large number of digits, with the one computed carefully by finite difference, or by automatic differentiation on a small size problem [61]. Here is one possible organization of the adjoint calculations:

Step 0: Forward Map and Objective Function

Identify

- The map $\varphi : x \in (I\!R^n, \langle\ ,\ \rangle_E) \rightsquigarrow v \in (I\!R^q, \langle\ ,\ \rangle_F)$ to be inverted
- The objective function $G(x, v)$ whose gradient is to be computed

The choice of G will depend on the derivative one wants to compute, see (2.10), (2.12), and (2.13) for examples. It is also necessary to specify the scalar products on $I\!R^n$ and $I\!R^q$ to make gradient and transposition well defined:

- The parameter space $E = I\!R^n$ is equipped with the Euclidean scalar product. If needed, the gradient with respect to calibrated parameters (Sect. 3.1.1 in next chapter) follows immediately by the chain rule.
- The scalar product $\langle\ ,\ \rangle_F$ on the data space $F = I\!R^q$ determines the norm $\|\cdot\|_F$ used to compare the model output $\varphi(x)$ to the data z, and to build up the data misfit objective function J. This norm can take into account the uncertainty on each component of the observation,

as in formula (1.8) in the Knott–Zoeppritz example of Sect. 1.1. For infinite dimensional problems where J involves an integral over space and/or time, as in (1.61), (1.71), and (1.75), the norm $\|\cdot\|_F$ is usually chosen so that the discrete objective function is an approximation of the continuous one (see also Sect. 3.1.2 in next chapter).

Step 1: State-Space Decomposition

Dispatch the equations defining φ between an affine state-space Y of dimension p, a set of p state equation $e(x,y) = 0$, and an observation operator $M(y)$. One may have to choose among different equivalent formulations of the state equation: for example, $y-1/x = 0$ and $xy-1 = 0$ are two equivalent state equations, and one has to decide which one to call $e(x,y) = 0$ and enter in Definition 2.25 of the Lagrangian. To get the simplest calculations, and usually the most efficient code, a good rule is to choose the formulation that is the easiest to differentiate (the second one in our example). As a result, the $\varphi : x \rightsquigarrow v$ mapping is given by

$$x \in I\!R^n \rightsquigarrow y \in Y \text{ solution of } e(x,y) = 0, \tag{2.23}$$

$$y \in Y \rightsquigarrow v = M(y) \in I\!R^q, \tag{2.24}$$

where M is explicit and the computational effort is in the resolution of $e(x,y) = 0$.

It is important to check that there are as many equations as unknowns: the number of equations $e(x,y) = 0$ should be equal to the dimension p of the state-space Y, that is, to the dimension of its tangent vector space $\delta Y = I\!R^p$.

Step 2: Lagrangian

Combine the objective function $G(x,v)$ chosen in step 0 and the decomposition $e(x,y) = 0$, $v = M(y)$ of φ chosen in step 2 to build up the Lagrangian:

$$\mathcal{L}(x,y,\lambda) = G(x, M(y)) + \langle e(x,y), \lambda \rangle_Z, \tag{2.25}$$

for any $x \in I\!R^n$, $y \in Y$, and $\lambda \in I\!R^P$.

The only thing left to choose in the above formula is the scalar product $\langle\, , \rangle_Z$ on the right-hand side space of the state equation. When the equation $e(x,y) = 0$ comes from the discretization of a continuous problem, this degree of freedom can be used to make the $\langle e(x,y),\lambda\rangle_Z$ term in (2.25) to mimic the corresponding term of the continuous Lagrangian. This will allow to interpret the vector λ as a discretization of the continuous adjoint variable.

A useful check is the following:

- Write down explicitly what the parameter vector x, the state vector y, the state-space Y, and its tangent space δY are in the problem under consideration
- Make sure that the Lagrangian $\mathcal{L}$ is an explicit function of its arguments: all quantities other than x, y, and λ appearing in the right-hand side of (2.25) should be known.

From this point on, there is no more decision to make all calculations follow from the formula (2.25) for the Lagrangian and the scalar product $\langle\,,\rangle_E$ on the parameter space chosen in step 0.

Step 3: Adjoint Equation

Differentiate the Lagrangian (2.25) at point x, y_x, λ, with respect to the state $y \in Y$ (so that $\delta y \in \delta Y = I\!R^p$) in order to obtain the "variational form" (2.19) of the adjoint equation, and reorganize the result by factorizing δy_j for all $j = 1 \dots p$:

$$\frac{\partial \mathcal{L}}{\partial y}(x, y_x, \lambda)\,\delta y = \sum_{j=1}^{p} h_j(x, y_x, \lambda)\,\delta y_j \quad \forall \delta y_j \in I\!R, \quad j = 1 \dots p. \tag{2.26}$$

This reorganization is the most delicate and tedious part. Once it is done, the "computational form" of the adjoint equations for $\lambda \in I\!R^p$ are obtained by equating to zero the coefficients of δy_j in (2.26):

$$h_j(x, y_x, \lambda) = 0 \quad \forall j = 1 \dots p. \tag{2.27}$$

We call λ_x the solution of the above adjoint equation.

Step 4: Gradient Equation

Differentiate the Lagrangian (2.25), at point x, y_x, λ_x, with respect to its first argument $x \in I\!R^n$, and reorganize the result by factorizing δx in the parameter scalar product $\langle\ ,\ \rangle_E$ in order to obtain the "variational form" (2.17) of the gradient equation, which we recall here:

$$\langle \nabla G, \delta x \rangle_E = \frac{\partial \mathcal{L}}{\partial x}(x, y_x, \lambda_x)\, \delta x \quad \forall \delta x \in I\!R^n. \tag{2.28}$$

When $I\!R^n$ is equipped with the usual Euclidean scalar product, the ith component of ∇G is simply the coefficient of δx_i in $\frac{\partial \mathcal{L}}{\partial x}(x, y_x, \lambda_x)\, \delta x$.

2.5 Example 1: The Adjoint Knott–Zoeppritz Equations

We compute here the derivatives of the forward map $\varphi : x \rightsquigarrow v$ described in Sect. 1.1: the parameter vector $x \in I\!R^4$ is made of the dimensionless contrast and background coefficients (e_ρ, e_P, e_S, χ) across the elastic interface, and the output vector $v \in I\!R^q$ is made of the q reflection coefficients $R_1, \ldots, R_q$ computed by the sequence of formula (1.2) for q different incidence angles $\theta_1, \ldots, \theta_q$.

We illustrate on this simple example the step-by-step adjoint approach of Sect. 2.3.

Step 0: Forward Map and Objective Function

Because of the independence of the calculations performed for each incidence angle, we only need to compute the derivative of the *forward map* $\psi : x = (e_\rho, e_P, e_S, \chi) \in I\!R^4 \rightsquigarrow v = R \in I\!R$ *for one given incidence angle* θ.

Hence the forward map here is $\psi : x \rightsquigarrow R$, and the objective function is $G(x, R) = R$.

We equip both parameter space $I\!R^4$ and output space $I\!R$ with the usual Euclidean scalar products, so that the Jacobian ψ' of ψ and its gradient $\nabla\psi$ are transposed matrices for the chosen scalar products.

Step 1: State-Space Decomposition

We use here the decomposition suggested in Sect. 1.1:

$$y = (e, f, S_1, S_2, \ \ldots\ , P, Q, R) \in I\!R^{19} \quad \text{(state vector)}, \tag{2.29}$$
$$M = [0 \ldots 0\, 1] \quad \text{(observation operator)}, \tag{2.30}$$

which has to be complemented by the state equation $e(x, y) = 0$. This requires to rewrite the formula (1.2) in the form of a sequence of equations. If we want to avoid the need to differentiate square roots or quotients, we can choose for $e(x, y) = 0$:

$$\left\{\begin{array}{ll} e - e_S - e_\rho = 0 & \\ f - 1 + e_\rho^2 = 0 & \\ S_1 - \chi(1 + e_P) = 0 & \\ S_2 - \chi(1 - e_P) = 0 & \\ (1 - e_S)T_1 - 2 = 0 & \\ (1 + e_S)T_2 - 2 = 0 & \\ q^2 - S_1 \sin^2\theta = 0 & \\ M_1^2 + q^2 - S_1 = 0 & (M_1 \geq 0) \\ M_2^2 + q^2 - S_2 = 0 & (M_2 \geq 0) \\ N_1^2 + q^2 - T_1 = 0 & (N_1 \geq 0) \\ N_2^2 + q^2 - T_2 = 0 & (N_2 \geq 0) \\ D - eq^2 = 0 & \\ A - e_\rho + D = 0 & \\ K - D + A = 0 & \\ B - 1 + K = 0 & \\ C - 1 - K = 0 & \\ P - M_1(B^2 N_1 + f N_2) - 4eDM_1M_2N_1N_2 = 0 & \\ Q - M_2(C^2 N_2 + f N_1) - 4q^2A^2 = 0 & \\ (P + Q)R - (P - Q) = 0. & \end{array}\right. \tag{2.31}$$

When the chosen equations have more than one solution, the condition in parenthesis indicate which one is to be chosen. As expected, there are 19 equations for 19 state unknowns.

Step 2: Lagrangian

With the objective function G of step 0 and the state-space decomposition of step 1, the Lagrangian reads

$$\begin{aligned} \mathcal{L}(x, y, \lambda) \quad &= \\ R + \lambda_1 \quad & (e - e_S - e_\rho) \\ +\lambda_2 \quad & (f - 1 + e_\rho^2) \\ +\lambda_3 \quad & (S_1 - \chi(1 + e_P)) \end{aligned} \tag{2.32}$$

$$
\begin{aligned}
&+\lambda_4 \quad (S_2 - \chi(1-e_{\mathrm{P}}))\\
&+\lambda_5 \quad ((1-e_{\mathrm{S}})T_1 - 2)\\
&+\lambda_6 \quad ((1+e_{\mathrm{S}})T_2 - 2)\\
&+\lambda_7 \quad (q^2 - S_1 \sin^2\theta)\\
&+\lambda_8 \quad (M_1^2 + q^2 - S_1)\\
&+\lambda_9 \quad (M_2^2 + q^2 - S_2)\\
&+\lambda_{10} \quad (N_1^2 + q^2 - T_1)\\
&+\lambda_{11} \quad (N_2^2 + q^2 - T_2)\\
&+\lambda_{12} \quad (D - eq^2)\\
&+\lambda_{13} \quad (A - e_\rho + D)\\
&+\lambda_{14} \quad (K - D + A)\\
&+\lambda_{15} \quad (B - 1 + K)\\
&+\lambda_{16} \quad (C - 1 - K)\\
&+\lambda_{17} \quad (P - M_1(B^2N_1 + fN_2) - 4eDM_1M_2N_1N_2)\\
&+\lambda_{18} \quad (Q - M_2(C^2N_2 + fN_1) - 4q^2A^2)\\
&+\lambda_{19} \quad ((P+Q)R - (P-Q)).
\end{aligned}
$$

At this point, it is useful to summarize the arguments of the Lagrangian and the spaces where they belong:

$$x = (e_\rho, e_{\mathrm{P}}, e_{\mathrm{S}}, \chi) \in I\!R^4 \qquad \text{(parameter)}, \tag{2.33}$$

$$y = (e, f, S_1, S_2, \ \ldots\ , P, Q, R) \in I\!R^{19} \qquad \text{(direct state)}, \tag{2.34}$$

$$\lambda = (\lambda_1, \ldots, \lambda_{19}) \in I\!R^{19} \qquad \text{(adjoint state)}, \tag{2.35}$$

and to check the following:

– The direct and adjoint state vectors have the same dimension

– The vectors x, y, and λ contain all the information necessary to compute the Lagrangian $\mathcal{L}$ as an explicit function (here the only quantity that appears in the right-hand side of (2.32) and is not in the arguments of $\mathcal{L}$ is the incidence angle θ, which is known).

Step 3: Adjoint Equation

The variational form (2.19) of the adjoint equation is obtained by differentiating the 20 terms of the Lagrangian (2.32) with respect to the state variables only, and equating the result to zero:

$$
\begin{aligned}
\frac{\partial \mathcal{L}}{\partial y}(x,y,\lambda)\,\delta y \;&= \qquad\qquad\qquad\qquad\qquad\qquad (2.36)\\
\delta R + \lambda_1 \;\; & \delta e\\
+ \lambda_2 \;\; & \delta f\\
+ \lambda_3 \;\; & \delta S_1\\
+ \lambda_4 \;\; & \delta S_2\\
+ \lambda_5 \;\; & (1 - e_{\mathrm{S}})\delta T_1\\
+ \lambda_6 \;\; & (1 + e_{\mathrm{S}})\delta T_2\\
+ \lambda_7 \;\; & (\delta(q^2) - \delta S_1 \sin^2\theta)\\
+ \lambda_8 \;\; & (2M_1\,\delta M_1 + \delta(q^2) - \delta S_1)\\
+ \lambda_9 \;\; & (2M_2\,\delta M_2 + \delta(q^2) - \delta S_2)\\
+ \lambda_{10} \; & (2N_1\,\delta N_1 + \delta(q^2) - \delta T_1)\\
+ \lambda_{11} \; & (2N_2\,\delta N_2 + \delta(q^2) - \delta T_2)\\
+ \lambda_{12} \; & (\delta D - \delta e\, q^2 - e\,\delta(q^2))\\
+ \lambda_{13} \; & (\delta A + \delta D)\\
+ \lambda_{14} \; & (\delta K - \delta D + \delta A)\\
+ \lambda_{15} \; & (\delta B + \delta K)\\
+ \lambda_{16} \; & (\delta C - \delta K)\\
+ \lambda_{17} \; & (\delta P - \delta M_1(B^2 N_1 + f N_2)\\
& \quad - M_1(2B\,\delta B\, N_1 + B^2\,\delta N_1 + \delta f\, N_2 + f\,\delta N_2)\\
& \quad - 4\delta e\, D M_1 M_2 N_1 N_2 - 4e\delta D\, M_1 M_2 N_1 N_2\\
& \quad - 4eD\delta M_1\, M_2 N_1 N_2 - 4eDM_1 \delta M_2\, N_1 N_2\\
& \quad - 4eDM_1 M_2 \delta N_1\, N_2 - 4eDM_1 M_2 N_1 \delta N_2)\\
+ \lambda_{18} \; & (\delta Q - \delta M_2\,(C^2 N_2 + f N_1)\\
& \quad - M_2(2C\,\delta C\, N_2 + C^2\,\delta N_2 + \delta f\, N_1 + f\,\delta N_1)\\
& \quad - 4\delta(q^2)\, A^2 - 8q^2 A\,\delta A)\\
+ \lambda_{19} \; & ((\delta P + \delta Q)R + (P+Q)\delta R - \delta P + \delta Q) \quad = 0\\
\forall \delta y = (\delta e, \delta f, \delta S_1, \delta S_2,\; & \ldots\; , \delta P, \delta Q, \delta R) \in I\!R^{19}.
\end{aligned}
$$

The computational form of the adjoint equations is obtained by equating to zero the coefficient of $\delta e, \delta f, \delta S_1, \delta S_2, \ \dots \ , \delta P, \delta Q, \delta R$ in (2.36):

$$\left\{\begin{array}{lll}
0 & = & \lambda_1 - q^2\,\lambda_{12} - 4DM_1M_2N_1N_2\,\lambda_{17} \\
0 & = & \lambda_2 - M_1N_2\,\lambda_{17} - M_2N_1\,\lambda_{18} \\
0 & = & \lambda_3 - \lambda_8 \\
0 & = & \lambda_4 - \lambda_9 \\
0 & = & (1-e_{\rm S})\,\lambda_5 - \lambda_{10} \\
0 & = & (1+e_{\rm S})\,\lambda_6 - \lambda_{11} \\
0 & = & \lambda_7 + \lambda_8 + \lambda_9 + \lambda_{10} + \lambda_{11} - e\,\lambda_{12} - 4A^2\,\lambda_{18} \\
0 & = & 2M_1\,\lambda_8 - 4eDM_2N_1N_2\,\lambda_{17} - (B^2N_1 + fN_2)\,\lambda_{17} \\
0 & = & 2M_2\,\lambda_9 - 4eDM_1N_1N_2\,\lambda_{17} - (C^2N_2 + fN_1)\,\lambda_{18} \\
0 & = & 2N_1\,\lambda_{10} - M_1B^2\,\lambda_{17} - 4eDM_1M_2N_2\,\lambda_{17} - M_2f\,\lambda_{18} \\
0 & = & 2N_2\,\lambda_{11} - M_1f\,\lambda_{17} - 4eDM_1M_2N_1\,\lambda_{17} - M_2C^2\,\lambda_{18} \\
0 & = & \lambda_{12} + \lambda_{13} - \lambda_{14} - 4eM_1M_2N_1N_2\,\lambda_{17} \\
0 & = & \lambda_{13} - \lambda_{14} - 8q^2A\,\lambda_{18} \\
0 & = & \lambda_{14} + \lambda_{15} - \lambda_{16} \\
0 & = & \lambda_{15} - 2M_1BN_1\,\lambda_{17} \\
0 & = & \lambda_{16} - 2M_2CN_2\,\lambda_{18} \\
0 & = & \lambda_{17} - (1-R)\,\lambda_{19} \\
0 & = & \lambda_{18} + (1+R)\,\lambda_{19} \\
0 & = & (P+Q)\,\lambda_{19} + 1.
\end{array}\right. \tag{2.37}$$

Equations in (2.37) are the *adjoint Knott–Zoeppritz equations.* They are solved easily backwards, computing first λ_{19}, then λ_{18}, etc....

Step 4: Gradient Equation

Differentiation of the Lagrangian (2.32) with respect to its first argument $x = (e_\rho, e_P, e_S, \chi)$ gives

$$\begin{array}{rl}
\dfrac{\partial \mathcal{L}}{\partial x}(x,y,\lambda) & = \\
-\lambda_1 & (\delta e_{\rm S} + \delta e_\rho) \\
+\lambda_2 & 2e_\rho\,\delta e_\rho \\
-\lambda_3 & (\delta\chi\,(1+e_{\rm P}) + \chi\,\delta e_{\rm P})
\end{array} \tag{2.38}$$

$$
\begin{array}{ll}
-\lambda_4 & (\delta\chi\,(1-e_{\rm P}) - \chi\,\delta e_{\rm P}) \\
-\lambda_5 & \delta e_{\rm S}\,T_1 \\
+\lambda_6 & \delta e_{\rm S}\,T_2 \\
-\lambda_{13} & \delta e_\rho .
\end{array}
$$

The gradient equation (2.28) gives then the vector $\nabla G = \nabla R$:

$$
\frac{\partial R}{\partial e_\rho}\delta e_\rho + \frac{\partial R}{\partial e_{\rm P}}\delta e_{\rm P} + \frac{\partial R}{\partial e_{\rm S}}\delta e_{\rm S} + \frac{\partial R}{\partial \chi}\delta\chi = \frac{\partial \mathcal{L}}{\partial x}(x, y_x, \lambda_x)\,\delta x \quad \forall \delta x \in I\!R^n. \tag{2.39}
$$

Comparing (2.38) and (2.39), we see that $\dfrac{\partial R}{\partial e_\rho}$ is the coefficient of δe_ρ in (2.38), etc..., which gives

$$
\frac{\partial R}{\partial e_\rho} = -\lambda_1 + 2\lambda_2 e_\rho - \lambda_{13} \tag{2.40}
$$

$$
\frac{\partial R}{\partial e_{\rm P}} = \chi(\lambda_4 - \lambda_3) \tag{2.41}
$$

$$
\frac{\partial R}{\partial e_{\rm S}} = -\lambda_1 - \lambda_5\,T_1 + \lambda_6\,T_2 \tag{2.42}
$$

$$
\frac{\partial R}{\partial \chi} = -\lambda_3(1+e_{\rm P}) - \lambda_4(1-e_{\rm P}) \tag{2.43}
$$

Remark 2.5.1 *We are here in the favorable situation where the forward map* φ *is the juxtaposition of* q *independent "component" maps* $\psi_1, \dots, \psi_q$*. So if we value to* 1 *the computational cost of* φ*, the cost of each* $\psi_i, i = 1, \dots, q$*, is* $1/q$*. Then each row* r_i *of* $D = \varphi'(x)$ *can be computed as earlier as* $\nabla\psi_i^T$ *at the cost of* $1/q$ *(one adjoint equation for* ψ_i*), so that the whole Jacobian* D *can be evaluated at an additional cost of* 1*, that is, at the same cost as the gradient* ∇J *in the general case.* ■

2.6 Examples 3 and 4: Discrete Adjoint Equations

The examples 3 and 4 of Sects. 1.5 and 1.6 are similar from the point of view of differentiation. So we consider here one problem that contains them both:

$$\begin{cases} -\nabla\cdot(a\nabla u) + k(u) \ = \ f \quad \text{in } \ \Omega, \\ u \ = \ u_e \quad \text{on a part } \ \partial\Omega_{\mathrm{D}} \quad \text{of } \ \partial\Omega, \\ a\,\dfrac{\partial u}{\partial \nu} \ = \ g \quad \text{on } \ \partial\Omega_{\mathrm{N}} \ = \ \partial\Omega \setminus \partial\Omega_{\mathrm{D}}, \end{cases} \tag{2.44}$$

where Ω is a domain of $I\!R^2$ with boundary $\partial\Omega$, $\partial\Omega_{\mathrm{D}}$, and $\partial\Omega_{\mathrm{N}}$ form a partition of $\partial\Omega$, and ν is the outer normal to Ω. We suppose that k, f, g, and Ω satisfy hypothesis (1.55)–(1.57), that u_e is in $H^{1/2}(\partial\Omega_{\mathrm{D}})$, and that a belongs to the admissible parameter set C defined in (1.66). Under these hypotheses, the state equation (2.44) (as well as its variational formulation (2.47) below) admits a unique solution $u \in H^1(\Omega)$, and so we can consider the following observations:

$$\begin{cases} Z \ \in \ I\!L^2(\Omega) & \text{measure of} \quad \nabla u \quad \text{in } \ \Omega, \\ z \ \in \ L^2(\Omega) & \text{measure of} \quad u \quad \text{in } \ \Omega, \\ z_N \ \in \ L^2(\partial\Omega_{\mathrm{N}}) & \text{measure of} \quad u \quad \text{on } \partial\Omega_{\mathrm{N}}, \end{cases} \tag{2.45}$$

which correspond to the observation operator:

$$M : u \in H^1(\Omega) \rightsquigarrow (\nabla u, u, u_{|\partial\Omega_{\mathrm{N}}}) \in I\!L^2(\Omega) \times L^2(\Omega) \times L^2(\partial\Omega_{\mathrm{N}})$$

and the objective function

$$J(a,k,f,g,u_e) = \int_\Omega \|Z - \nabla u\|^2 + \int_\Omega |z-u|^2 + \int_{\partial\Omega_N} |z_N - u|_{\partial\Omega_N}|^2. \tag{2.46}$$

Though it is probably not realistic from a practical point of view to estimate simultaneously a, k, f, g, and u_e by minimization of J, the above formulation is convenient as it will allow for the simultaneous determination of the partial derivatives of J with respect to its five arguments by the adjoint approach.

To compute a *discrete gradient*, as explained in the introduction of this chapter, we follow the rule *first discretize, then differentiate*: we first reduce the problem to finite dimension, and only then apply the adjoint state approach to compute the gradient.

2.6.1 Discretization Step 1: Choice of a Discretized Forward Map

We choose for this example a standard finite element approximations of the state equation (2.44), which we describe now. It is based on the variational formulation of (2.44):

$$\begin{cases} \text{find } u \in H^1(\Omega) \text{ with } u_{|\partial\Omega_D} = u_e \text{ such that} \\ \int_\Omega a\nabla u \nabla w + \int_\Omega k(u)w = \int_\Omega fw + \int_{\partial\Omega_N} gw \\ \text{for all } w \in H^1(\Omega) \text{ with } w_{|\partial\Omega_D} = 0, \end{cases} \tag{2.47}$$

and consists in choosing a finite dimensional subspace W_h of $H^1(\Omega)$, where the approximated solution u_h and the test function w_h will reside, quadrature formula to approximate the integrals which appear in (2.47), and finite dimensional approximations of the parameters a, k, f, g, and u_e.

1. *The space* W_h. We suppose that Ω is a polyhedron of $I\!R^2$, and we cover Ω by a triangulation $\mathcal{T}_h$, that is, a family of nondegenerated triangles K called *elements*, whose union is Ω, and such that two distinct elements of $\mathcal{T}_h$ are either disjoints, or share only an edge or a vertex. $\mathcal{T}_h$ is called the *simulation mesh*. The index h refers to the size of the elements of $\mathcal{T}_h$:

$$h = \max_{K \in \mathcal{T}_h} \text{diam}(K),$$

but it will more generally be used to indicate the discretized quantity associated to a continuous one. The triangulation $\mathcal{T}_h$ is chosen *adapted* to the boundary conditions, that is, such that $\partial\Omega_D$ and $\partial\Omega_N$ are made of edges of elements of $\mathcal{T}_h$. We denote by Ω_h the set of nodes of $\mathcal{T}_h$, that is, the collection of all vertices M of the elements K.

We choose for W_h the lowest order Lagrange finite element space:

$$W_h = \{ w_h \in \mathcal{C}(\overline{\Omega}) \mid w|_K \in \mathcal{P}_1 \}, \tag{2.48}$$

where $\mathcal{C}(\overline{\Omega})$ is the space of continuous functions over the closure $\overline{\Omega}$ of Ω, and $\mathcal{P}_1$ is the space of polynomials of degree one in two variables. The degrees of freedom of W_h are the values u_M of u_h at the nodes Ω_h of $\mathcal{T}_h$:

$$u_h = (u_M \in I\!R,\ M \in \Omega_h). \tag{2.49}$$

They are linked to the function u_h by

$$u_h(x) = \sum_{M \in \Omega_h} u_M\, w_M(x) \quad \forall x \in \Omega, \tag{2.50}$$

where the basis functions w_M are defined by

$$w_M \in W_h, \; w_M(P) = \begin{cases} 0 & \text{if } P \neq M \\ 1 & \text{if } P = M \end{cases} \quad \forall M, P \in \Omega_h. \tag{2.51}$$

The dimension of W_h is the number N_h of nodes in Ω_h. So we have to find N_h (nonlinear) equations to determine u_h.

2. *The quadrature formula.* We use the "trapezoidal rule" to approximate the integral of a regular function over an element K or one edge A of $\mathcal{T}_h$. If we denote by $a_i, i = 1, 2, 3$, the vertices of K and by $a_i, i = 1, 2$, the endpoints of an edge A, we obtain, for any function ψ continuous on K,

$$\begin{cases} \int_K \psi \approx I_K(\psi) = \frac{|K|}{3} \sum_{i=1}^{3} \psi(a_i), \\ \int_A \psi \approx I_A(\psi) = \frac{|A|}{2} \sum_{i=1}^{2} \psi(a_i). \end{cases} \tag{2.52}$$

The integrals over Ω and $\partial\Omega_{\mathrm{N}}$ are then approximated by

$$\begin{cases} \int_\Omega \psi \approx I_\Omega(\psi) = \sum_{K \in \mathcal{T}_h} I_K(\psi|_K), \\ \int_{\partial\Omega_{\mathrm{N}}} \psi \approx I_{\partial\Omega_{\mathrm{N}}}(\psi) = \sum_{A \in \partial\Omega_{\mathrm{N}}} I_A(\psi|_A), \end{cases} \tag{2.53}$$

where ψ is any function on Ω whose restriction to the interior $\overset{\circ}{K}$ of an element K of $\mathcal{T}_h$ is continuous on $\overset{\circ}{K}$, and has a limit on the edges and vertices of K. Hence the restrictions $\psi|_K$ to K and $\psi|_A$ to a boundary edge A are well defined at the vertices of K and the endpoints of A. This ensures that $I_K(\psi|_K)$ and $I_{\partial\Omega_{\mathrm{N}}}(\psi)$ in the right-hand sides of the Definition (2.53) make sense.

We shall use in the sequel the quadrature formula (2.53) for functions ψ, which are either continuous on $\overline{\Omega}$ or are piecewise constant on $\mathcal{T}_h$.

3. *The discretized parameters.* We have to choose the finite dimensional inputs corresponding to a, k, f, g, and u_e. Because the computational cost in the adjoint approach is independent of the number of parameters, one computes the gradient with respect to the *simulation parameters*, which correspond to the most comprehensive discretization of the parameters compatible with the chosen approximation of the state equation (see Sect. 3.3 in the next chapter). Their number exceeds usually by far the number of parameters that could actually be identified

without further regularization. So the optimization has to be performed with respect to a (usually smaller) number of *optimization parameters.* The choice of these optimization parameters is addressed in the Chap. 3 on parameterization, and the calculation of the gradient with respect to the optimization parameters from the gradient with respect to the simulation parameters is discussed in Sects. 3.3 and 3.8.2.

We describe now a possible choice of simulation parameters approximating a, k, f, g, u_e.

In the variational formulation (2.47), a is the coefficient of $\nabla u \nabla w$ in an integral over Ω. As u and w are approximated in the space W_h of functions which are piecewise linear on $\mathcal{T}_h$, their gradients are constant vector fields on each element K, so that any variation of a *inside* an element K is not going to be seen, only its mean value over K plays a role. This suggests to choose for a_h a piecewise constant parameter on $\mathcal{T}_h$:

$$a_h = (\ a_K \in I\!R,\ K \in \mathcal{T}_h\ ). \tag{2.54}$$

The nonlinearity $u \rightsquigarrow k(u)$ can be approximated by a function k_h depending on a finite number n_k of coefficients: polynomial, continuous piecewise linear function, closed form formula, etc.

The right-hand side f can be approximated by a function f_h, which is either piecewise constant on the triangles of $\mathcal{T}_h$ or piecewise linear in the same space W_h as u_h. We choose the second solution, which leads to slightly simpler formulas. The degrees of freedom of f_h are then its values on the nodes M of Ω_h:

$$f_h = (f_M \in I\!R,\ M \in \Omega_h). \tag{2.55}$$

Similarly, the right-hand side g on the boundary $\partial\Omega_{\rm N}$ can be approximated by a function g_h, which is either piecewise constant on the edges A of $\partial\Omega_{\rm N}$, or continuous on $\partial\Omega_{\rm N}$ and linear on each of its edge A. We choose the second solution, so the degrees of freedom of g_h are its values on the subset $\partial\Omega_{{\rm N},h}$ of nodes M of Ω_h, which are located on $\partial\Omega_{\rm N}$ (we do not include in Ω_h the endpoints of $\partial\Omega_{\rm N}$, when they exist, as it will turn out that the value of g_M at such points has no influence on the solution):

$$g_h = (\ g_M \in I\!R,\ M \in \partial\Omega_{{\rm N},h}). \tag{2.56}$$

There is no choice for the discretization $u_{e,h}$ of u_e: if one wants to satisfy exactly the boundary condition $u_h = u_{e,h}$ on $\partial\Omega_{\rm D}$, then $u_{e,h}$ is necessarily in the trace of W_h on $\partial\Omega_{\rm D}$, that is, in the space of continuous functions on $\partial\Omega_{\rm D}$, which are linear over each edge A of $\partial\Omega_{\rm D}$. The degrees of freedom of $u_{e,h}$ are then its values on the subset $\partial\Omega_{{\rm D},h}$ of nodes M of Ω_h, which are located on $\partial\Omega_{\rm D}$, including the endpoints of $\partial\Omega_{\rm D}$ when they exist:

$$u_{e,h} = (\ u_{e,M} \in I\!\!R,\ \ M \in \partial\Omega_{{\rm D},h}). \tag{2.57}$$

The sets of node $\partial\Omega_{{\rm N},h}$ and $\partial\Omega_{{\rm D},h}$ form a partition of the subset $\partial\Omega_h$ of Ω_h made of nodes located on the boundary $\partial\Omega$. All the terms in the variational formulation (2.47) have now a finite dimensional counterpart, and so we can define the *finite dimensional variational formulation*:

$$\left\{\begin{array}{l} \text{find } u_h \in W_h \text{ with } u_{h|\partial\Omega_{\rm D}} = u_{e,h} \text{ such that} \\ I_\Omega(a_h \nabla u_h \nabla w_h) + I_\Omega(k_h(u_h) w_h) = I_\Omega(f_h w_h) + I_{\partial\Omega_{\rm N}}(g_h w_h) \\ \text{for all } w_h \in W_h \text{ with } w_{h|\partial\Omega_{\rm D}} = 0. \end{array}\right. \tag{2.58}$$

As the node of the quadrature formulas I_Ω and $I_{\partial\Omega_{\rm N}}$ coincide with the nodes $M \in \Omega_h$ of the degrees of freedom of u_h, it is a simple matter to deduce from the above equation, where $u_h \in W_h$ is still a *function*, a system of nonlinear equations for the *vector* of degrees of freedom $u_M \in \Omega_h$:

$$\left\{\begin{array}{ll} u_M = u_{e,M} & \forall M \in \partial\Omega_{{\rm D},h}, \\ \sum_{P\in\Omega_h\setminus\partial\Omega_{{\rm D},h}} A_{M,P}\, u_P + \alpha_M\, k_h(u_M) = & - \sum_{P\in\partial\Omega_{{\rm D},h}} A_{M,P}\, u_{e,P} \\ & +\alpha_M f_M \\ & (+\partial\alpha_M\, g_M \text{ if } M \in \partial\Omega_{{\rm N},h}), \\ & \forall M \in \Omega_h\setminus\partial\Omega_{{\rm D},h}. \end{array}\right. \tag{2.59}$$

where

$$A_{M,P} = I_\Omega(a_h\, \nabla w_M \nabla w_P) \quad \forall M,P \in \Omega_h\setminus\partial\Omega_{{\rm D},h}, \tag{2.60}$$

and where α_M and $\partial\alpha_M$ are geometric coefficients related to the triangulation $\mathcal{T}_h$:

$$\alpha_M = \frac{1}{3} \sum_{K\in\mathcal{T}_h, K\ni M} |K|, \qquad \partial\alpha_M = \frac{1}{2} \sum_{A\subset\partial\Omega_N, A\ni M} |A|. \tag{2.61}$$

The nonlinear state equations (2.59) have to be solved on the computer by some iterative algorithm. However, one does not consider this algorithm as being a part of the process to be differentiated: the vector u_h produced by the code is considered to be the unique solution of equation (2.59). Of course, this is acceptable if the algorithm has been run up to a satisfactory convergence.

At this point, the "parameter-to-state" map $a_h,\ k_h,\ f_h,\ g_h,\ u_{e,h} \rightsquigarrow u_h$ is perfectly defined from a computational point of view.

2.6.2 Discretization Step 2: Choice of a Discretized Objective Function

We choose here an approximation J_h to the continuous objective function J defined in (2.46). Using the approximation of u and the approximation of the integrals over Ω and $\partial\Omega_N$ of the previous section, it is natural to define J_h by

$$\left\{ \begin{array}{rcl} J_h(a_h, k_h, f_h, g_h, u_{e,h}) & = & I_\Omega(\|Z_h - \nabla u_h\|^2) \\ & + & I_\Omega(|z_h - u_h|^2) \\ & + & I_{\partial\Omega_\mathrm{N}}(|z_{\mathrm{N},h} - u_{h|\partial\Omega_\mathrm{N}}|^2), \end{array} \right. \tag{2.62}$$

where the data $Z = (Z_h, z_h, z_{N,h})$ satisfy the following:

- Z_h is a piecewise constant vector field on $\mathcal{T}_h$
- z_h is a function of W_h, that is, continuous piecewise linear on Ω
- $z_{\mathrm{N},h}$ is a continuous piecewise linear function on $\partial\Omega_\mathrm{N}$

The corresponding degrees of freedom are

$$\left\{ \begin{array}{lll} Z_h & = (Z_K \in I\!R^2,\ K \in \mathcal{T}_h\ ) & \text{(measure of } \nabla u\text{)}, \\ z_h & = (z_M \in I\!R,\ M \in \Omega_h\ ) & \text{(measure of } u\text{)}, \\ z_{\mathrm{N},h} & = (z_{\mathrm{N},h,M} \in I\!R, M \in \partial\Omega_{\mathrm{N},h}) & \text{(measure of } u \text{ on } \partial\Omega_\mathrm{N}\text{)}. \end{array} \right. \tag{2.63}$$

The discretization of the continuous problem is now completed, and so we can move to the determination of the gradient of the discrete objective function J_h with respect to all its arguments by the adjoint approach.

2.6.3 Derivation Step 0: Forward Map and Objective Function

Our objective is to compute the partial derivatives of J_h with respect to the

$$n = \mathrm{Card}\mathcal{T}_h + n_k + \mathrm{Card}\Omega_h + \mathrm{Card}\partial\Omega_h$$

degrees of freedom of the parameter vector $x = (a_h, k_h, f_h, g_h, u_{e,h})$, so we equip the parameter space $E = I\!R^n$ with the usual Euclidean scalar product.

The discrete objective function (2.62) is of the form (2.11), with a *forward map* φ_h and an *objective function* G_h given by:

$$\varphi_h : x = (a_h, k_h, f_h, g_h, u_{e,h}) \rightsquigarrow v = (\nabla u_h, u_h, u_{h|\partial\Omega_N}), \tag{2.64}$$

where u_h is solution of (2.58), and:

$$\left\{ \begin{array}{rcl} G_h(x,v) & = & I_\Omega(\|Z_h - V_h\|^2) \\ & + & I_\Omega(|z_h - v_h|^2) \\ & + & I_{\partial\Omega_N}(|z_{N,h} - v_{N,h}|^2), \end{array} \right. \tag{2.65}$$

where $v = (V_h, v_h, v_{N,h})$ is in the data space defined by (2.63). This data space is equipped with the scalar product associated to the norm in the right-hand side of (2.65). As required, G_h is an explicit function of x and v (it even does not depend on x !), and $G_h(x, \varphi_h(x))$ coincides with $J_h(x)$.

2.6.4 Derivation Step 1: State-Space Decomposition

We have to now dispatch (2.58) defining u_h into a possibly affine state-space Y and a set of state equations $e(x, u_h) = 0$, which defines $u_h \in Y$. There are two options here, depending on whether or not we include the Dirichlet boundary condition in the state-space Y:

Option 1: Y does not include the boundary condition. We consider here the *vector* state-space:

$$Y = \delta Y = W_h.$$

The state equations $e(x,y) = 0$ which determine $y = u_h \in Y$ are then (see (2.58))

$$u_{h|\partial\Omega_D} = u_{e,h}, \tag{2.66}$$

$$I_\Omega(a_h \nabla u_h \nabla w_h) + I_\Omega(k_h(u_h) w_h) = I_\Omega(f_h w_h) + I_{\partial\Omega_N}(g_h w_h) \tag{2.67}$$

$$\forall w_h \in W_h \text{ s.t. } w_{h|\partial\Omega_D} = 0,$$

which, together with the observation operator

$$M : u_h \rightsquigarrow (\nabla u_h, u_h, u_{h|\partial\Omega_N}), \tag{2.68}$$

defines the forward map φ_h chosen in (2.64).

Option 2: Y includes the boundary condition. This option is available when one does not need to compute the gradient of J_h with respect to the Dirichlet data $u_{e,h}$. So we eliminate $u_{e,h}$ from the parameter vector

$$x = (a_h, k_h, f_h, g_h),$$

and introduce it in an *affine* state-space

$$Y = \{w_h \in W_h \mid w_{h|\partial\Omega_D} = u_{e,h}\},$$

whose tangent vector space is

$$\delta Y = \{w_h \in W_h \mid w_{h|\partial\Omega_D} = 0\},$$

which happens now to coincide with the space of test functions of the variational formulation for u_h. The state equation $e(x, y) = 0$ which determine $y = u_h \in Y$ is then (compare with (2.66) and (2.67))

$$\begin{aligned} I_\Omega(a_h \nabla u_h \nabla w_h) + I_\Omega(k_h(u_h)w_h) &= I_\Omega(f_h w_h) \\ &\quad + I_{\partial\Omega_N}(g_h w_h) \quad \forall w_h \in \delta Y, \end{aligned} \tag{2.69}$$

which, together with the same observation operator M as in (2.68), defines the forward map φ_h chosen in (2.64).

2.6.5 Derivation Step 2: Lagrangian

We take advantage here of the fact that the *state equation* $e(x, y) = 0$ is partly or totally under *variational form*: the Lagrange multiplier associated with a variational equation *can always be chosen in the space of the test functions of the variational formulation*, in which case the corresponding term $\langle e(x, y), \lambda\rangle_Z$ of the Lagrangian is immediately obtained by replacing the test function by the Lagrange multiplier λ.

Option 1: In this case, the state is

$$u_h \in Y = W_h$$

and there are two Lagrange multipliers:

– $\lambda_{D,h}$ associated with the Dirichlet boundary condition (2.66), which specifies the value of u_h at each node $M \in \partial\Omega_{D,h}$. So we chose

$$\lambda_{D,h} = (\ \lambda_{D,M} \in I\!R,\ \ M \in \partial\Omega_{D,h}),$$

and we define the scalar product between $\lambda_{\mathrm{D},h}$ and $u_{e,h} - u_h|_{\partial\Omega_{\mathrm{D}}}$ via a quadrature formula $I_{\partial\Omega_{\mathrm{D}}}$ defined in the same way as $I_{\partial\Omega_{\mathrm{N}}}$ as in (2.53).

– λ_h associated with the variational formulation (2.67), which one can simply take in the space of the test functions

$$\lambda_h \in \{w_h \in W_h \mid w_h|_{\partial\Omega_{\mathrm{D}}} = 0\}. \tag{2.70}$$

With the objective function G_h of (2.65) and the state equations (2.66) and (2.67), the Lagrangian for the first option is then

$$\begin{aligned} &\mathcal{L}(a_h, k_h, f_h, g_h, u_{e,h}; u_h; \lambda_{D,h}, \lambda_h) = \\ &I_\Omega(\|Z_h - \nabla u_h\|^2) + I_\Omega(|z_h - u_h|^2) + I_{\partial\Omega_{\mathrm{N}}}(|z_{\mathrm{N},h} - u_h|^2) \\ &+ I_{\partial\Omega_{\mathrm{D}}}((u_{e,h} - u_h)\lambda_{\mathrm{D},h}) \\ &+ I_\Omega(a_h \nabla u_h \cdot \nabla\lambda_h) + I_\Omega(k_h(u_h)\lambda_h) - I_\Omega(f_h\lambda_h) - I_{\partial\Omega_N}(g_h\lambda_h). \end{aligned} \tag{2.71}$$

By construction, the dimension of the vector $\lambda_{\mathrm{D},h}, \lambda_h$ is equal to the number of equations in (2.66) and (2.67), and $\mathcal{L}$ is an explicit function of all its arguments, as one can check on (2.71).

Option 2: The state satisfies now

$$u_h \in Y = \{w_h \in W_h \mid w_{h|_{\partial\Omega_{\mathrm{D}}}} = u_{e,h}\},$$

and there is only one Lagrange multiplier λ_h associated with the variational state equation (2.69), which lives in its test function space

$$\lambda_h \in \delta Y = \{w_h \in W_h \mid w_h|_{\partial\Omega_{\mathrm{D}}} = 0\},$$

which is the same space as in (2.70). The Lagrangian for the second option uses the same function G_h as option 1, and the single state equation (2.69)

$$\begin{aligned} &\mathcal{L}(a_h, k_h, f_h, g_h; u_h; \lambda_h) = \\ &I_\Omega(\|Z_h - \nabla u_h\|^2) + I_\Omega(|z_h - u_h|^2) + I_{\partial\Omega_{\mathrm{N}}}(|z_{N,h} - u_h|^2) \\ &+ I_\Omega(a_h \nabla u_h \cdot \nabla\lambda_h) + I_\Omega(k_h(u_h)\lambda_h) - I_\Omega(f_h\lambda_h) - I_{\partial\Omega_{\mathrm{N}}}(g_h\lambda_h). \end{aligned} \tag{2.72}$$

The introduction of the boundary condition in the state-space leads hence to a Lagrangian function with fewer arguments and fewer terms, so that the adjoint state determination will be slightly simpler, but the price to pay is that this approach will not give the gradient with respect to the boundary condition $u_{e,h}$.

2.6.6 Derivation Step 3: Adjoint Equation

Option 1: The equation for $\lambda_h \in \{w_h \in W_h \mid w_h|_{\partial\Omega_D} = 0\}$ is obtained by equating to zero the differential of (2.71) with respect to u_h:

$$\begin{aligned}
\frac{\partial \mathcal{L}}{\partial u_h}(a_h, k_h, f_h, g_h, u_{e,h}; u_h; \lambda_{D,h}, \lambda_h) &= \\
& -2I_\Omega((Z_h - \nabla u_h)\cdot \nabla \delta u_h) \\
& -2I_\Omega((z_h - u_h)\delta u_h) \\
& -2I_{\partial\Omega_\mathrm{N}}((z_{\mathrm{N},h} - u_h)\delta u_h) \\
& \quad -I_{\partial\Omega_\mathrm{D}}(\delta u_h \lambda_{\mathrm{D},h}) \\
& \quad +I_\Omega(a_h \nabla \delta u_h \cdot \nabla \lambda_h) + I_\Omega(k_h'(u_h)\delta u_h \lambda_h) \\
&= 0 \quad \forall \delta u_h \in \delta Y,
\end{aligned} \tag{2.73}$$

where

$$\delta Y = W_h.$$

Equations (2.70) and (2.73) define uniquely the adjoint states $\lambda_{\mathrm{D},h}$ and λ_h. Choosing successively $\delta u_h \in \{w_h \in W_h \mid w_h|_{\partial\Omega_\mathrm{D}} = 0\}$ and $\delta u_h = w_M \ \forall M \in \partial\Omega_{h,\mathrm{D}}$, where w_M is the basis function of W_h associated with node M, we obtain the following decoupled adjoint equations for λ_h and $\lambda_{\mathrm{D},h}$:

$$\left\{\begin{array}{l}
\text{find } \lambda_h \in W_h \text{ with } \lambda_h|_{\partial\Omega_\mathrm{D}} = 0 \text{ such that} \\
I_\Omega(a_h \nabla \lambda_h \cdot \nabla w_h) + I_\Omega(k_h'(u_h)\lambda_h w_h) = \\
\qquad 2I_\Omega((Z_h - \nabla u_h)\cdot \nabla w_h) \\
\qquad +2I_\Omega((z_h - u_h) w_h) \\
\qquad +2I_{\partial\Omega_\mathrm{N}}((z_{\mathrm{N},h} - u_h) w_h) \\
\text{for all } w_h \in W_h \text{ with } w_{h|_{\partial\Omega_\mathrm{D}}} = 0,
\end{array}\right. \tag{2.74}$$

$$\left\{\begin{array}{l}
\text{find } \lambda_{\mathrm{D},h} \in \{\lambda_{\mathrm{D},M} \in I\!R, M \in \partial\Omega_{\mathrm{D},h}\} \text{ such that} \\
I_{\partial\Omega_\mathrm{D}}(\lambda_{\mathrm{D},h} w_M) = \\
\qquad +I_\Omega(a_h \nabla \lambda_h \cdot \nabla w_M) + I_\Omega(k_h'(u_h)\lambda_h w_M) \\
\qquad -2I_\Omega((Z_h - \nabla u_h)\cdot \nabla w_M) \\
\qquad -2I_\Omega((z_h - u_h) w_M) \\
\qquad -2I_{\partial\Omega_\mathrm{N}}((z_{\mathrm{N},h} - u_h) w_M) \\
\text{for all basis functions } w_M \in W_h, M \in \partial\Omega_{\mathrm{D},h}.
\end{array}\right. \tag{2.75}$$

Equation (2.74) is very similar to the variational formulation (2.58) for u_h, but with different right-hand sides: in the case where one does not observe the gradient of u, there is no $2I_\Omega((Z_h - \nabla u_h)\cdot\nabla w_h)$ term in (2.74), and we obtain immediately that λ_h is an approximation to the solution λ of an equation similar to the elliptic equation (2.44) defining u, but with right-hand sides $2(z-u)$ instead of f in Ω and $2(z_{\mathrm{N}}-u)$ instead of g on $\partial\Omega_{\mathrm{N}}$.

We postpone the interpretation of (2.74) for λ_h in the general case, and of (2.74) for $\lambda_{\mathrm{D},h}$, after we have established the continuous adjoint equations at the end of Sect. 2.6.

The system of linear equations for the node values $\lambda_M, M\in\Omega_h$ and $\lambda_{\mathrm{D},M}, M\in\partial\Omega_{D,h}$ resulting from (2.74) and (2.75) is (compare with (2.59))

$$\left\{\begin{array}{l} \lambda_M = 0 \qquad\qquad\qquad\qquad \forall M\in\partial\Omega_{\mathrm{D},h} \\ \sum_{P\in\Omega_h\setminus\partial\Omega_{\mathrm{D},h}} A_{M,P}\,\lambda_P + \alpha_M\, k_h'(u_M)\lambda_M \;= \\ \qquad +2\sum_{K\ni M}|K|\,(Z_K-\nabla u_h|_K)\cdot\nabla w_M|_K \\ \qquad +2\alpha_M(z_M-u_M) \\ \qquad (+2\,\partial\alpha_M(z_{\mathrm{N},M}-u_M) \quad \text{if } M\in\partial\Omega_{\mathrm{N},h}) \\ \qquad\qquad\qquad\qquad \forall M\in\Omega_h\setminus\partial\Omega_{\mathrm{D},h}, \end{array}\right.$$

$$\left\{\begin{array}{l} \partial\alpha_M\lambda_{\mathrm{D},M} = \;+\sum_{P\in\Omega_h\setminus\partial\Omega_{\mathrm{D},h}} A_{M,P}\,\lambda_P + \alpha_M\, k_h'(u_M)\lambda_M \\ \qquad -2\sum_{K\ni M}|K|\,(Z_K-\nabla u_h|_K)\cdot\nabla w_M|_K \\ \qquad -2\alpha_M(z_M-u_M) \\ \qquad (-2\,\partial\alpha_M(z_{\mathrm{N},M}-u_M) \quad \text{if } M\in\partial\Omega_{\mathrm{N},h}) \\ \qquad\qquad\qquad\qquad \forall M\in\partial\Omega_{\mathrm{D},h}. \end{array}\right.$$

Option 2: Similarly, we obtain the equation for $\lambda_h\in\delta Y$ by equating to zero the differential of (2.72) with respect to u_h:

$$\begin{aligned} \frac{\partial\mathcal{L}}{\partial u_h}(a_h,k_h,f_h,g_h;u_h;\lambda_h) \;&= \qquad\qquad (2.76)\\ &-2I_\Omega((Z_h-\nabla u_h)\cdot\nabla\delta u_h) \\ &-2I_\Omega((z_h-u_h)\delta u_h) \\ &-2I_{\partial\Omega_{\mathrm{N}}}((z_{\mathrm{N},h}-u_h)\delta u_h) \\ &\quad +I_\Omega(a_h\nabla\delta u_h\cdot\nabla\lambda_h)+I_\Omega(k_h'(u_h)\delta u_h\lambda_h) \\ &\qquad = \;0 \qquad\qquad \forall\delta u_h\in\delta Y, \end{aligned}$$

where now

$$\delta Y = \{w_h\in W_h \mid w_h|_{\partial\Omega_{\mathrm{D}}}=0\},$$

so that (2.76) coincides with (2.74) of option 1.

Hence the two options for the boundary condition define the same adjoint state λ_h, but $\lambda_{D,h}$ is defined only in option 1.

2.6.7 Derivation Step 4: Gradient Equation

Option 1: Differentiation of the Lagrangian (2.71) with respect to all parameters $a_h, k_h, f_h, g_h, u_{e,h}$ for fixed direct and adjoint states u_h, $\lambda_{D,h}$ and λ_h gives

$$\begin{aligned} \delta J_h \;=\; & I_{\partial\Omega_D}(\delta u_{e,h}\lambda_{D,h}) \\ & + I_\Omega(\delta a_h \nabla u_h \cdot \nabla\lambda_h) \\ & + I_\Omega(\delta k_h(u_h)\lambda_h) \\ & - I_\Omega(\delta f_h\, \lambda_h) \\ & - I_{\partial\Omega_N}(\delta g_h\, \lambda_h). \end{aligned} \tag{2.77}$$

If we expand the term $I_{\partial\Omega_D}(\delta u_{e,h}\lambda_{D,h})$ using the definition of $I_{\partial\Omega_D}$, and pick the coefficient of $\delta u_{e,M}$ in the resulting formula, we see that

$$\frac{\partial J_h}{\partial u_{e,M}} = \partial\alpha_M \lambda_{D,M} \qquad \forall M \in \partial\Omega_{D,h}. \tag{2.78}$$

We obtain similarly from the $I_\Omega(\delta a_h \nabla u_h \cdot \nabla\lambda_h)$ term

$$\frac{\partial J_h}{\partial a_K} = |K|(\nabla u_h \cdot \nabla\lambda_h)|_K. \tag{2.79}$$

In order to determine the *gradient with respect to the nonlinearity* k_h, let us denote by $\kappa_1, \ldots, \kappa_{n_k}$ the coefficients that define the $u \rightsquigarrow k_h(u)$ function. The differential of k_h is then

$$\delta k_h(u_h) = \sum_{j=1,\ldots,n_k} \frac{\partial k_h}{\partial \kappa_j}(u_h)\delta\kappa_j. \tag{2.80}$$

We can now substitute into (2.77) the value of δk_h given by (2.80) and pick the coefficients of $\delta\kappa_j$, which gives the following expressions for the partial derivatives of J_h with respect to $\kappa_1, \ldots, \kappa_{n_k}$:

$$\frac{\partial J_h}{\partial \kappa_h} = I_\Omega\Big(\frac{\partial k_h}{\partial \kappa_j}(u_h)\,\lambda_h\Big). \tag{2.81}$$

In the case where k_h is a polynomial of degree $n_k - 1$ in the variable u with coefficients $\kappa_0, \ldots, \kappa_{n_k-1}$, we obtain from (2.80)

$$\frac{\partial J_h}{\partial \kappa_j} = I_\Omega(u_h^j \lambda_h), \qquad j = 0, \ldots, n_k - 1.$$

But k_h can also be searched for as a continuous piecewise linear function of u; in this case, the coefficients $\kappa_j, j = 1, \ldots, n_k$, are the values of k_h at a given set of values $u_j, j = 1, \ldots, n_k$, of u, so that

$$\frac{\partial J_h}{\partial \kappa_j} = I_\Omega(w_j(u_h)\lambda_h),$$

where w_j is the continuous piecewise linear function of u defined by $w_j(u_i) = \delta_{i,j}$, with $\delta_{i,j} = 0$ if $i \neq j$ or 1 if $i = j$.

Finally, the two last terms of (2.77) give the gradient with respect to the right-hand sides f_h and g_h:

$$\frac{\partial J_h}{\partial f_M} = \alpha_M \lambda_M \qquad \forall M \in \Omega_h, \tag{2.82}$$

$$\frac{\partial J_h}{\partial g_M} = \partial\alpha_M \lambda_M \qquad \forall M \in \partial\Omega_{N,h}. \tag{2.83}$$

Option 2: The only difference with option 1 is that the $I_{\partial\Omega_D}((u_{e,h}-u_h)\lambda_{D,h})$ term is missing in the Lagrangian (2.72) for option 2. Hence, the formula (2.78) for the derivative with respect to the boundary condition $u_{e,h}$ is not available in this option, but the formula (2.79), (2.81)–(2.83) for the other derivatives are the same.

2.7 Examples 3 and 4: Continuous Adjoint Equations

It is in fact possible to compute the derivative of the objective function (2.46) of Examples 3 and 4 for the original infinite dimensional problem, before any reduction to finite dimension is done. As we shall see in this section, this determination is formally simpler than the determination of the discrete gradient, as one does not need to consider all the formulas used for the discretization. But it requires a good understanding of functional analysis, distribution theory, and Green formulas, and from this point of view the determination of the discrete gradient, which works only with real numbers, is more elementary.

We present purposely the continuous gradient *after* the discrete one, to emphasize the fact that the continuous gradient is *not* a preliminary step to the discrete gradient. However, if they are determined first, the formulas for the continuous Lagrangian can be used as a guideline for the choice of the scalar products for the discrete Lagrangian, in order to ensure that the discrete adjoint state is an approximation of the continuous one.

The forward map is (compare with (2.64))

$$\varphi : x = (a, k, f, g, u_e) \rightsquigarrow v = (\nabla u, u, u_{|\partial\Omega_N}), \tag{2.84}$$

where $u \in H^1(\Omega)$ is the solution of (2.47), and according to (2.46), the objective function $G(x, v)$ to be differentiated is (compare with (2.65))

$$G(x,v) = \int_\Omega \|Z - \nabla v\|^2 + \int_\Omega |z - v|^2 + \int_{\partial\Omega_N} |z_N - v|^2.$$

This completes the step 0 of derivation. One can then choose the state-space decomposition corresponding to option 1 above with the vector state-space:

$$Y = \delta Y = H^1(\Omega),$$

so that the state equations (2.47) rewrite

$$u_{|\partial\Omega_D} = u_e \text{ in } H^{1/2}(\partial\Omega_D), \tag{2.85}$$

$$\int_\Omega a\nabla u \cdot \nabla w + \int_\Omega k(u)w = \int_\Omega fw + \int_{\partial\Omega_N} gw \tag{2.86}$$
$$\forall w \in H^1(\Omega) \text{ such that } w_{|\partial\Omega_D} = 0.$$

This, together with the observation operator

$$M : u \rightsquigarrow (\nabla u, u, u_{|\partial\Omega_N}),$$

defines the forward map φ chosen in (2.84) and completes the step 1 of derivation. In step 2, one introduces first the two Lagrange multipliers:

– λ_D associated with the Dirichlet boundary condition (2.85). The choice of the function space for λ_D is a little technical: as λ_D is expected to define a linear functional on the dense subspace $H^{1/2}(\partial\Omega_{\rm D})$ of $L^2(\partial\Omega_{\rm D})$, where (2.85) holds, it is natural to require that

$$\lambda_{\rm D} \in H^{-1/2}(\partial\Omega_{\rm D}),$$

where $H^{-1/2}(\partial\Omega_{\rm D}) \supset L^2(\partial\Omega_{\rm D})$ is the dual space of $H^{1/2}(\partial\Omega_{\rm D}) \subset L^2(\partial\Omega_{\rm D})$. For any $\lambda_{\rm D} \in H^{-1/2}(\partial\Omega_{\rm D})$ and $\mu \in H^{1/2}(\partial\Omega_{\rm D})$, we denote by $\langle\lambda_{\rm D},\mu\rangle_{H^{-1/2},H^{1/2}}$ the value of the linear functional $\lambda_{\rm D}$ on the function μ. In the case where $\lambda_{\rm D}$ happens to be in the dense subset $L^2(\Omega)$ of $H^{-1/2}(\partial\Omega_{\rm D})$, one has simply

$$\langle\lambda_{\rm D},\mu\rangle_{H^{-1/2},H^{1/2}} = \int_{\partial\Omega_{\rm D}} \lambda_{\rm D}\,\mu \tag{2.87}$$

– λ associated to the variational formulation (2.86), which one can simply take in the space of the test functions of (2.86):

$$\lambda \in \{w \in H^1(\Omega) \mid w|_{\partial\Omega_{\rm D}} = 0\}.$$

The corresponding Lagrangian function is then defined by (compare with (2.71))

$$\begin{aligned} \mathcal{L}(a,k,f,g,u_e;u;\lambda_{\rm D},\lambda) = \\ \int_\Omega \|Z-\nabla u\|^2 + \int_\Omega |z-u|^2 + \int_{\partial\Omega_{\rm N}} |z_{\rm N}-u|^2 \\ +\langle u_e - u|_{\partial\Omega_{\rm D}}, \lambda_{\rm D}\rangle_{H^{1/2},H^{-1/2}} \\ +\int_\Omega a\nabla u\cdot\nabla\lambda + \int_\Omega k(u)\lambda - \int_\Omega f\lambda - \int_{\partial\Omega_{\rm N}} g\lambda\,. \end{aligned} \tag{2.88}$$

Differentiation of the Lagrangian with respect to u gives, as above, two decoupled equations for λ and $\lambda_{\rm D}$ (compare with (2.74) and (2.75)):

$$\left\{\begin{array}{rcl} \text{find } \lambda \in H^1(\Omega) \text{ with } \lambda|_{\partial\Omega_{\rm D}} = 0 \text{ such that} && \\ \int_\Omega a\nabla\lambda\cdot\nabla w + \int_\Omega k'(u)\,\lambda w &=& +2\int_\Omega (Z-\nabla u)\cdot\nabla w \\ && +2\int_\Omega (z-u)w \\ && +2\int_{\partial\Omega_N}(z_N-u)w \\ \text{for all } w \in H^1(\Omega) \text{ with } w_{|\partial\Omega_{\rm D}} = 0, && \end{array}\right. \tag{2.89}$$

$$\left\{\begin{array}{rcl} \text{find } \lambda_{\rm D} \in H^{-1/2}(\partial\Omega_{\rm D}) \text{ such that} && \\ \langle\lambda_{\rm D}, w_{|\partial\Omega_{\rm D}}\rangle_{H^{-1/2},H^{1/2}} &=& +\int_\Omega a\nabla\lambda\cdot\nabla w + \int_\Omega k'(u)\,\lambda w \\ && -2\int_\Omega (Z-\nabla u)\cdot\nabla w \\ && -2\int_\Omega (z-u)w \\ && -2\int_{\partial\Omega_{\rm N}}(z_{\rm N}-u)w \\ \text{for all } w \in H^1(\Omega). && \end{array}\right. \tag{2.90}$$

Proposition 2.7.1 *The* continuous adjoint equations *(2.89) and (2.90) have necessarily unique solutions* λ *and* $\lambda_{\rm D}$. *These equations are a* weak formulation *of the following set of partial differential equations:*

$$\begin{cases} -\nabla\cdot(a\nabla\lambda) + k'(u)\lambda \;=\; 2(z-u) - 2\nabla\cdot(Z-\nabla u) \quad \text{in } \Omega, \\ \lambda \;=\; 0 \quad \text{on } \partial\Omega_{\rm D}, \\ a\,\dfrac{\partial\lambda}{\partial\nu} \;=\; 2(z_{\rm N}-u) + 2(Z-\nabla u)\cdot\nu \quad \text{on } \partial\Omega_{\rm N}, \end{cases} \tag{2.91}$$

$$\lambda_{\rm D} = a\,\frac{\partial\lambda}{\partial\nu} - 2(Z-\nabla u)\cdot\nu \quad \text{on } \partial\Omega_{\rm D}. \tag{2.92}$$

Proof. The existence and uniqueness of the solution λ and $\lambda_{\rm D}$ of the continuous adjoint equations (2.89) and (2.90) follow immediately from Proposition 2.3.3.

We recall first the concept of weak formulation: let us call *weak solution* a solution λ, $\lambda_{\rm D}$ of (2.89), (2.90), and *classical solution* a solution λ, $\lambda_{\rm D}$ of (2.91), (2.92) in the usual sense of *functions.* The system of equations (2.89) and (2.90) is then a *weak formulation* of the set of partial differential equations (2.91) and (2.92) if and only if:

– Any *classical solution* is a *weak solution*
– Any *regular weak solution* is a *classical solution*

Without the regularity assumption, there is no hope for a weak solution to be a classical solution, as the solution of (2.89) and (2.90) is not smooth enough for all terms in (2.91) and (2.92) to make sense as functions: when λ is in $H^1(\Omega)$ and a is in $L^\infty(\Omega)$, the vector field $a\nabla u$ is in $L^2(\Omega)\times L^2(\Omega)$, so that, for example,

– The term $\nabla\cdot(a\nabla\lambda)$ does not make sense as a function on Ω, but only as a distribution

– The term $a\,\frac{\partial\lambda}{\partial\nu}$, which is by definition $a\nabla u\cdot\nu$, does not makes sense as a function on $\partial\Omega$ (functions of $L^2(\Omega)$ have no trace on $\partial\Omega$)

We check first that any classical solution is a weak one. Let λ, $\lambda_{\rm D}$ be a classical solution of (2.91) and (2.92). Multiplying the first equation of (2.91) by $w\in H^1(\Omega)$, integrating over Ω and using the Green formula

$$\begin{cases} -\int_\Omega \nabla\cdot(a\nabla\lambda - 2(Z-\nabla u))\,w = +\int_\Omega (a\nabla\lambda - 2(Z-\nabla u))\cdot\nabla w \\ \qquad\qquad\qquad\qquad\qquad\qquad - \int_{\partial\Omega}(a\nabla\lambda - 2(Z-\nabla u))\cdot\nu\, w, \end{cases} \tag{2.93}$$

we obtain that

$$\left\{\begin{array}{ll} \int_\Omega a\nabla\lambda\cdot\nabla w + \int_\Omega k'(u)\,\lambda\, w = & 2\int_\Omega(Z-\nabla u)\cdot\nabla w \\ & +2\int_\Omega(z-u)w \\ & +\int_{\partial\Omega_{\rm D}}(a\nabla\lambda - 2(Z-\nabla u))\cdot\nu\, w \\ & +\int_{\partial\Omega_{\rm N}}(a\nabla\lambda - 2(Z-\nabla u))\cdot\nu\, w \\ \text{for all } w\in H^1(\Omega), & \end{array}\right. \tag{2.94}$$

and, using (2.92) and the third equation of (2.91)

$$\left\{\begin{array}{ll} \int_\Omega a\nabla\lambda\cdot\nabla w + \int_\Omega k'(u)\,\lambda\, w = & 2\int_\Omega(Z-\nabla u)\cdot\nabla w \\ & +2\int_\Omega(z-u)w \\ & +\int_{\partial\Omega_{\rm D}}\lambda_{\rm D}\, w \\ & +2\int_{\partial\Omega_{\rm N}}(z_{\rm N}-u)w \\ \text{for all } w\in H^1(\Omega). & \end{array}\right. \tag{2.95}$$

This equation is to be compared with the weak formulation (2.89) and (2.90):
– It reduces to (2.89) when the test function w is chosen such that $w=0$ on $\partial\Omega_{\rm D}$. Hence, the classical solution λ satisfies (2.89)
– Because $\lambda_{\rm D}$ is by hypothesis a function, formula (2.87) holds, so that (2.95) coincides with (2.90). This shows that the classical solution $\lambda_{\rm D}$ satisfies (2.90)

We conclude the proof by checking that any regular weak solution is a classical solution. So let λ, $\lambda_{\rm D}$ be a regular solution of (2.89) and (2.90). Let us choose w in (2.89) in the space $\mathcal{D}(\Omega)$ of test functions of distributions. This space is made of infinitely differentiable functions, which vanish over some neighborhood of $\partial\Omega$. Hence the integral over $\partial\Omega_{\rm N}$ disappears in (2.89), which now reads, in the sense of distributions,

$$-\nabla\cdot(a\nabla\lambda) + k'(u)\lambda \;=\; 2(z-u) - 2\nabla\cdot(Z-\nabla u) \;\text{ in } \mathcal{D}'(\Omega). \tag{2.96}$$

But we have supposed that λ is a smooth function, so that (2.96) holds in the sense of functions, everywhere on Ω, and hence coincides with the first equation of (2.91). So we see that λ satisfies the first equation of (2.91). It satisfies also trivially the second equation of (2.91) by definition of the weak solution.

We prove now that λ and $\lambda_{\rm D}$ also satisfy the third equation of (2.91) as well as equation (2.92). To do that, we multiply (2.96) – which now holds everywhere on Ω – by a test function $w\in H^1(\Omega)$, integrate over Ω, and use the Green formula (2.93), which gives (2.94) as above. Subtracting (2.94) from (2.89) for a w which vanishes over $\partial\Omega_{\rm D}$ gives

$$\begin{cases} 0 = 2\int_{\partial\Omega_{\rm N}}(z_{\rm N}-u)w - \int_{\partial\Omega_{\rm N}}(a\nabla\lambda - 2(Z-\nabla u))\cdot\nu\, w \\ \text{for all } w \in H^1(\Omega) \text{ with } w_{|\partial\Omega_{\rm D}} = 0. \end{cases} \tag{2.97}$$

Because of hypothesis (1.57), when w spans the subspace of $H^1(\Omega)$ made of functions that vanish over $\partial\Omega_{\rm D}$, its trace on $\partial\Omega_{\rm N}$ spans the dense subspace $H^{1/2}(\partial\Omega_{\rm N})$ of $L^2(\partial\Omega_{\rm N})$, and (2.97) implies that the coefficient of w is zero, which shows that λ satisfies the third equation of (2.91).

Then adding (2.94) and (2.90) gives, using the third equation of (2.91),

$$\begin{cases} \langle\lambda_{\rm D}, w_{|\partial\Omega_{\rm D}}\rangle_{H^{-1/2},H^{1/2}} = \int_{\partial\Omega_{\rm D}}(a\nabla\lambda - 2(Z-\nabla u))\cdot\nu\, w \\ \text{for all } w \in H^1(\Omega). \end{cases} \tag{2.98}$$

Combining (2.98) with formula (2.87), which holds because of the smoothness hypothesis made on $\lambda_{\rm D}$, we obtain

$$\begin{cases} \int_{\partial\Omega_{\rm D}}\lambda_{\rm D}\, w = \int_{\partial\Omega_{\rm D}}(a\nabla\lambda - 2(Z-\nabla u))\cdot\nu\, w \\ \text{for all } w \in H^1(\Omega). \end{cases}$$

Once again, because of hypothesis (1.57), when w spans $H^1(\Omega)$, its trace on $\partial\Omega_{\rm D}$ spans the dense subspace $H^{1/2}(\partial\Omega_{\rm D})$ of $L^2(\partial\Omega_{\rm D})$, which shows that $\lambda_{\rm D}$ satisfies (2.92). This ends the proof of Proposition 2.7.1. ■

Now that the direct state u and the adjoint state $\lambda_{\rm D}, \lambda$ are known, we simply have to differentiate the Lagrangian (2.88) with respect to $x = (a, k, f, g, u_e)$ for fixed $u, \lambda_{\rm D}, \lambda$ to obtain the differential of the least squares objective function J defined in (2.46)

$$\begin{aligned} \delta J &= \langle\delta u_e, \lambda_{\rm D}\rangle_{H^{1/2},H^{-1/2}} \\ &\quad + \int_\Omega \delta a\nabla u\cdot\nabla\lambda + \int_\Omega \delta k(u)\,\lambda - \int_\Omega \delta f\,\lambda - \int_{\partial\Omega_{\rm N}} \delta g\,\lambda. \end{aligned} \tag{2.99}$$

This formula is the continuous equivalent of the "gradient equations" of the discrete case: it shows that $\lambda_{\rm D}$ is the derivative with respect to the Dirichlet boundary condition u_e, that $\nabla u\cdot\nabla\lambda$ is the derivative with respect to the diffusion coefficient a, that λ is the gradient with respect to the right-hand side f, and that the trace of λ on $\partial\Omega_{\rm N}$ is the gradient with respect to the

Neumann condition g. When the adjoint state λ is regular enough so that its level sets

$$C_v = \{x \in I\!R^2 \mid u(x) = v\}, \qquad u_{\min} \leq v \leq u_{\max}$$

are regular curves, the differential with respect to k can be written as

$$\delta J = \int_{u_{\min}}^{u_{\max}} \delta k(v) \int_{C_v} \lambda,$$

so that the derivative of J with respect to the nonlinearity k for a value v of u is the integral of λ along the level line C_v of u.

Remark 2.7.2 *This example gives a first illustration of the difference between* discrete *and* discretized *gradient approaches mentioned in the introduction of the chapter: they both require to choose a discretization of the direct state equation (2.44) and the objective function (2.46), for example, the ones described in Sects. 2.6.1 and 2.6.2. But once this is done, the discrete adjoint equations (2.74) and (2.75) and the discrete derivative formulas (2.77) follow unambiguously, whereas determination of the discretized adjoint equations and gradient formula would require further to discretize the adjoint equations (2.91) and (2.92) and the derivative formulas (2.99). There are usually many different ways to do this, so that there is no guarantee that the discretized adjoint equations will coincide with the discrete adjoint equations (2.74) and (2.75), which are the only one that lead to the exact gradient of* J_h. *For the problem under consideration, it is reasonable to think that a seasoned numerical analyst would choose the discrete adjoint equation (2.74) as an approximation to (2.91), but it is most unlikely that he would have chosen the intricate equation (2.75) as an approximation to the simple equation (2.92), in which case the discretized gradient will be only an approximation to the exact gradient of the discrete objective function.* ■

2.8 Example 5: Differential Equations, Discretized Versus Discrete Gradient

Consider the system of differential equations (state equation):

$$\frac{\mathrm{d}u}{\mathrm{d}t} = f(u(t), a) \quad \text{for } t > 0, \qquad u(0) = u_0, \tag{2.100}$$

where $t \rightsquigarrow u(t) \in I\!R^m$ is the (infinite dimensional) state variable, and where the parameter vector $a \in I\!R^n$ and the initial data $u_0 \in I\!R^m$ are to be estimated.

Consider also that a measure $z \in I\!R^m$ of the state $u(T)$ at a given time T is available for this (observation operator: $u \rightsquigarrow u(T)$). The least squares objective function is then

$$J(a, u_0) = \frac{1}{2}\|z - u(T)\|^2_{I\!R^m}. \tag{2.101}$$

It is now an exercise to determine the continuous adjoint state and derivative formulas as we did for the elliptic problem in Sect. 2.7 (the Green formula is replaced by integration by part). To compute the gradient with respect to the initial condition u_0, one decides first not to include the initial condition in an affine state-space, but rather to consider it as a state equation (this corresponds exactly to the **Option 1** choice for the state space Y in Sect. 2.6). The starting point is hence the following Lagrangian:

$$\left\{ \begin{array}{rcl} \mathcal{L}(a, u_0, u, \lambda, \lambda_0) & = & \frac{1}{2}\|z - u(T)\|^2_{I\!R^m} \\ & + & \int_0^T (f(u(t), a) - \frac{\mathrm{d}u}{\mathrm{d}t}) \cdot \lambda \\ & + & (u_0 - u(0)) \cdot \lambda_0, \end{array} \right. \tag{2.102}$$

the resulting adjoint equations are

$$-\frac{\mathrm{d}\lambda}{dt} = \left(\frac{\partial f}{\partial u}\right)^T (u(t), a)\ \lambda(t) \text{ for } t > 0, \quad \lambda(T) = u(T) - z, \tag{2.103}$$

$$\lambda_0 = \lambda(0), \tag{2.104}$$

and the formulas for the derivatives ("gradient equations") is

$$\delta J = \int_0^T \frac{\partial f}{\partial a}(u(t), a)\delta a \cdot \lambda(t) + \lambda_0 \cdot \delta u_0.$$

The gradients of J with respect to a and u_0 are then

$$\nabla_a J = \int_0^T \left(\frac{\partial f}{\partial a}\right)^T (u(t), a)\ \lambda(t), \tag{2.105}$$

$$\nabla_{u_0} J = \lambda_0 = \lambda(0). \tag{2.106}$$

To solve this problem on the computer, we discretize the differential equations (2.100) and the objective function (2.101). We introduce the times $t^k, k = 0 \dots K$, such that

$$0 = t^0 < \cdots < t^k < \cdots < t^K = T,$$

and the corresponding time steps

$$h^{k+1/2} = t^{k+1} - t^k.$$

Then we can decide, for example, to replace (2.100) by the discrete state equation

$$\frac{u^{k+1} - u^k}{h^{k+1/2}} = f(u^{k+\theta}, a) \ , k = 1 \dots K \ , \quad u^0 = u_0 \tag{2.107}$$

for some $\theta \in [0, 1]$, where we have used the convenient notation

$$u^{k+\theta} = (1 - \theta)u^k + \theta u^{k+1}.$$

The scheme is explicit for $\theta = 0$, and implicit for $0 < \theta \leq 1$. We denote by

$$u_h = (u^k, k = 0 \dots K) \in I\!R^{(K+1)m}$$

the solution of (2.107), which is supposed to exist. The vector u_h is the *state* and $I\!R^{(K+1)m}$ *the state-space* of the system.

The *observation operator* associated with final time data is $M : u_h \rightsquigarrow u^K$, and we can decide, for example, to approximate the objective function J by

$$J_h(a, u_0) = \frac{1}{2} \|z - u^K\|^2_{I\!R^m}$$

At this point, we have the choice between

– Either take advantage of the fact that we have already computed the continuous adjoint, and go for the discretized adjoint approach

– Or use the continuous adjoint only as a guideline for the choice of the scalar products, and go for the discrete adjoint approach

We investigate now these two possibilities.

2.8.1 Implementing the Discretized Gradient

There are **two more decisions** to make in this approach:

– One has to choose one discretization of the adjoint equations (2.103) and (2.104). By analogy with (2.107), it is natural to define the approximation

$$\lambda_h = (\lambda^k, k = K \dots 0) \in \mathbb{R}^{(K+1)m}$$

of the solution $t \rightsquigarrow \lambda(t)$ of (2.103) by

$$\frac{\lambda^{k-1} - \lambda^k}{h^{k-1/2}} = \left(\frac{\partial f}{\partial u}\right)^{\mathrm{T}}(u^{k-\theta}, a)\, \lambda^{k-\theta}, \;\; k = K \dots 1, \;\; \lambda^K = u^K - z. \tag{2.108}$$

It is then also natural to replace (2.104) for the multiplier λ_0 associated with the initial condition by

$$\lambda_{0,h} = \lambda^0. \tag{2.109}$$

– One has to discretize the formulas (2.105) and (2.106) for the gradients of J. Replacing the integral by the trapezoidal rule, we obtain

$$(\nabla_h)_a\, J_h = \sum_1^K h^{k-1/2} \left\{ \left(\frac{\partial f}{\partial a}\right)^{\mathrm{T}}(u^{k-1}, a)\;\; \lambda^{k-1} + \left(\frac{\partial f}{\partial a}\right)^{\mathrm{T}}(u^k, a)\;\; \lambda^k \right\} \tag{2.110}$$

$$(\nabla_h)_{u_0}\, J_h = \lambda_0 = \lambda^0. \tag{2.111}$$

The index h in ∇_h remembers us that this quantity is only in general an **approximation to the gradient**.

2.8.2 Implementing the Discrete Gradient

The good side here is that there is **no more decision** to make, and that we shall obtain the **exact gradient** of J_h. But the dark side is that we need to redo the **adjoint calculation at the discrete level**. This usually leads to cumbersome calculations, which we detail now.

We write first the discrete Lagrangian $\mathcal{L}_h(a, u_0; u_h; \lambda_h, \lambda_{0,h})$. We need to choose a name for the Lagrange multiplier associated with the kth equation of (2.107): as this equation computes u^{k+1} from u^k, its multiplier is associated with the $[k, k+1]$ interval. A natural solution would be to call it $\lambda^{k+1/2}$ in absence of further information. But it will appear on the final discrete adjoint equations that this multiplier can be interpreted more precisely as an approximation of $\lambda(t)$ at the time $t^{k+1-\theta} = (1-\theta)t^{k+1} + \theta t^k$, so that we shall call it $\lambda^{k+1-\theta}$. So we define the discrete Lagrange multipliers by

$$\lambda_h = (\lambda^{1-\theta} \dots \lambda^{K-\theta}) \in I\!R^{Km}, \quad \lambda_{0,h} \in I\!R^m. \tag{2.112}$$

Then we need to choose a scalar product on $I\!R^{Km}$ and $I\!R^m$. By analogy with the continuous formula (2.102), we choose on $I\!R^{Km}$, a scalar product that mimics the integral from 0 to T of the scalar product in $I\!R^m$, and we equip $I\!R^m$ with the usual scalar product. The discrete Lagrangian is then

$$\begin{cases} \mathcal{L}_h(a, u_0; u_h; \lambda_h, \lambda_{0,h}) \;=\; \frac{1}{2}\|z - u^K\|^2_{I\!R^m} \\ +\sum_{k=0}^{K-1} h^{k+1/2}\left(f(u^{k+\theta}, a) - (u^{k+1} - u^k)/h^{k+1/2}\right) \cdot \lambda^{k+1-\theta} \\ +(u_0 - u^0) \cdot \lambda_{0,h}, \end{cases}$$

where as previously $u^{k+\theta}$ is a notation for $(1-\theta)u^k + \theta u^{k+1}$.

The discrete adjoint equation is then obtained as usual by equating to zero the partial derivative of $\mathcal{L}_h$ with respect to u_h:

$$\begin{cases} \dfrac{\partial \mathcal{L}_h}{\partial u_h}\delta u_h \;=\; (u^K - z) \cdot \delta u^K \\ +\sum_{k=0}^{K-1}\left(h^{k+1/2}\dfrac{\partial f}{\partial u}(u^{k+\theta}, a)\,\delta u^{k+\theta} - \delta u^{k+1} + \delta u^k\right) \cdot \lambda^{k+1-\theta} \\ -\delta u^0 \cdot \lambda_{0,h} \;=\; 0 \qquad \forall \delta u_h = (\delta u^k, k = 0 \dots K) \in I\!R^{(K+1)m}. \end{cases} \tag{2.113}$$

The $K+1$ formulas for the computation of λ_h and $\lambda_{0,h}$ will be obtained by equating to zero the coefficients of δu_k in (2.113) for $k = 0 \dots K$. We perform for this purpose a *discrete integration by parts*, which consists in rewriting (2.113) in the form $\sum_{k=0}^{K}(\dots)\delta u^k = 0$. We call A, B, C, D, E the five terms of (2.113), and reorganize them into the desired form. We notice first that δu^K is missing in the D term, but present in A. So we *define* $\lambda^{K+1-\theta} \in I\!R^m$ by

$$\lambda^{K+1-\theta} = u^K - z, \tag{2.114}$$

and rewrite the $A + D$ terms as

$$A + D = \sum_{k=0}^{K} \delta u^k \cdot \lambda^{k+1-\theta},$$

which is of the desired form. Similarly, we see that δu^0 is missing in the C term, but present in E, so we *define* $\lambda^{-\theta} \in I\!R^m$ by

$$\lambda^{-\theta} = \lambda_{0,h}, \tag{2.115}$$

and rewrite the $C + E$ terms as

$$C + E = -\sum_{k=0}^{K-1} \delta u^{k+1} \cdot \lambda^{k+1-\theta} - \delta u^0 \cdot \lambda^{-\theta},$$

or, with an index shift

$$C + E = -\sum_{k=0}^{K} \delta u^k \cdot \lambda^{k-\theta},$$

which is also of the desired form. We work now on the B term:

$$B = \sum_{k=0}^{K-1} h^{k+1/2} \frac{\partial f}{\partial u}(u^{k+\theta}, a)\delta u^{k+\theta} \cdot \lambda^{k+1-\theta},$$

or, remembering that $\delta u^{k+\theta} = (1-\theta)\delta u^k + \theta \delta u^{k+1}$,

$$\begin{aligned} B = (1-\theta) &\sum_{k=0}^{K-1} h^{k+1/2} \frac{\partial f}{\partial u}(u^{k+\theta}, a)\delta u^k \cdot \lambda^{k+1-\theta} \\ +\theta &\sum_{k=0}^{K-1} h^{k+1/2} \frac{\partial f}{\partial u}(u^{k+\theta}, a)\delta u^{k+1} \cdot \lambda^{k+1-\theta}. \end{aligned}$$

We see that δu^K is missing in the sum of the first line. But if we define

$$t^{K+1} = t^K, \qquad h^{K+1/2} = t^{K+1} - t^K = 0,$$

we can extend this sum up to index $k = K$, as this adds only a zero term. Similarly, δu^0 is missing in the sum of the second line, but if we define

$$t^{-1} = t^0, \qquad h^{-1/2} = t^0 - t^{-1} = 0,$$

we can extend the sum to $k = -1$. This gives, after an index shift,

$$B = (1-\theta) \sum_{k=0}^{K} h^{k+1/2} \frac{\partial f}{\partial u}(u^{k+\theta}, a)\delta u^k \cdot \lambda^{k+1-\theta}$$

$$+\theta \sum_{k=0}^{K} h^{k-1/2} \frac{\partial f}{\partial u}(u^{k-1+\theta}, a)\delta u^k \cdot \lambda^{k-\theta}.$$

If we define h^k and θ^k for $k = 0 \dots K$ by

$$\begin{cases} h^k &= t^{k+1-\theta} - t^{k-\theta} = (1-\theta)h^{k+1/2} + \theta h^{k-1/2} \\ h^k\theta^k &= \theta h^{k-1/2}, \qquad h^k(1-\theta^k) = (1-\theta)h^{k+1/2}. \end{cases}$$

we see that

$$t^k = (1-\theta^k)t^{k-\theta} + \theta^k t^{k+1-\theta},$$

and we can rewrite B as follows:

$$B = \sum_{k=0}^{K} h^k(1-\theta^k) \left(\frac{\partial f}{\partial u}(u^{k+\theta}, a)\right)^T \lambda^{k+1-\theta} \cdot \delta u^k$$

$$+\sum_{k=0}^{K} h^k\theta^k \left(\frac{\partial f}{\partial u}(u^{k-1+\theta}, a)\right)^T \lambda^{k-\theta} \cdot \delta u^k,$$

which is of the desired form.

The final **discrete adjoint equations** for the determination of $\lambda_h = (\lambda^{1-\theta} \dots \lambda^{K-\theta})$ defined in (2.112), $\lambda^{K+1-\theta}$ and $\lambda^{-\theta}$ defined in (2.114) and (2.115), and $\lambda_{0,h}$ are then

$$\frac{\lambda^{k-\theta} - \lambda^{k+1-\theta}}{h^k} = (1-\theta^k)\left(\frac{\partial f}{\partial u}(u^{k+\theta}, a)\right)^T \lambda^{k+1-\theta} \tag{2.116}$$

$$+\theta^k\left(\frac{\partial f}{\partial u}(u^{k-1+\theta}, a)\right)^T \lambda^{k-\theta}, \qquad k = K \dots 0,$$

$$\lambda^{K+1-\theta} = u^K - z \tag{2.117}$$

$$\lambda_{0,h} = \lambda^{-\theta} \tag{2.118}$$

Differentiation of the Lagrangian $\mathcal{L}_h$ with respect to the parameters a and u_0

$$\delta\mathcal{L}_h = \sum_{k=0}^{K-1} h^{k+1/2} \left(\frac{\partial f}{\partial a}(u^{k+\theta}, a)\, \delta a\right) \cdot \lambda^{k+1-\theta} - \delta u_0 \cdot \lambda_{0,h}$$

gives then the **discrete gradient equations** by picking up the coefficients of δa and δu_0:

$$\begin{cases} \nabla_a J_h = \sum_{k=0}^{K-1} h^{k+1/2} \left(\dfrac{\partial f}{\partial a}(u^{k+\theta}, a) \right)^T \lambda^{k+1-\theta}, \\ \nabla_{u_0} J_h = -\lambda_{0,h}. \end{cases} \tag{2.119}$$

Comparison of Discretized and Discrete Gradients

Once again, we see that the discrete adjoint equations (2.116)–(2.118) and the formulas in (2.119) for the discrete gradient differ substantially from their discretized counterparts (2.108)–(2.111). Moreover, the discrete formulas are far from being the most natural ones, which makes it very unlikely for the discrete formulas to be chosen by chance when using the discretized approach. The author remembers of a stiff parabolic problem, where the discrete approach had been used for a time discretization with a constant time step, and the correctness of the gradient produced by the code had been validated by comparison with finite differences. But when it came to application to real data, a variable time step was used to limit the computational cost, and the code was simply modified by replacing everywhere the fixed time step by the variable one, which amounted to a discretized approach. At that point, the code blew up during the computations, and a lot of time was spent looking up for a coding error; but the problem disappeared only after the discrete adjoint and gradient formulas were implemented for the variable time steps.

Adaptive Time Steps

Most simulators for time-dependent problems use adaptive time stepping to adapt to the stiffness of the problem and reduce the computational costs. When this happens, the time steps depend on the parameters, and should be differentiated in the discrete gradient approach. But time stepping procedures are often not differentiable and poorly documented, and taking into account this dependance would lead to still more tedious calculations. Luckily enough, if the time steps are correctly estimated, small perturbations of these time steps should have a minor influence on the solution, and it is the author's experience that this influence can be usually neglected. Hence the first reasonable thing to do is to calculate the discrete gradient with the variable, but fixed time steps determined by the time stepping procedure for the current parameter value. Whenever the objective function has to be evaluated for a different value of the parameters, for example, during the line search, the time steps are determined again by the time stepping procedure.

2.9 Example 6: Discrete Marching Problems

It is sometime convenient to forget the continuous differential equation underlying a marching problem. This can be the case when a simulation code already exists for the marching problem of interest, and one wants to perform some optimization based on this code (see "Sensitivity versus adjoint" below). Without loss of generality, the state equations $e(a, u) = 0$ for such a problem can be written as

$$E^{k-1/2}(a, u^k, u^{k-1}) = 0, \quad k = 1 \dots K, \tag{2.120}$$

$$u^0 = u_0, \tag{2.121}$$

where $a \in I\!R^n$ is a vector of parameters, and $u_0 \in I\!R^m$ is the initial value. For sake of simplicity, we shall suppose here that u_0 is known, and that a is the parameter vector to be estimated, so that the parameter space is $E = I\!R^n$. But there is no difficulty in handling the case where u_0 is unknown (see **Option 1** in Sect. 2.6).

When (2.120) is nonlinear with respect to u^k, it has to be solved only approximately on the computer using an iterative scheme (a Newton algorithm, for example). Such algorithms are governed by tests, and hence are not differentiable. So it is practically impossible to include them in the definition of the forward map, and, for the sake of discrete gradient computations, one usually considers that (2.120) *is* the discrete equation, and that it is solved exactly by the computer. This acceptable if the equation is solved precisely enough.

As for the observation, let us consider, for example, the case where, at each "time" index k, a measurement z^k of

$$v^k = M^k(u^k) \tag{2.122}$$

is available, where M^k is the observation operator at index k:

$$M^k : I\!R^m \rightsquigarrow I\!R^{m_k}.$$

The observation space is then

$$F = I\!R^q, \qquad \text{where } q = m_1 + \cdots + m_K,$$

and the forward map to be inverted is

$$\varphi : a \in E = I\!R^n \rightsquigarrow (v^1 \dots v^k) \in F = I\!R^q. \tag{2.123}$$

Now, given a data vector

$$z = (z^1 \dots z^K) \in I\!R^q,$$

the data misfit function for the estimation of a is

$$J(a) = \frac{1}{2} \sum_{k=1}^{K} \|z^k - M^k(u^k)\|^2_{I\!R^{m_k}}. \tag{2.124}$$

A necessary step for the estimation of a knowing z is to compute the gradient of J with respect to the parameter a. We compare now on this problem the sensitivity functions and adjoint approaches. We recall that $\partial E^{k-1/2}/\partial u^k$ and $\partial E^{k-1/2}/\partial u^{k-1}$ denote the partial derivatives of $E^{k-1/2}$ with respect to its second and third arguments.

Sensitivity Functions Approach

According to Sect. 2.2, we differentiate (2.120)–(2.122) with respect to a_j for $j = 1 \dots N$. This gives the following formula for the jth column s_j of the Jacobian D of the forward map $\varphi : a \rightsquigarrow (v^1 \dots v^k)$ (jth *output sensitivity function*):

$$s_j^k = \frac{\partial M(u^k)}{\partial a_j} = (M^k)'(u^k)\,\frac{\partial u^k}{\partial a_j}, \qquad k = 1 \dots K,$$

where the *state sensitivity functions* $\partial u^k/\partial a_j$ are given by (all partial derivatives of $E^{k-1/2}$ are evaluated at a, u^k, u^{k-1})

$$\frac{\partial E^{k-1/2}}{\partial u^k}\frac{\partial u^k}{\partial a_j} + \frac{\partial E^{k-1/2}}{\partial u^{k-1}}\frac{\partial u^{k-1}}{\partial a_j} + \frac{\partial E^{k-1/2}}{\partial a_j} = 0, \tag{2.125}$$

$$\frac{\partial u^0}{\partial a_j} = 0. \tag{2.126}$$

The gradient of J is then given by

$$\frac{\partial J}{\partial a_j} = \sum_{k=1}^{K} (M^k(u^k) - z^k) \cdot s_j^k \quad j = 1 \dots N.$$

Adjoint Approach

We follow the step-by-step approach of Sect. 2.4:
Step 0: The forward map φ is already defined in (2.123), and, as we want to compute the gradient of J defined in (2.124), the objective function $G(a, v)$ is

$$G(a,v) = \frac{1}{2}\sum_{k=1}^{K} \|z^k - v^k\|^2_{I\!R^{m_k}} \qquad \text{(independant of } a\text{)}. \quad (2.127)$$

In absence of information on the need of parameter scaling, on the nature of uncertainty on the data z, and on any continuous analogon of the problem, we simply equip the parameter space $I\!R^n$ and the data space $I\!R^q$ with the usual Euclidean scalar products.
Step 1: We do not need to compute the gradient with respect to the initial condition u_0, as u_0 is known, and so we can include this condition in an affine state-space Y, according to **Option 2** of Sect. 2.6:

$$Y = \{u = (u^0 \dots u^K) \in I\!R^{(K+1)m} \mid u^0 = u_0\},$$

with the associated vector space:

$$\delta Y = \{\delta u = (0, \delta u^1 \dots \delta u^K) \in I\!R^{(K+1)m}\} = I\!R^p \text{ with } p = Km.$$

The corresponding observation operator is

$$M : u = (u^0 \dots u^K) \in Y \rightsquigarrow v = (M^1(u^1) \dots M^K(u^K)) \in I\!R^q. \quad (2.128)$$

Step 2: The Lagrangian is then, with G, is defined in (2.127) and M in (2.128):

$$\mathcal{L}(a,u,\lambda) = G(a, M(u)) + \sum_{k=1}^{K} E^{k-1/2}(a, u^k, u^{k-1}) \cdot \lambda^{k-1/2}, \qquad (2.129)$$

where

$$\begin{cases} a \in & I\!R^n \\ u = & (u^0, u^1 \dots u^K) \quad \in Y \\ \lambda = & (\lambda^{1/2} \dots \lambda^{K-1/2}) \quad \in I\!R^p. \end{cases}$$

Step 3: Differentiation of the Lagrangian with respect to the state u gives the variational form of the adjoint equation:

$$\frac{\partial \mathcal{L}}{\partial u}\delta u = A + B + C = 0 \quad \forall \delta u = (0, \delta u^1 \dots \delta u^K) \in \delta Y,$$

where

$$A = \sum_{k=1}^{K} (M^k(u^k) - z^k) \cdot (M^k)'(u^k)\delta u^k,$$

$$B = \sum_{k=1}^{K} \frac{\partial E^{k-1/2}}{\partial u^k}(a, u^k, u^{k-1})\, \delta u^k \cdot \lambda^{k-1/2},$$

$$C = \sum_{k=1}^{K} \frac{\partial E^{k-1/2}}{\partial u^{k-1}}(a, u^k, u^{k-1})\, \delta u^{k-1} \cdot \lambda^{k-1/2}.$$

We perform now a *discrete integration by part* to factorize the δu^k terms: one notices that the sum in the C term starts in fact at $k = 2$ (remember that $\delta u^0 = 0$), and that it can be extended to $K + 1$, provided we define $\lambda^{K+1/2} \in I\!R^p$ by $\lambda^{K+1/2} = 0$. After an index shift, the C term becomes

$$C = \sum_{k=1}^{K} \frac{\partial E^{k+1/2}}{\partial u^k}(a, u^{k+1}, u^k)\, \delta u^k \cdot \lambda^{k+1/2}.$$

Equating to zero successively the coefficients of $\delta u^1 \ldots \delta u^K$ gives the computational form of the adjoint equations (all partial derivatives of $E^{k-1/2}$ are evaluated at a, u^k, u^{k-1}, those of $E^{k+1/2}$ are evaluated at a, u^{k+1}, u^k)

$$\left(\frac{\partial E^{k-1/2}}{\partial u^k}\right)^{\mathrm{T}} \lambda^{k-1/2} + \left(\frac{\partial E^{k+1/2}}{\partial u^k}\right)^{\mathrm{T}} \lambda^{k+1/2} \qquad (2.130)$$

$$+ \left((M^k)'(u^k)\right)^T (M^k(u^k) - z^k) = 0 \quad \text{for } k = K \ldots 1,$$

which can be solved backwards starting from the final condition

$$\lambda^{K+1/2} = 0. \qquad (2.131)$$

Step 4: We differentiate now the Lagrangian (2.129) with respect to the parameter vector a (partial derivatives of $E^{k-1/2}$ are evaluated at a, u^k, u^{k-1}):

$$\delta J = \frac{\partial \mathcal{L}}{\partial a}\delta a = \sum_{k=1}^{K} \frac{\partial E^{k-1/2}}{\partial a}\, \delta a \cdot \lambda^{k-1/2},$$

and pick the coefficient of δa_j, which gives the gradient equations

$$\frac{\partial J}{\partial a_j} = \sum_{k=1}^{K} \frac{\partial E^{k-1/2}}{\partial a_j} \cdot \lambda^{k-1/2} \qquad j = 1 \ldots N.$$

Sensitivity Versus Adjoint

The general pros and cons of sensitivity function and adjoint approaches for the minimization of J apply here:

- The computational cost of the sensitivity function approach increases with the number of parameters (there are as many linearized equations (2.125) and (2.126) to solve as parameters), and changing the parameters is not easy (it requires to recode the linearized equations). But it gives not only the gradient of J, but also the Jacobian D of the forward map φ, and allows to use Gauss–Newton or Levenberg–Marquardt optimization algorithms. It also does not require a large memory, as the linearized equations can be solved along with the original marching scheme.

- The computational cost of the adjoint approach is independent of the number of parameters (one adjoint equation (2.130) and (2.131) in all cases), which makes it well suited for inverse problems with large number of parameters; changing the parameterization using the chain rule is easy once the gradient has been computed with respect to the (possibly large) set of *simulation parameters* (see (2.15) and Sect. 3.3 below). But it does not compute the Jacobian, and so one is limited to optimization algorithms of Quasi-Newton type, which usually require a larger number of iterations. Also, the memory requirement can be extremely large, as one needs to store or recompute the direct state $u^1 \dots u^K$ before solving backwards for the adjoint state $\lambda^{K-1/2} \dots \lambda^{1/2}$ (see [40]) for an optimal compromise between computation and storage).

The adjoint approach is the method of choice for the sensitivity analysis study, which is to be performed *before* the inversion itself to find out the number of parameters that can be retrieved from the data (Sect. 3.2 of Chap. 3): it allows to compute the Jacobian D with respect to the (usually large) number of *simulation parameters*, before any choice is made concerning the parameterization and the optimization parameters. This is possible because in the adjoint approach the Jacobian is computed row by row, with a cost proportional to the number of observations, but independent of the number of parameter (of course the function $G(a, v)$ to be used there is no more (2.127), but rather (2.12)).

We conclude this example with some implementation remarks.

First, given a forward modeling code that solves the marching problem (2.120) and (2.121), the matrix $\partial E^{k-1/2}/\partial u^k$ is necessarily formed somewhere in the code to perform the corresponding Newton iterations. But this matrix is also the matrix of the linearized equation (2.125) in the sensitivity approach. Hence the additional work to implement the sensitivity approach consists in forming the matrices $\partial E^{k-1/2}/\partial u^{k-1}$ and $\partial E^{k-1/2}/\partial a$, and in solving the same system as in Newton's iterations, but with n different right-hand sides. As the matrix of the system is already formed, the computational effort for the resolution of each linearized equation (2.125) is much less than the one required for one simulation. This approach has been used to implement sensitivity equations in a complex nonlinear reservoir simulation code [6].

Second, comparison of the sensitivity equations (2.125) and the adjoint equation (2.130) shows that the same matrices $\partial E^{k-1/2}/\partial u^k$, $\partial E^{k-1/2}/\partial u^{k-1}$, and $\partial E^{k-1/2}/\partial a$ appear at both places. So when given a modeling code with sensitivity equations capabilities, one can consider developing an adjoint code by identifying these matrices in the code, and recombining them into the desired adjoint equation. This approach has been used for the same reservoir simulation code as above in [79].

Chapter 3

Choosing a Parameterization

We address in this chapter practical aspects of parameterization for the same finite dimensional inverse problem (2.5) as in Chap. 2, where the forward map $x \in I\!R^n \rightsquigarrow v = \varphi(x) \in I\!R^q$ is defined by the state-space decomposition (2.1) and (2.2):

– Because of their possibly different physical nature, it is necessary to calibrate (Sect. 3.1) the simulation parameters by adimensionalization, to avoid artifacts in the conditioning of the problem

– The singular value decomposition of the Jacobian $D = \varphi'(x)$ allows then to estimate the number of independent parameters that can be retrieved for a given level of uncertainty on the data (Sect. 3.2)

– In the (usual...) case where the above number of retrievable parameters is smaller than the number of simulation parameters, it is necessary to add a priori information on the parameters to restore wellposedness – or better Q-wellposedness, see Chap. 4 – of the least squares problem. The regularization methods available to this purpose have been presented in Sect. 1.3.4. In any case, one has to make a clear distinction (Sect. 3.3) between simulation parameters (which are coefficients or source terms in the numerical simulation model) and optimization parameters (the ones that are seen by the optimizer, and will be actually estimated)

– The regularization by size reduction will be discussed in Chap. 4, and the Levenberg–Marquardt–Tychonov and the state-space regularization in Chap. 5. We present in this chapter the *regularization by parameterization*, which proceeds by reduction of the number of unknowns: the information is added by choosing a set of formulas to compute the simulation parameters from a smaller number of optimization parameters. A good parameterization

G. Chavent, *Nonlinear Least Squares for Inverse Problems: Theoretical Foundations and Step-by-Step Guide for Applications*, Scientific Computation,
DOI 10.1007/978-90-481-2785-6_3,

should allow to explain the data up to the noise level *and* should not lead to overparameterization (the dimension of the optimization vector should be smaller than the number of retrievable parameters determined above). We evaluate in this chapter four parameterization against these objectives: closed form formula (Sect. 3.4), singular vector basis (Sect. 3.5), multiscale approximation (Sect. 3.6), and adaptive parameterization (Sect. 3.7)

– Finally, we discuss in Sect. 3.8 some implementation issues:

• How to organize the inversion code to allow an easy experimentation with various choices of optimization parameters (Sect. 3.8.1)

• How to compute the gradient with respect to optimization parameters once the gradient with respect to numerical parameters is known (Sect. 3.8.2)

• And, in Sect. 3.9, we describe the maximum projected curvature (MPC) descent step, which is specially designed to enhance the performance of descent algorithms used for the resolution of nonlinear least squares problems.

3.1 Calibration

The unknown parameters correspond often to different physical quantities: for example, hydraulic conductivities, porosities, acoustic impedances, etc. Similarly, the available observations can involve temperatures, pressures, concentrations, etc.

Before trying to recover (part of) the parameters from the data, it is necessary to eliminate the poor conditioning that can be caused by the different orders of magnitude associated with the different physical quantities in the parameter and the data vectors. So one first has to *calibrate* the finite dimensional inverse problem (2.5) by using *dimensionless parameters* and *data.*

3.1.1 On the Parameter Side

Let X denote the unknown parameter as described by the physics. The principle is to first adimensionalize X by choosing a reference value X_{ref} for each physical parameter, and then to choose the *calibrated parameter vector* $x \in I\!R^n$ representing X in such a way that

1. The Euclidean norm of x represents a mean value of the adimensionalized parameters

2. And, in the case where X is a distributed parameter on a domain Ω, the Euclidean scalar product on $\mathbb{R}^n$ is proportional to the $L^2(\Omega)$ scalar product of the functions X_h approaching X

The first property is a matter of convenience, as it allows simply a more direct interpretation of the results. The second property is meant to ensure that the discretization does not change (in particular does not worsen...) the conditioning of the least squares optimization problem. Hence,

- *When X is a finite dimensional vector of scalar coefficients*, one chooses first, for each component X_i, a *reference value* $X_{i,\text{ref}} > 0$, which incorporates the best available knowledge on its magnitude ($X_{i,\text{ref}}$ can be the same for a group of components corresponding to the same physical quantity). The *calibrated* parameter vector x is then defined by

$$x_i = \frac{1}{\sqrt{n}} \frac{X_i}{X_{i,\text{ref}}}, \qquad i = 1 \dots n,$$

where the coefficient $\sqrt{n}$ ensures that

$$\|x\|_{\mathbb{R}^n} \stackrel{\text{def}}{=} \Big(\sum_{i=0}^{n} x_i^2 \Big)^{\frac{1}{2}} = \text{ mean value of } X/X_{\mathbf{ref}}.$$

- *When X is a function of a variable ξ defined over some domain Ω*, the reduction to a finite number n of scalar parameters is usually done by discretization (notations of Sect. 2.6.1): the domain Ω is covered by a simulation mesh $\mathcal{T}_h$ made of a finite number of cells or elements K, where the index $h > 0$ denotes the largest cell size in $\mathcal{T}_h$. A finite dimensional approximation space $\boldsymbol{E}_h$ is then chosen for the functions X; a function $\xi \rightsquigarrow X_h(\xi)$ of $\boldsymbol{E}_h$ is characterized by n "natural" degrees of freedom $X_1 \dots X_n$ (cell or node values).

  1. *For a discontinuous piecewise constant approximation:* The function $\xi \rightsquigarrow X_h$ takes a constant value X_i on the ith cell K_i. The L^2 scalar product of functions of $\boldsymbol{E}_h$ is given by

$$\int_\Omega X_{0,h} X_{1,h} = \sum_{i=1\dots n} |K_i| \, X_{0,i} X_{1,i},$$

where $|K_i|$ is the measure (length, area, or volume) of K_i. To ensure that the L^2 scalar product of the functions X_h is proportional to the Euclidean scalar product in $I\!R^n$ of the calibrated parameter vectors x, we define x_i on the cell K_i by

$$x_i = \Big(\frac{|K_i|}{|\Omega|}\Big)^{\frac{1}{2}} \frac{X_i}{X_{\text{ref}}}, \qquad i = 1 \ldots n, \tag{3.1}$$

where $|\Omega|$ is the measure (length, area, or volume) of Ω, and where $X_{\text{ref}} > 0$ is the *reference value* used to adimensionalize X. The coefficient $|\Omega|$ in (3.1) ensures, as above, that the Euclidean norm of x represents a mean value of the adimensionalized parameters:

$$\|x\|_{I\!R^n} \overset{\text{def}}{=} \Big(\sum_{i=1}^{n} x_i^2\Big)^{\frac{1}{2}} = \Big(\frac{1}{|\Omega|}\int_\Omega \frac{X_h(\xi)^2}{X_{\text{ref}}^2}\,\mathrm{d}\xi\Big)^{\frac{1}{2}}. \tag{3.2}$$

2. *For a continuous piecewise linear approximation:* The function $\xi \rightsquigarrow X_h$ is now continuous, and linear over each cell, as described in Sect. 2.6.1. The degrees of freedom of X_h are its values X_M at the nodes M of $\mathcal{T}_h$. One approximate scalar product in $L^2(\Omega)$ is then, with the notations (2.52) and (2.53),

$$I_\Omega(X_{0,h}X_{1,h}) \overset{\text{def}}{=} \sum_{M \text{ node of } \mathcal{T}_h} \alpha_M X_{0,M} X_{1,M} \approx \int_\Omega X_{0,h}X_{1,h}, \tag{3.3}$$

where the coefficients α_M, defined in (2.61), satisfy

$$\sum_{M \text{ node of } \mathcal{T}_h} \alpha_M = |\Omega|. \tag{3.4}$$

So we can still define the calibrated parameters by (compare with (3.1))

$$x_i = \Big(\frac{\alpha_{M_i}}{|\Omega|}\Big)^{\frac{1}{2}} \frac{X_{M_i}}{X_{\text{ref}}}, \qquad i = 1 \ldots n. \tag{3.5}$$

As in the case of discontinuous approximation, this ensures that the Euclidean norm of x is an (approximate) mean value of the adimensionalized parameter (compare with (3.2))

$$\|x\|_{I\!R^n} \overset{\text{def}}{=} \Big(\sum_{i=1}^{n} x_i^2\Big)^{\frac{1}{2}} = \Big(\frac{1}{|\Omega|} I_\Omega\Big(\frac{X_h(\xi)^2}{X_{\text{ref}}^2}\Big)\Big)^{\frac{1}{2}}, \tag{3.6}$$

With the above calibrations, the Euclidean scalar product in $I\!R^n$ corresponds, up to the multiplicative constant $1/(|\Omega|X_{\rm ref}^2)$, to the (possibly approximate) scalar product in $L^2(\Omega)$ for the physical parameters

$$\langle x, y\rangle_{I\!R^n} \stackrel{\rm def}{=} \sum_{i=1}^{n} x_i y_i = \frac{1}{|\Omega|X_{\rm ref}^2} \begin{cases} \int_\Omega X_h(\xi)Y_h(\xi)\,{\rm d}\xi, \\ \text{or} \\ I_\Omega(X_h Y_h), \end{cases} \tag{3.7}$$

according to the type of approximation chosen for the parameter (discontinuous piecewise constant or continuous piecewise linear).

Remark 3.1.1 *In the case where a continuously derivable approximation of* X *is required, then necessarily some of the degrees of freedom will be derivatives (instead of values) of* X_h*, and it will be necessary, to satisfy (3.2), (3.6), and (3.7), to use a nondiagonal scalar product on the parameter space* $I\!R^n$*. It will then be necessary to exercise some caution for the gradient calculation, as explained in Remark 3.1.2 below.* ■

3.1.2 On the Data Side

Similarly, let Z denote the data vector given by the physics, and ΔZ_j denote the uncertainty on its jth component. It is convenient to use this uncertainty to define the *calibrated* data z_j:

$$z_j = \frac{1}{\sqrt{q}} \frac{Z_j}{\Delta Z_j}, \qquad j = 1 \dots q.$$

This amounts to use ΔZ_j as unit to measure the discrepancy between the output of the model and the corresponding data x (as in (1.8), for example). So if $\delta Z \in I\!R^q$ is a vector of data perturbations such that $|\delta Z_j| \le \Delta Z_j$ for $j = 1 \dots q$, the vector δz of corresponding calibrated data perturbation satisfies

$$|\delta z_j| \le 1/\sqrt{q} \quad \text{for} \quad j = 1 \dots q \quad \text{and hence} \qquad \|\delta z\| \le \Delta z \stackrel{\rm def}{=} 1,$$

so that the uncertainty on the calibrated data vector z is $\Delta z = 1$.

When the Z_j's are all independent Gaussian variables with standard deviation σ_j, one can take $\Delta Z_j = \sigma_j$, in which case problem (2.5) coincides with the maximum likelihood estimator.

3.1.3 Conclusion

For the rest of the chapter, the parameter x and data z that appear in the NLS inverse problem (2.5) will be the *calibrated parameter and data* defined earlier, with the parameter space E and data space F equipped with the Euclidean norms

$$\|x\|_E = \|x\| = \Big(\sum_{i=1}^{n} x_i^2\Big)^{1/2}, \qquad \|v\|_F = \|v\| = \Big(\sum_{j=1}^{q} v_j^2\Big)^{1/2},$$

which represent mean values of the adimensionalized parameters and data.

Remark 3.1.2 *The calibration could have been taken into account as well by introducing weighted scalar products and norms on the parameter space $I\!R^n$ and the data space $I\!R^q$. But this approach is error prone from a practical point of view, as the transpose of a matrix is no more simply obtained by exchanging rows and columns! For example, if $\langle\cdot,\cdot\rangle_\Lambda$ is the scalar product on $I\!R^q$ associated with a symmetric positive definite matrix Λ, the transposed M^{T_Λ}, for this scalar product, of a $q{\times}q$ matrix M is $\Lambda^{-1}M^{T}\Lambda$, where M^{T} is the usual transposed. This can be easily overseen in the numerical calculations.*

This approach, however, cannot be avoided in the case where a full (i.e., not diagonal) covariance matrix of the data is available – but all the material below can be easily adapted. ■

3.2 How Many Parameters Can be Retrieved from the Data?

Before any attempt is made to minimize the least squares objective function (2.5), the first thing to do is determine the largest number of independent parameters that can be retrieved from the sole observations at a given noise level on the data. This has to be done before any a-priori information is added into the system: the parameters to be used for this determination are the *simulation parameters* (Sect. 3.3 below), which are input to the simulation code before any parameterization is chosen. In most cases, the original unregularized problem is underdetermined, with a number n of simulation parameters much larger than the number q of observations.

This analysis is performed on the linearized problem at a few (usually one) nominal value x_{nom}. It is based on the knowledge of the "noise level" on the data, and on the singular values decomposition of the Jacobian $\varphi'(x)$. There is a large amount of literature on this subject, which usually takes into account the statistical properties of the errors on the data, see for instance [74]. We give below an elementary approach based on uncertainty analysis, but which is sufficient in the large number of cases where the statistical information on the data is lacking or scarce.

So one chooses a nominal parameter value $x_{\text{nom}} \in C$, and replaces the forward map φ by its linearization $\varphi^{\text{lin}}_{\text{nom}}$:

$$\forall \delta x \in I\!R^n, \qquad \varphi^{\text{lin}}_{\text{nom}}(\delta x) = \varphi(x_{\text{nom}}) + \varphi'(x_{\text{nom}})\,\delta x.$$

Let $z_{\text{nom}} = \varphi(x_{\text{nom}}) = \varphi^{\text{lin}}_{\text{nom}}(0)$ be the exact data corresponding to x_{nom}, $\Delta z > 0$ the uncertainty level on the data, δz with $\|\delta z\|_F \leq \Delta z$ a data error vector, $z = z_{\text{nom}} + \delta z$ the corresponding noise corrupted data, and $\tilde{x} = x_{\text{nom}} + \widetilde{\delta x}$ the corresponding solution of the linearized unconstrained inverse problem:

$$\widetilde{\delta x} \text{ minimizes } \frac{1}{2}\|\varphi^{\text{lin}}_{\text{nom}}(\delta x) - z\|_F^2 = \frac{1}{2}\|\varphi'(x_{\text{nom}})\,\delta x - \delta z\|_F^2 \text{ over } I\!R^n. \tag{3.8}$$

We evaluate the size of the error $\widetilde{\delta x}$ induced by the error δz on the data z_{nom} at the nominal value z_{nom}:

- *Absolute uncertainty:* One performs a *singular value decomposition* (SVD) of the $q \times n$ matrix $\varphi'(x_{\text{nom}})$. This produces two orthonormal bases

$$\begin{cases} e_1, \dots, e_n &= \text{basis of parameter space } I\!R^n, \\ \epsilon_1, \dots, \epsilon_q &= \text{basis of data space } I\!R^q, \end{cases} \tag{3.9}$$

  and a sequence of $r \leq \min\{n, q\}$ strictly positive numbers:

$$\mu_1 \geq \mu_2 \geq \cdots \geq \mu_r > 0 \tag{3.10}$$

  called the *singular values* such that

$$\begin{cases} \varphi'(x_{\text{nom}})e_i &= \mu_i \epsilon_i, \qquad i = 1, \cdots, r, \\ \varphi'(x_{\text{nom}})e_i &= 0, \qquad i = r+1, \cdots n. \end{cases} \tag{3.11}$$

  It is then convenient to *complement* the above singular values $\mu_i, i = 1 \cdots r$, by defining:

$$\mu_{r+1} = \mu_{r+2} = \cdots = \mu_n = 0.$$

The solution of (3.8) is then given by

$$\mu_i \, \widetilde{\delta x_i} = \delta z_i, \qquad i = 1 \ldots r, \tag{3.12}$$

where $\widetilde{\delta x_i}$ and δz_i are the coefficients of $\widetilde{\delta x}$ and δz on the singular bases:

$$\begin{cases} \widetilde{\delta x} &= \sum_{i=1}^{n} \widetilde{\delta x_i} \, e_i, \\ \delta z &= \sum_{j=1}^{q} \delta z_j \, \epsilon_j. \end{cases}$$

This shows first that the perturbations $\widetilde{\delta x_i}, i = r+1 \cdots n$, cannot be retrieved from the available data, as they do not appear in (3.12), which give the solutions of (3.8). Hence the uncertainty on the components x_i of $x_{\rm nom}$ on singular vectors corresponding to zero singular values is infinite:

$$\Delta x_i = +\infty, \qquad i = r+1 \cdots n.$$

Then for $i = 1 \cdots r$, (3.12) gives, as $|\delta z_i| \leq \|\delta z\|_F \leq \Delta z$,

$$\mu_i \, |\widetilde{\delta x_i}| \leq \Delta z, \qquad i = 1 \ldots r,$$

so that the uncertainty Δx_i on the component x_i of $x_{\rm nom}$ on singular vectors corresponding to nonzero singular values is given by

$$\Delta x_i = \frac{\Delta z}{\mu_i}, \qquad i = 1 \ldots r. \tag{3.13}$$

Summing up, we see that the components of $x_{\rm nom}$ on singular vectors associated with zero singular values cannot be retrieved (the corresponding uncertainty Δx_i is infinite). For the components on the other singular vectors, the larger the singular value, the smaller the uncertainty! To determine which of these components can be retrieved in a stable way, one has to estimate the *relative uncertainty*.

- *Relative uncertainty:* We suppose that the forward map φ satisfies

$$\varphi \text{ is defined at } 0 \quad \text{and} \quad \varphi(0) = 0$$

(this condition can always be satisfied by translating the origin in the parameter and data spaces).

Then, if the linearized model at x_{nom} is valid, one has

$$0 = \varphi(0) \simeq \underbrace{\varphi(x_{\text{nom}})}_{z_{\text{nom}}} + \varphi'(x_{\text{nom}})(0 - x_{\text{nom}}),$$

so that

$$z_{\text{nom}} \simeq \varphi'(x_{\text{nom}})x_{\text{nom}}.$$

But the largest singular value μ_1 is the norm of the Jacobian $\varphi'(x_{\text{nom}})$ associated with the usual Euclidian norm, hence,

$$\|z_{\text{nom}}\| \lesssim \mu_1 \|x_{\text{nom}}\|. \tag{3.14}$$

Combining (3.13) and (3.14) gives

$$\frac{\Delta x_i}{\|x_{\text{nom}}\|} \lesssim \frac{\mu_1}{\mu_i} \frac{\Delta z}{\|z_{\text{nom}}\|}, \qquad i = 1 \ldots r. \tag{3.15}$$

So we see that the relative uncertainty on the component x_i of x_{nom} on one of the first r singular vectors is amplified at most by a factor μ_1/μ_i from the relative uncertainty on the data. For the $n-r$ remaining components x_i corresponding to singular values $\mu_i = 0$, one has seen that $\Delta x_i = +\infty$, so that formula (3.15) remains valid in that case.

- *Number of retrievable parameters:* The component of x_{nom} on the ith singular vector e_i is said to be *retrievable* if it is *above the noise level*, that is, if its relative uncertainty is *smaller than one.*

  Hence the *number* $n_r \leq n$ *of retrievable parameters* for an uncertainty level Δz on the data is given by

$$n_r = \text{ number of singular values } \mu_i \text{ s.t.: } \frac{\mu_i}{\mu_1} \geq \frac{\Delta z}{\|z_{\text{nom}}\|}.$$

  The determination of the number n_r of retrievable parameter is the first numerical computation to perform for the solution of an inverse problem. It gives an upper bound to the number of independent parameters that can be estimated from the data in a stable way.

  Its practical determination requires the computation of the Jacobian $\varphi'(x_{\text{nom}})$. Because the number n of (simulation) parameters is usually much larger than the number q of data, this is most efficiently done row-by-row by the adjoint state technique (Sect. 2.3 of Chap. 2). Of course, in the rare case where $n \leq q$, a column-by-column determination of the Jacobian using the sensitivity functions approach (Sect. 2.2 of Chap. 2) is the computationally less expensive approach. Examples of application of this uncertainty analysis can be found in [50, 2, 80].

Remark 3.2.1 *It can happen that the computation of the Jacobian is not feasible when there are large number of parameters* **and** *data, so that n_r cannot be determined before the inversion is actually performed. However, the determination of the gradient $\nabla J(x)$ is always feasible by adjoint state. It is advisable in this case to use an* adaptive parameterization *(Sect. 3.7 below), which can use the information given by $\nabla J(x)$ to take care automatically of the tradeoff between data fitting and overparameterization.* ■

3.3 Simulation Versus Optimization Parameters

It is important to distinguish two (usually distinct) parameter vectors:

- The *simulation parameters* x_{sim} (denoted by x in Sect. 1.3.4) are the ones that are input to the simulation code. They should contain all model parameters one might dream to invert for, and provide the most comprehensive description of the numerical forward model used.

  In particular, when the unknown parameter is a function, the numerical computation of the forward model φ requires its discretization over some mesh $\mathcal{T}_h$, and the simulation parameter x_{sim} will be the (large) vector made of all the calibrated degrees of freedom of this function over $\mathcal{T}_h$. Its dimension n_{sim} will hence in general be much larger than the dimension q of the data vector z.

  For example, if the unknown parameter is the diffusion coefficient a in the inverse problem of Sect. 1.6, and if the resolution of the elliptic equation is made, as proposed in Sect. 2.6.1, on a mesh $\mathcal{T}_h$ of triangular finite elements K, the natural choice for x_{sim} is the vector of the *calibrated values* of the discretized parameter a_h on $\mathcal{T}_h$, where a_h is defined in (2.54). The simulation parameter is hence given by (3.1), where now X_i represents the (constant) value a_{K_i} of a_h over the ith element of $\mathcal{T}_h$, and X_{ref} the reference value chosen to adimensionalize a.

Not surprisingly, the analysis of the previous section, when applied to the estimation of x_{sim} from the data z, usually shows that the number n_r of retrievable parameters is strictly smaller than the dimension n_{sim} of x_{sim}. Hence solving the optimization problem (2.5) with the parameters $x = x_{\text{sim}}$ is purposeless, as it would be underdetermined. It is then necessary to regularize the problem by adding a-priori information, as discussed in Sect. 1.3.4, until all parameters are "above the noise level" induced by the uncertainty on the data. We focus in the present chapter on the *regularization by parameterization* approach, which proceeds by reduction of the number of unknowns.

- The *optimization parameters* x_{opt} are the ones that are input to the optimization code, and hence which are actually estimated by minimization of the objective function. The vector x_{opt} is used to parameterize the simulation parameter x_{sim} (c.f. (1.16) of Sect. 1.3.4):

$$x_{\text{sim}} = \psi(x_{\text{opt}}), \qquad x_{\text{opt}} \in C_{\text{opt}} \quad \text{with} \quad \psi(C_{\text{opt}}) \subset C, \tag{3.16}$$

where ψ is the *parameterization routine.*

The dimension n_{opt} of x_{opt} has to be taken small enough to avoid over-parameterization, that is, the Jacobian of φ as a function from $I\!R^{n_{\text{opt}}}$ to $I\!R^q$ should have all its singular values above the noise level (Fig. 3.1), and large enough for the solution $\widehat{x_{\text{opt}}}$ of (2.5) to explain the data up to the noise level. Moreover, the chosen parameterization should if possible enhance the conditioning of the optimization problem (2.5).

The choice of optimization parameters is a critical, but delicate step of the inversion, and one is very often led to experiment with it. It is hence practically important to organize the code in such a way that *changing optimization parameters* incurs a *minimum amount of changes* in the code. When the information flux between the *optimization routine* and the *forward+adjoint modeling code* is organized as in the flowchart of Fig. 3.8 below, changing optimization parameters requires only to replace the two subroutines that compute $x_{\text{sim}} = \psi(x_{\text{opt}})$ (*parameterization routine*) and $g_{\text{opt}} = \psi'(x_{\text{opt}})^T . g_{\text{sim}}$ (*adjoint parameterization routine*). This suggest that these two subroutine should be developed and tested at the same time, prior to being inserted in the inversion code. The necessary mathematical tools have been developed in Chap. 2, and are applied in Sect. 3.8.2 below to an example.

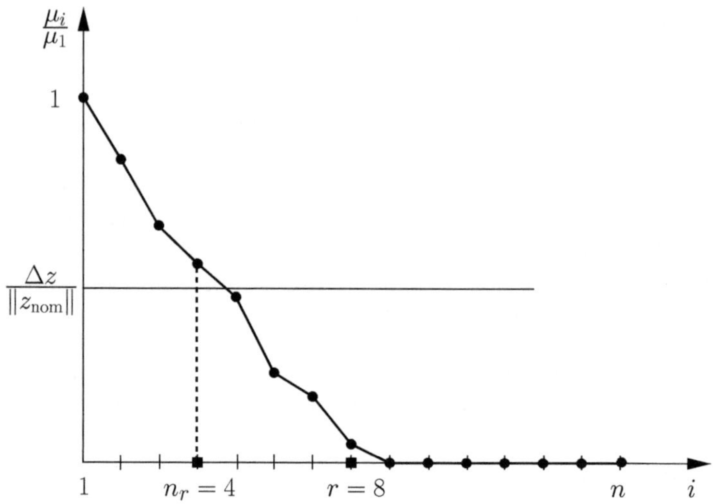

Figure 3.1: Graphical determination of the number of retrievable parameters: $\mu_1, \ldots, \mu_n =$ singular values, in decreasing order

Remark 3.3.1 *Even in the case where all parameters in* x_{sim} *are retrievable, the choice* $x_{\text{opt}} = x_{\text{sim}}$ *does not necessarily produce the best conditioning of the optimization problem, and one can be led to choose a parameter* x_{opt} *distinct from* x_{sim}*, see* adapted multiscale basis *in Sect. 3.6.* ■

We describe below four approaches to reduce the number of unknown parameters.

3.4 Parameterization by a Closed Form Formula

When the information content of the data is very small, and the number of simulation parameters is large, one solution is to express the simulation parameters as a function of a small number of coefficients through a closed form formula.

For example, suppose that the unknown parameter is a one-dimensional temperature profile T on the $[0, 1]$ interval. A natural simulation parameter is then

$$x_{\text{sim}} = (T_1/T_{\text{ref}}, \ldots, T_{n_{\text{sim}}}/T_{\text{ref}}),$$

where T_{ref} is a reference temperature used for calibration purpose (Sect. 3.1), and where T_i, $i = 1 \ldots, n_{\text{sim}}$, represents the values of the temperature on a discretization of $[0, 1]$ into n_{sim} intervals of length $h = 1/n_{\text{sim}}$. Suppose that the singular value analysis of Sect. 3.2 shows that only a few number $n_{r,\text{sim}}$ of independent simulation parameters can be recovered from the data, and that one expects from the physics of the problem that the temperature profile has the shape of a bump. One can then incorporate this information into our formulation by searching (for example) for a Gaussian temperature profile:

$$x_{\text{sim},i} = \frac{T_{\max}}{T_{\text{ref}}} \exp\Big(-\frac{1}{2}\Big(\frac{(i-1/2)h - \xi_{\max}}{\Delta\xi}\Big)^2\Big), \; i = 1 \ldots n_{\text{sim}},$$

where the optimization parameter is

$$x_{\text{opt}} = \Big(\frac{T_{\max}}{T_{\text{ref}}}, \xi_{\max}, \Delta\xi\Big) \in I\!R^3.$$

Even if $n_{r,\text{sim}}$ was larger than three, it is necessary to perform again the analysis of Sect. 3.1, this time with the x_{opt} parameter, as there is no guarantee that the directions of $I\!R^{n_{\text{sim}}}$ "spanned" (the quotes come from the fact that $x_{\text{opt}} \rightsquigarrow x_{\text{sim}}$ is not linear) by the chosen parameterization are in the subspace of $I\!R^{n_{\text{sim}}}$ spanned by the singular vectors associated with the largest singular values μ_{sim}.

The effect on conditioning of parameterizing by a closed form formula is unpredictable. However the author's experience is that it tends rather to deteriorate, this being partly compensated by the very small number of parameters.

In the present situation where the simulation parameters x_{sim} are expressed by a simple formula in term of the optimization parameters x_{opt}, the determination of $\nabla_{x_{\text{opt}}} J$ from $\nabla_{x_{\text{sim}}} J$ is very easy (Sect. 3.8.2 below).

3.5 Decomposition on the Singular Basis

It is the first natural idea after one has performed the singular value decomposition of the Jacobian, at some nominal value $x_{\text{sim}}^{\text{nom}}$, in order to determine the number n_r of retrievable parameters (Sect. 3.2): why not reduce the search to the n_r coefficients of x_{sim} on the singular basis vectors associated with singular values that are above the noise level? With the notations of Sect. 3.2, this amounts to choose for optimization parameters and parameterization map:

$$\left\{ \begin{array}{lcll} n_{opt} & = & n_r \leq n_{sim}, & \\ x_{opt,i} & = & x_i^{SVD} \stackrel{\text{def}}{=} \langle x_{sim}, e_i \rangle_{I\!R^{n_{sim}}}, & i = 1 \ldots n_r, \\ x_{sim} & = & \psi(x_{opt}) = \sum_{i=1}^{n_r} x_{opt,i}\, e_i + \sum_{i=n_r+1}^{n_{sim}} (x_i^{SVD})_{nom}\, e_i. & \end{array} \right. \tag{3.17}$$

There is no need, with this choice, to redo the analysis of Sect. 3.2 with the parameters x_{opt}, as by construction they are all above the noise level!

This parameterization is interesting for the analysis of the problem: the directions of the singular vectors $\epsilon_1 \ldots \epsilon_{n_r}$ indicate which part of the data space actually carries useful information on the parameter x. Similarly the singular vectors $e_1 \ldots \epsilon_{n_r}$ indicate which combination of the simulation parameters can be best estimated.

These nice properties are counterbalanced by the fact that the corresponding physical interpretation is not always easy (as, e.g., in [36] for a large size geophysical inverse problem). A nice exception is when the singular vectors associated with the largest singular values happen to point in the direction of axes of the simulation parameter space $I\!R^{n_{sim}}$. It is then possible to order the simulation parameters according to decreasing singular values, so that the n_r first simulation parameters are retrievable, each with its own uncertainty level, with the remaining ones having to be fixed (at their nominal values, for example). This is the case for the inversion of the Knott–Zoeppriz equations of Sect. 1.1, see [50].

Another difficulty comes from the fact that the Jacobian – and hence its singular vectors – depends on the choice of the nominal parameter x_{sim}^{nom} where it is evaluated. So the nice properties of the singular vector can get lost when the current parameter changes during the course of the optimization, which makes their use purposeless for uncertainty analysis.

For all these reasons, the choice (3.17) is seldom used to define the optimization parameters – but performing a SVD at a nominal value is nevertheless useful to estimate n_r and gain insight into the problem.

3.6 Multiscale Parameterization

Multiscale parameterization concerns the case of distributed parameters, where the unknown is a function $\xi \in \Omega \rightsquigarrow X(\xi) \in I\!R$: most often ξ is the space variable, in which case X describes a heterogeneity, but not necessarily: it can also be a dependent variable (as temperature, concentration, saturation, etc.), in which case X represents a nonlinearity.

3.6.1 Simulation Parameters for a Distributed Parameter

According to Sect. 3.3, the natural simulation parameter $x_{\rm sim}$ for a distributed physical parameter X is the vector of the *calibrated values* x_i of its approximation $X_h \in \boldsymbol{E}_h$ over the simulation mesh $\mathcal{T}_h$:

$$x_{{\rm sim},i} = x_i, \; i = 1 \dots n_{\rm sim}. \tag{3.18}$$

These calibrated values x_i, defined in Sect. 3.1.1, are the coefficients of the decomposition of X_h on a set of *calibrated local basis functions* e_i, $i = 1 \dots n_{\rm sim}$,

$$X_h = \sum_{i=1}^{n_{\rm sim}} x_{{\rm sim},i}\, e_i.$$

The formulas for x_i given in Sect. 3.1.1 show the following:

1. *For a discontinuous piecewise constant approximation*, x_i is given by (3.1) and e_i by

$$\begin{cases} e_i = X_{\rm ref} \left(\frac{|\Omega|}{|K_i|}\right)^{\frac{1}{2}} \chi_{K_i}, \quad i = 1 \cdots n_{\rm sim}, \\ |e_i|^2_{L^2(\Omega)} = X^2_{\rm ref}|\Omega|, \end{cases} \tag{3.19}$$

   where χ_{K_i} denotes the characteristic function of the ith cell K_i of the simulation mesh $\mathcal{T}_h$.

2. *For a continuous piecewise linear approximation*, x_i is given by (3.5) and e_i by

$$\begin{cases} e_i = X_{\rm ref} \left(\frac{|\Omega|}{\alpha_{M_i}}\right)^{\frac{1}{2}} \omega_{M_i}, \quad i = 1 \cdots n_{\rm sim}, \\ |e_i|^2_{h,L^2(\Omega)} = X^2_{\rm ref}|\Omega| = 2|e_i|^2_{L^2(\Omega)}, \end{cases} \tag{3.20}$$

where ω_{M_i} denotes the local basis function of $\boldsymbol{E}_h$ associated to node M_i (its value is 1 at the ith node M_i and 0 at all other nodes of $\mathcal{T}_h$), and where $|\cdot|_{h,L^2(\Omega)}$ is the norm on $\boldsymbol{E}_h$ associated to the scalar product

$$\langle X_h, Y_h \rangle_{h,L^2(\Omega)} \stackrel{\text{def}}{=} I_\Omega(X_h Y_h),$$

with I_Ω defined in (3.3).

Finally, for both approximations,

• The simulation parameters $x_{\text{sim},i}$ represent calibrated *local* (cell or node) *values* of the unknown parameter function over the mesh $\mathcal{T}_h$

• The calibrated basis $e_1 \dots e_{n_{\text{sim}}}$ of $\boldsymbol{E}_h$ given by (3.19) or (3.20) is *normalized* (to the value $X_{\text{ref}}|\Omega|^{1/2}$) and *orthogonal* for the scalar product $\langle\cdot,\cdot\rangle_{L^2(\Omega)}$ (discontinuous piecewise constant functions) or $\langle\cdot,\cdot\rangle_{h,L^2(\Omega)}$ (continuous piecewise linear functions)

• The Euclidean scalar product of $\mathbb{R}^{n_{\text{sim}}}$ corresponds (up to the coefficient $X^2_{\text{ref}}|\Omega|$) – see (3.7) to the scalar product $\langle\cdot,\cdot\rangle_{L^2(\Omega)}$ or $\langle\cdot,\cdot\rangle_{h,L^2(\Omega)}$.

Once the simulation parameters x_{sim} have been chosen, the maximum number of retrievable parameters n_{r} can be determined as explained in Sect. 3.2.

3.6.2 Optimization Parameters at Scale k

A mesh $\mathcal{T}'$ is a sub-mesh of the mesh $\mathcal{T}$ (one writes $\mathcal{T}' \supset \mathcal{T}$), if $\mathcal{T}'$ is a refinement of $\mathcal{T}$, that is, if any element K of $\mathcal{T}$ is the union of elements K' of $\mathcal{T}'$. In a scale-by-scale approach, the parameter is searched for successively over a sequence of embedded meshes $\mathcal{T}^0 \subset \mathcal{T}^1 \subset \cdots \subset \mathcal{T}^K$, which allow to represent more and more details.

In Sect. 3.6.1, we have used as *simulation parameter space* the approximation space $\boldsymbol{E}_h$ introduced in Sect. 3.1.1 over the *simulation mesh* $\mathcal{T}_h$. The same approximation can now be used over each *coarse mesh* $\mathcal{T}^k$ to define a sequence of *embedded function spaces* $\boldsymbol{E}^k$:

$$\boldsymbol{E}^0 \subset \boldsymbol{E}^1 \cdots \subset \boldsymbol{E}^K \quad \text{with } \dim \boldsymbol{E}^k = n_k\ ,\ k = 0 \dots K.$$

Each space $\boldsymbol{E}^k$ is the *approximation space at scale* k, with $k = 0$ corresponding to the *coarse* scale, and $k = K$ to the *fine* scale. The finer mesh $\mathcal{T}^K$ is chosen fine enough so that

$$n_k \geq n_{\text{r}},$$

where n_r is the number n_r of retrievable parameters determined at the end of Sect. 3.6.1. If n_r is not known, one can choose $\mathcal{T}^k$ of a size similar to $\mathcal{T}_h$ – but the optimization will stop at a scale k usually smaller than K.

Let $e_i^k \in \boldsymbol{E}^k$, $i = 1 \dots n_k$, be a collection of basis functions of $\boldsymbol{E}^k$ at each scale $k = 0 \dots K$. The choice of this basis (local or multiscale) will be discussed in paragraph 4 of Sect. 3.6.3. The *optimization parameter* $x_{\text{opt}}^k \in I\!R^{n_k}$ at scale k is then made of the coefficients on this basis of the approximation x^k of x in $\boldsymbol{E}^k$:

$$x^k = \sum_{i=1}^{n_k} x_{\text{opt},i}^k \, e_i^k.$$

Given an optimization parameter $x_{\text{opt}}^k \in I\!R^{n_k}$, the simulation parameter $x_{\text{sim}} \in I\!R^{n_{\text{sim}}}$ is computed (see (3.16)) by

$$x_{\text{sim}} = \psi^k \, x_{\text{opt}}^k, \tag{3.21}$$

where the *parameterization matrix* ψ^k represents the *interpolation* from the optimization parameter space $\boldsymbol{E}^k$ equipped with the chosen basis e_i^k, $i = 1 \dots n_k$, to the simulation parameter space $\boldsymbol{E}_h$ equipped with the local basis e_i, $i = 1 \dots n_{\text{sim}}$.

Choosing $\mathcal{T}^K = \mathcal{T}_h$ for the fine scale mesh will simplify somewhat the interpolation matrix ψ^k, as the nodes of a coarse mesh $\mathcal{T}^k$ are now also nodes of the fine and simulation mesh $\mathcal{T}^K = \mathcal{T}_h$ over which one interpolates. But this is possible only in certain situations, for example, when $\mathcal{T}_h$ is a regular two-dimensional $N_1 \times N_2$ rectangular mesh with $N_1 = 2^K N_1^0$ and $N_2 = 2^K N_2^0$, in which case $\mathcal{T}^0$ is a $N_1^0 \times N_2^0$ mesh.

3.6.3 Scale-By-Scale Optimization

The admissible sets at each scale are defined by

$$C^k = \{x_{\text{opt}}^k \in I\!R^{n_k} \mid \sum_{i=1}^{n_k} x_{\text{opt},i}^k \, e_i^k \in C\}.$$

Then, given an initial guess $x^{\text{init}} \in C^0$, the optimization problem (2.5) is solved successively at scales $k = 0, 1, \dots,$ the result of the optimization at scale $k-1$ being used (after interpolation over the mesh $\mathcal{T}^k$) as initial value for the optimization at scale k, until the data are explained up to the uncertainty level

$$\begin{cases} \text{set: } k = 0 \quad \text{and} \quad \hat{x}_{\text{opt}}^{-1} = x^{\text{init}}, \\ \text{starting from } \hat{x}_{\text{opt}}^{k-1} \in C^k \text{ , find } \hat{x}_{\text{opt}}^{k} = argmin_{x_{\text{opt}} \in C^k} J(x_{\text{opt}}), \\ \text{increment } k \text{ until the fit is down to uncertainty level.} \end{cases} \tag{3.22}$$

In this approach, the final number of parameters that will be estimated is not known before the end of the scale-by-scale optimization.

It was observed in a certain number of cases [57, 58, 15, 64, 69] that this approach, which searches for the features of the unknown parameter successively from coarse to fine scale.

– Allows local optimization algorithms to converge to (or near to) the global minimum of the objective function, instead of stopping in a possibly remote local minimum or stationary point

– And can enhance the conditioning of the optimization problems

There is to date no rigorous mathematical explanation of this phenomenon, though it should not be out of reach to formalize the following sketch of analysis:

– We first introduce a class of "nicely nonlinear" inverse problems, for which scale-by-scale optimization is expected to perform well

– Then we consider a prototype "nicely nonlinear" problem with low dimensions (two parameters and three data!) whose attainable set can be represented on a picture

– We explain on this picture the properties of multiscale approach with respect to local minima or stationary points and conditioning

– And finally we discuss the influence of the choice of basis functions at each scale on the conditioning of the optimization problem.

1. *Nicely nonlinear problems:* At a point $x_{\text{sim}} \in C$ and in a unit direction $u \in I\!R^{n_{\text{sim}}}$, we define

   - $s = \|V\| =$ **sensitivity** of the forward map φ, where $V = D_u\varphi(x_{\text{sim}})$
   - $\kappa = \|A\|/\|V\|^2 =$ (upper bound to the) **curvature** of the forward map φ at x_{sim} in the direction u, where $A = D^2_{u,u}\varphi(x_{\text{sim}})$

   By construction, V is the *velocity* and A the *acceleration* at $t = 0$ along the curve $t \rightsquigarrow \varphi(x_{\text{sim}} + tu)$ drawn on the attainable set $\varphi(C)$. These quantities play an important role in the analysis of the nonlinear inverse problem (2.5) given in Chap. 4, where they are discussed in detail.

Let then $(e_i^k, i = 1 \dots n_k)$ denote, for each scale k, a *normalized* and *local* basis of $\boldsymbol{E}^k$. This basis can be defined, over the mesh $\mathcal{T}^k$, in the same way the basis (3.19) or (3.20) of $\boldsymbol{E}_h$ was defined over $\mathcal{T}_h$.

We can introduce now a rather loosely defined class of problems, where the scale-by-scale decomposition of the parameter space structures the sensitivity and curvature levels:

Definition 3.6.1 *The NLS inverse problem (2.5) is "nicely nonlinear" if the* sensitivity decreases *and the* curvature increases *in the directions of the local basis function* e_i^k *when* k *moves* from coarse to fine *scales.*

In short, "nicely nonlinear" problems are more sensitive but less nonlinear at coarse scales, and less sensitive but more nonlinear at fine scales. Hence the scale-by-scale resolution (3.22), which solves the coarse scale problem first, can be expected to perform well.

An example of "nicely nonlinear" problem (see [57, 58]) is the estimation of a one-dimensional diffusion parameter, defined in Sect. 1.4 and analyzed for wellposedness in Sect. 4.8. One can also conjecture that the estimation of the two-dimensional diffusion parameter (defined in Sect. 1.6 and analyzed in Sects. 4.9 and 5.4) is "nicely nonlinear," as suggested in [30] and in Remark 4.9.5.

At the opposite end of the spectrum, one finds the problem of estimating the sound velocity from wavefield measurements: large scale perturbations of the sound velocity produce phase-shifts ($\simeq$ translation in time) in the observed wavefield, which correspond to a high curvature because of the high frequency content of the geophysical signals – a very difficult problem, which cannot benefit from the multiscale approach described here [24].

2. *A prototype nicely nonlinear problem:* Consider a forward map $\varphi : \mathbb{R}^2 \rightsquigarrow \mathbb{R}^3$ and an admissible set $C = [-1, +1] \times [-1, +1] \subset \mathbb{R}^2$, where the x_1-axis corresponds to coarse scale (the space $\boldsymbol{E}^0$ above), and the x_2-axis to the fine details (a supplementary subspace $\boldsymbol{W}^1$ of

$\boldsymbol{E}^0$ in $\boldsymbol{E}^1$, see (3.25) below). Let us suppose that this problem is "nicely nonlinear":

- On the subset $C^0 = C \cap \{x|x_2 = 0\}$ of C, the curvature of φ is small enough for the attainable set $\varphi(C^0)$ to be the piece of curve shown in Fig. 3.2, top
- On the segments of $C^1 = C$ parallel to the second axis $\{x|x_1 = 0\}$, the curvature of φ increases significantly, but at the same time its sensitivity also decreases significantly, so that the attainable set $\varphi(C^1)$ might look like the piece of surface in Fig. 3.2, bottom.

3. *Stationary points and scale-by-scale resolution:* Let the data $z \in I\!R^3$ be given as in Fig. 3.3, and $x^{\text{init}} = (-1, 0) \in C^0 \subset C$ be the given initial guess of x. We see the following in Figs. 3.3 and 3.4:

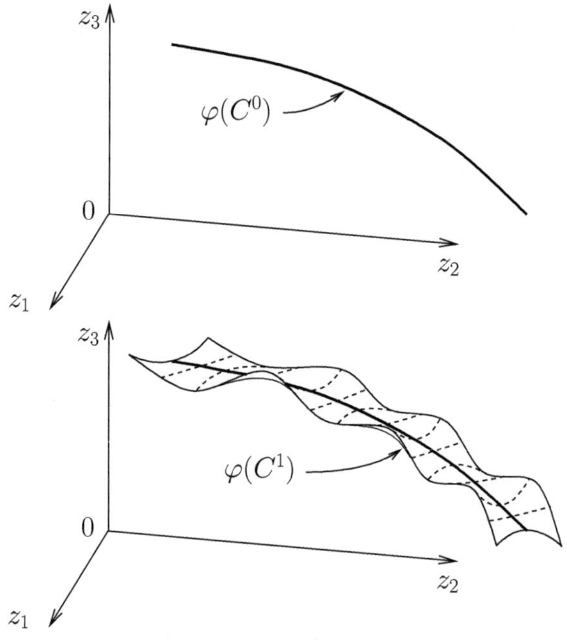

Figure 3.2: A representation of the attainable set for the prototype example at scale 0 and 1

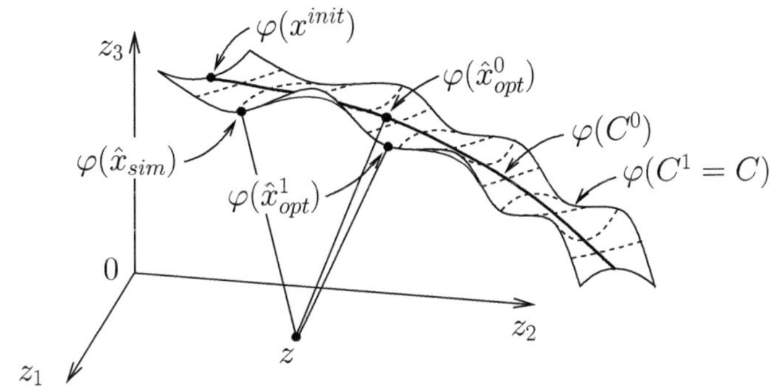

Figure 3.3: Comparison of local vs. multiscale parameterization for the convergence of gradient algorithms

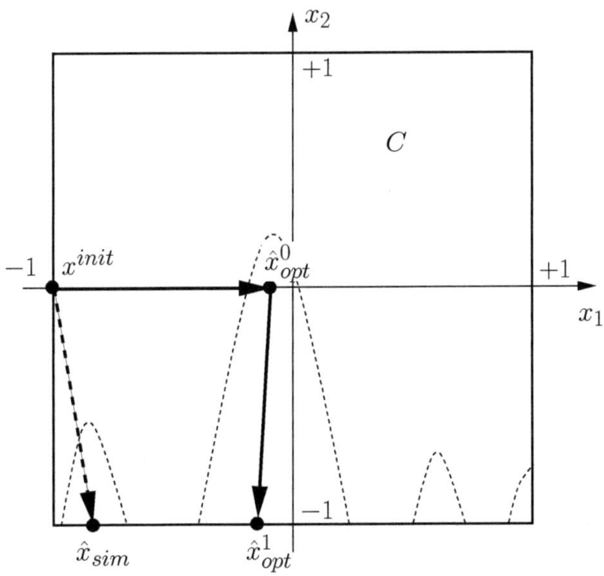

Figure 3.4: Optimization paths in C for the local (*dashed line*) and the multiscale (*full line*) resolution. The *thin dashed lines* correspond to one level line of the objective function

- If the inversion is performed by solving the usual formulation (2.5) by a local gradient method with respect to the local variables $x_{\text{sim}} = (x_1, x_2)$, the algorithm will stop, because of the oscillations of the attainable set $\varphi(C)$, at a local minimizer $\hat{x}_{\text{sim}}$ far from the global one.
- Let now the inversion be performed according to the scale-by-scale formulation (3.22). At the scale $k = 0$, any local algorithm will reach the global minimum $\hat{x}^0_{\text{opt}}$, as z is at a distance of $\varphi(C^1)$ smaller than its smallest radius of curvature. Then at scale $k = 1$, the local algorithm will start at $\hat{x}^0_{\text{opt}}$, and stop at one nearby local minimum $\hat{x}^1_{\text{opt}}$, which by construction is much closer to the global minimum – it coincides with it on the chosen example.

We see on this example that the ability of the multiscale resolution to overcome the problem of local minima or stationary points for a nicely enough nonlinear problem relies solely on a correct resolution in $\mathbb{R}^{n_k}$ of the optimization problems (3.22), in the sense that, at each scale k, the optimization algorithm converges to the nearest local minimum or stationary point.

4. *Choosing optimization parameters at scale k:* To perform the scale-by-scale optimization (3.22), it is necessary to decide whether the optimization vector x^k_{opt} at scale k is made of the coefficients of X_h on a *local* or a *multiscale* basis of $\boldsymbol{E}_k$:

   (A) *On a local basis*: In most cases, using at each scale $k = 0 \ldots K$ the simplest *local* basis $(e^k_i, i = 1 \ldots n_k)$ will produce satisfactory results – but maybe not with the best computational efficiency:

   (a) *For a discontinuous piecewise constant approximation:* $\boldsymbol{E}^k$ is made of piecewise constant functions over the elements of $\mathcal{T}^k$, and the optimization parameters and the associated local basis functions at scale k are given by

$$\begin{cases} x^k_{\text{opt},i} &= \left(\dfrac{|K_i|}{|\Omega|}\right)^{\frac{1}{2}} \dfrac{X_i}{X_{\text{ref}}} & i = 1 \cdots n_k, \\ e^k_i &= X_{\text{ref}} \left(\dfrac{|\Omega|}{|K_i|}\right)^{\frac{1}{2}} \chi_{K_i} & i = 1 \cdots n_k, \\ |e_i|_{L^2(\Omega)} &= X_{\text{ref}}\, |\Omega|^{\frac{1}{2}} & i = 1 \cdots n_k, \end{cases} \tag{3.23}$$

where now K_i is the ith cell of the *coarse* mesh $\mathcal{T}^k$ (compare with the definitions (3.1), (3.18), and (3.19) of the simulation parameters and basis defined over the fine *simulation* mesh) $\mathcal{T}_h$.
Because of the $L^2(\Omega)$ orthogonality and normalization of the functions e_i^k in (3.23), the Euclidean scalar product for x_{opt}^k in $I\!R^{n_k}$ coincides (up to the coefficient $X_{\text{ref}}^2|\Omega|$) with the $L^2(\Omega)$ scalar product in $\boldsymbol{E}_k$.

(b) *For a continuous piecewise linear approximation:* Here $\boldsymbol{E}^k$ is made of continuous piecewise linear functions over coarse triangular elements of $\mathcal{T}^k$. The optimization parameters and the local basis functions at scale k are then given by

$$\left\{\begin{array}{lll} x_i & = \left(\dfrac{\alpha_{M_i}}{|\Omega|}\right)^{\frac{1}{2}} \dfrac{X_{M_i}}{X_{\text{ref}}} & i=1\cdots n_k, \\ e_i^k & = X_{\text{ref}}\left(\dfrac{|\Omega|}{\alpha_{M_i}}\right)^{\frac{1}{2}} \omega_{M_i}^k & i=1\cdots n_k, \\ |e_i^k|_{k,L^2(\Omega)}^2 & = X_{\text{ref}}^2\,|\Omega| \;=\; 2|e_i^k|_{L^2(\Omega)}^2 & i=1\cdots n_k, \end{array}\right. \tag{3.24}$$

where now ω_M^k is the function of $\boldsymbol{E}^k$ with value 1 at the ith node M_i and value 0 at all other nodes of the *coarse* mesh $\mathcal{T}^k$, and where $|\cdot|_{k,L^2(\Omega)}$ denotes the approximate $L^2(\Omega)$-norm on $\boldsymbol{E}^k$ evaluated using the quadrature formula (3.3) over the triangulation $\mathcal{T}^k$ instead of $\mathcal{T}_h$ (compare with the definitions (3.5), (3.18), and (3.20) of the simulation parameters and basis).
As in (3.20), the basis (3.24) is not any more orthogonal in $L^2(\Omega)$, but it is orthogonal for the approximate scalar product $\langle\cdot,\cdot\rangle_{k,L^2(\Omega)}$. Hence the interpretation of the Euclidean scalar product in $I\!R^{n_k}$ as the scalar product in $L^2(\Omega)$ is retained for this approximate scalar product.

The implementation of the scale-by-scale optimization with a local basis at each scale is relatively simple. Because of the proportionality of the Euclidean scalar product in $I\!R^{n_k}$ and the $L^2(\Omega)$ scalar product in $\boldsymbol{E}^k$, the optimization problem retains at each scale the conditioning of the continuous problem.

(B) *On an adapted multiscale basis*: When computation time is at stakes, one can take advantage of the link between scale and sensitivity to enhance the conditioning of the problem at scale k by using for x^k_{opt} an *adapted multiscale basis* $(\tilde{\epsilon}^\ell_j, j = 1 \dots p_\ell, \ell = 0 \dots k)$ of $\boldsymbol{E}^k$, which we define now.

Let $\boldsymbol{W}^k$, $k = 1 \dots K$ denote a supplementary space of $\boldsymbol{E}^{k-1}$ in $\boldsymbol{E}^k$. A function of $\boldsymbol{W}^k$ represents details that are seen at scale k but not at scale $k-1$ – hence the name *detail space* given to $\boldsymbol{W}^k$. The corresponding *multiscale decomposition* of $\boldsymbol{E}^K$ into a direct sum of subspaces is

$$\boldsymbol{E}^K = \boldsymbol{E}^0 \oplus \boldsymbol{W}^1 \oplus \cdots \oplus \boldsymbol{W}^K. \tag{3.25}$$

The space $\boldsymbol{E}^0$ is sometimes called the *background space*, as it corresponds to the absence of any details. Depending on the choice made for $\boldsymbol{W}^k$, the decomposition (3.25) can be *orthogonal* (if $\boldsymbol{W}^k \perp \boldsymbol{E}^{k-1}$ for $k = 1 \dots K$), or *oblique* in the other cases. Define

$$\begin{cases} \left(\epsilon^0_j\ ,\ j = 1 \cdots p_0 \overset{\text{def}}{=} n_0\right) & = \quad \text{normalized local basis of } \boldsymbol{E}^0, \\ \text{and, for } k = 1 \cdots K, & \\ \left(\epsilon^k_j\ ,\ j = 1 \cdots p_k\right) & = \quad \text{normalized local basis of } \boldsymbol{W}^k, \end{cases} \tag{3.26}$$

where p_0 is the dimension n_0 of $\boldsymbol{E}^0$ and $p_k = n_k - n_{k-1}$ is the dimension of the detail space $\boldsymbol{W}^k$. A normalized *multiscale basis* e^k_i, $i = 1 \dots n_k$, of $\boldsymbol{E}^k$ is then obtained by retaining only the first n_k vectors in the list (3.26) above. However, most optimization algorithms are invariant with respect to a change of orthonormal basis. There is hence nothing to expect from choosing for $\boldsymbol{E}^k$ an orthonormal multiscale basis instead of the simple local basis! But one can take advantage of the properties of "nicely nonlinear" problems and use an *adapted multiscale basis*, which boosts the sensitivity of the variables according to the size of the details they represent:

- For the determination of the background ($k = 0$), the natural choice is to use the optimization parameters x_{opt} and the local basis $\epsilon^0_j = e^0_j$ defined by (3.23) or (3.24). So the first step of a scale-by-scale resolution is the same for an adapted multiscale basis or a local basis.

- For the determination of the details at a scale $k \in \{1 \dots K\}$, let s_j^k denote the sensitivity in the directions of the multiscale basis (3.26) at some nominal parameter $x_{\text{sim}}^{\text{nom}}$:

$$s_j^k = \|V_j^k\| = \|D_{\epsilon_j^k}\varphi(x_{\text{sim}}^{\text{nom}})\|,$$

and define a *mean relative sensitivity* of details at scale k:

$$s_r^k = \Big(\frac{1}{p_k}\sum_{i=1\dots p_k}(s_i^k)^2\Big)^{1/2} \Big/ \Big(\frac{1}{n_0}\sum_{i=1\dots n_0}(s_i^0)^2\Big)^{1/2}. \qquad (3.27)$$

When k increases, the support of the basis functions ϵ_j^k becomes smaller, and so one can expect that, according to Definition 3.6.1 of a "nicely nonlinear" problem,

$$1 > s_r^1 > \dots > s_r^K.$$

So one can define an *adapted* multiscale basis for $\boldsymbol{E}^K$ (and hence for each $\boldsymbol{E}^k$!) by

$$\tilde{\epsilon}_j^k = \epsilon_j^k / s_r^k, \qquad k = 1 \dots K, \qquad (3.28)$$

where the coefficient $1/s_r^k$ tries to compensates for the loss of sensitivity of the forward model in the directions of finer details. Using this adapted multiscale basis tends to spherize the level lines of the objective function at scale k, and hence speeds up the resolution. If one does not want to estimate numerically s_r^k by (3.27), one can experiment with $s_r^k = (h_k/h_0)^\alpha$ for various $\alpha > 0$, where h_k is the size of the cells of $\mathcal{T}^k$.

Remark 3.6.2 *When* $\mathbf{W}^k \perp \mathbf{E}^{k-1}$ *(the decomposition (3.25) is* orthogonal*), basis functions* ϵ_j^k *of* $\mathbf{W}^k$ *are orthogonal to the local basis functions* e_i^k *of* $\mathbf{E}^{k-1}$ *defined in (3.23) or (3.24). Hence allowing the parameter to have a nonzero component on* ϵ_j^k does not change *its mean values at scale* $k-1$*. In many applications (as the estimation of a diffusion coefficient for example), the model output is sensitive to the mean value of the parameter, so that opening of new degrees of freedom in* $\mathbf{W}^k$ *will not deteriorate, at first order, the quality of the fit to the data obtained at scale* $k-1$*. For such problems, the optimization routine, when searching for* $\hat{x}_{\text{opt}}^k$ *in* $\mathbf{E}^k$*, will have to work* *out its way in practice in the much smaller space* $\mathbf{W}^k$, *as the coefficients on* $\mathbf{E}^{k-1}$ *are already almost at their optimal value.*

Oppositely, when the multiscale basis is associated with an oblique decomposition, allowing the parameter to have a nonzero component on a detail direction ϵ_j^k *of* $\mathbf{W}^k$ does change *its mean values at scale* $k-1$. *Hence it will be necessary, during the course of the optimization at scale* k, *to change coordinates* both *in* $\mathbf{W}^k$ and *in* $\mathbf{E}^{k-1}$ *to maintain the correct mean values determined at scale* $k-1$.

Multiscale bases associated to orthogonal decompositions should hence be preferred, as they tend to ensure a faster resolution of the optimization problem at each scale k, *by concentrating on the smaller problem of determining the details in* $\mathbf{W}^k$. *But – with the exception of the Haar basis, which is limited to piecewise constant functions on rectangular meshes – they are difficult to construct, as the basis vectors* ϵ_j^k *of* $\mathbf{W}^k$ *cannot in general be determined explicitly over the finite domain* Ω. *There exists a large amount of literature on wavelets [48], which are finite support orthogonal multiscale bases – but they are defined over an infinite domain, and cannot be used directly on bounded domains. Another approach is to use a numerical orthogonalization procedure as the Gram–Schmitt procedure, but this is computationally expensive, and the finite support property of the basis functions is lost.* ■

Remark 3.6.3 *Instead of performing the scale-by-scale optimization (3.22), one can minimize simultaneously with respect to the whole set of multiscale parameters, starting from a first initial guess* $x^{\text{init}} \in C^0$:

$$\textit{starting from } x^{\text{init}} \in C^0, \textit{ find } \hat{x}^K_{\text{opt}} = argmin_{x_{\text{opt}} \in C^K} J(x_{\text{opt}}).$$

Because of the invariance of optimization algorithms with respect to orthonormal change of basis, no improvement is to be expected concerning the stationary points and conditioning problems if an orthonormal multiscale basis is used.

Using the same adapted basis (3.28) as in the scale-by-scale resolution is not a good idea: boosting the sensitivity of fine scales to enhance conditioning will at the same time increase the nonlinearity, and hence worsen the stationary points problem instead of relieving it!

But it was observed in [57, 58] that going the other way, that is, using a multiscale basis that damps the sensitivity of fine scales, can allow to overcome the stationary points problem – at the price of a worse conditioning.

This is the case, for example, of the Haar basis (Remark 3.6.4 below), or of the adapted basis:

$$\tilde{\epsilon}_j^k = s_r^k \epsilon_j^k, \qquad k = 1 \dots K.$$

When such a basis is used, the optimization algorithm works first in the subspace of the most sensitive coordinates, that is, those corresponding to coarse scales, thus allowing to overcome the stationary points problems. It is only when these coordinates have been adjusted that the algorithm feels the need to adjust the coordinates for the finer scales, which have been made much less sensitive. The graph of the objective function as a function of the iteration number has then a stair shape, where going down one stair corresponds to grasping to the adjustment of the next finer scale. ■

3.6.4 Examples of Multiscale Bases

We describe now the simplest multiscale bases for discontinuous piecewise constant and continuous piecewise linear parameters.

Approximation by a discontinuous piecewise constant function.

We consider the case of parameters that are, at scale k, piecewise constant functions over a two-dimensional rectangular mesh $\mathcal{T}^k$ obtained by k divisions of a rectangular background mesh $\mathcal{T}^0$. The dimension of the corresponding parameter space $\boldsymbol{E}^k$ is then $n_k = 2^k \times 2^k\, n_0$, and the dimension of a supplementary space W^k is $p_k = 4^k n_0 - 4^{k-1} n_0 = 3n_{k-1}$. The mesh $\mathcal{T}^k$ is hence obtained by dividing each rectangular cell L of $\mathcal{T}^{k-1}$ into four smaller rectangular cells A, B, C, D. The kth refinement is called *regular* if all cells L of $\mathcal{T}^k$ are divided into four *equal* rectangles.

We associate to L the three functions $\epsilon_{L,\mathrm{lr}}^k$, $\epsilon_{L,\mathrm{tb}}^k$, and $\epsilon_{L,\mathrm{sp}}^k$ defined in Fig. 3.5 (the indexes *lr*, *tb*, *sp* stand, respectively, for left-right, top-bottom, and saddle-point), with

$$a = m\frac{|L|}{4|A|}, \qquad b = m\frac{|L|}{4|B|}, \qquad c = m\frac{|L|}{4|C|}, \qquad d = m\frac{|L|}{4|D|},$$

where $m > 0$ is a normalization coefficient.

One checks easily that:

$$\int_L \epsilon_{L,\mathrm{lr}}^k = 0, \qquad \int_L \epsilon_{L,\mathrm{tb}}^k = 0, \qquad \int_L \epsilon_{L,\mathrm{sp}}^k = 0,$$

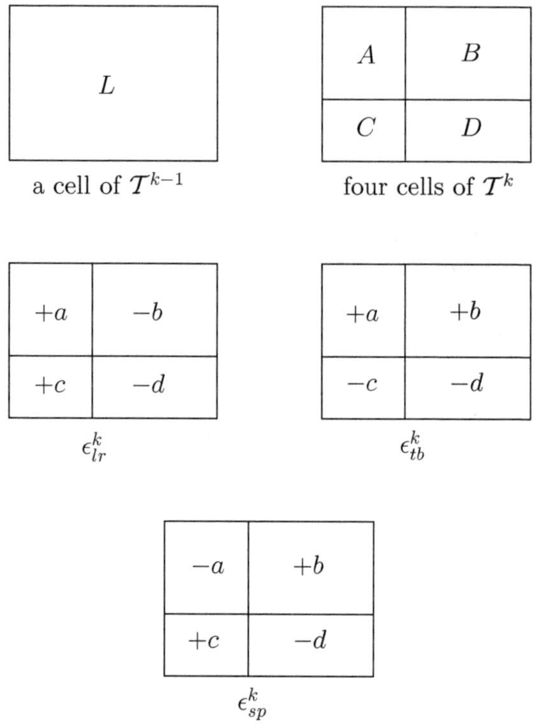

Figure 3.5: Definition of the three basis functions of $\boldsymbol{W}^k$ associated to the cell L of $\mathcal{T}^{k-1}$

and

$$|\epsilon^k_{L,\mathrm{lr}}|_{L^2(L)} = |\epsilon^k_{L,\mathrm{tb}}|_{L^2(L)} = |\epsilon^k_{L,\mathrm{sp}}|_{L^2(L)} = m\frac{|L|}{\bar{\ell}^{\frac{1}{2}}},$$

where $\bar{\ell}$ is equal to four times the harmonic mean of the areas of rectangles A, B, C, and D ($\bar{\ell} = |L|$ if the kth refinement is regular!):

$$\frac{4}{\bar{\ell}} = \frac{1}{4}\Big(\frac{1}{|A|} + \frac{1}{|B|} + \frac{1}{|C|} + \frac{1}{|D|}\Big).$$

The functions $\epsilon^k_{L,\mathrm{lr}}$, $\epsilon^k_{L,\mathrm{tb}}$, $\epsilon^k_{L,\mathrm{sp}}$ are linearly independent and orthogonal to $\boldsymbol{E}^{k-1}$. So the *orthogonal* supplementary space $\boldsymbol{W}^k$ to $\boldsymbol{E}^{k-1}$ in $\boldsymbol{E}^k$ is

$$\boldsymbol{W}^k = \mathrm{span}\{\epsilon^k_{L,\mathrm{lr}}, \epsilon^k_{L,\mathrm{tb}}, \epsilon^k_{L,\mathrm{sp}},\ L \in \mathcal{T}^{k-1}\}.$$

The basis functions $\epsilon^k_{L,\mathrm{lr}}$, $\epsilon^k_{L,\mathrm{tb}}$, $\epsilon^k_{L,\mathrm{sp}}$ of $\boldsymbol{W}^k$ associated to the same cell L are not orthogonal in general (except for a regular refinement, see Remark 3.6.4). But basis functions associated to different cells (at same or different scales) are orthogonal. This basis of $\boldsymbol{W}^k$ can be normalized independently of the scale by a proper choice of m over each cell L:

$$\begin{cases} m = X_{\mathrm{ref}} \dfrac{\bar{\ell}^{\frac{1}{2}} |\Omega|^{\frac{1}{2}}}{|L|} \quad \text{implies (compare with (3.23)):} \\ |\epsilon^k_{L,\mathrm{lr}}|_{L^2(L)} = |\epsilon^k_{L,\mathrm{tb}}|_{L^2(L)} = |\epsilon^k_{L,\mathrm{sp}}|_{L^2(L)} = X_{\mathrm{ref}} |\Omega|^{\frac{1}{2}}. \end{cases}$$

With this normalization, the multiscale basis functions (3.26) of $\boldsymbol{E}^k$ have the same norm $X_{\mathrm{ref}}|\Omega|^{\frac{1}{2}}$ as the local basis functions (3.23). The local basis is orthogonal, and the multiscale basis is orthogonal by blocks, one block being made of the three basis functions associated to one cell L of a meshes $\mathcal{T}^k, k = 1 \cdots K$.

Remark 3.6.4 *When the refinement is regular, one has $a = b = c = d = m$ and $\bar{\ell} = |L|$, and the functions $\epsilon^k_{L,\mathrm{lr}}$, $\epsilon^k_{L,\mathrm{tb}}$, $\epsilon^k_{L,\mathrm{sp}}$ are orthogonal. The case $m = 1$ corresponds then to the classical Haar basis – notice that the norm of the basis functions of the Haar basis decreases as $|L|^{1/2}$ when the scale is refined.* ■

Approximation by a continuous piecewise linear function.

We consider now the case where the parameter space $\boldsymbol{E}^k$ at scale k is made of continuous piecewise linear (on triangles) or bilinear (on rectangles) functions, over a mesh $\mathcal{T}^k$ obtained by k refinements of a background mesh $\mathcal{T}^0$ made of triangles and rectangles. It is supposed that the initial mesh and its subsequent refinements are done in such a way that all meshes $\mathcal{T}^k$ are regular meshes in the sense of finite elements (the intersection of two cells of such a mesh is either void, or an edge, or a vertex). Let

$$\mathcal{V}^k = \{M \text{ such that } M \text{ is a node of } \mathcal{T}^k \text{ but not of } \mathcal{T}^{k-1}\}.$$

By construction, $\mathcal{V}^k$ contains $p_k = n_k - n_{k-1}$ nodes M.

So a first choice is to choose for ϵ^k_M the local basis functions (3.24) of $\boldsymbol{E}^k$ associated to the nodes M of $\mathcal{V}^k$:

$$\epsilon^k_M = e^k_M, \qquad M \in \mathcal{V}^k.$$

With this choice, the space

$$\boldsymbol{W}^k = \text{span}\{\epsilon_M^k, M \in \mathcal{V}^k\}$$

is an *oblique* supplementary space of E^{k-1} in E^k. Hence the corresponding multiscale basis $\epsilon_M^k, k = 0 \cdots K, M \in \mathcal{V}^k$ is not orthogonal, neither between scales, nor inside a given supplementary space. But it is normalized, as its elements have the same L^2-norm $\frac{\sqrt{2}}{2} X_{\text{ref}} |\Omega|^{\frac{1}{2}}$ as the local basis functions e_i^k, $i = 1 \cdots n_K$ of $\boldsymbol{E}^k$ (see (3.24)).

The corresponding parameterization matrix ψ^k (see 3.21) involves a lot of interpolation, so that the change of variable can be computationally expensive; its implementation and that of its transposed are relatively delicate and error-prone, though following the procedure described in Sect. 2.4 with the forward map $\varphi = \psi^k$ and the objective function G defined in (2.13) should limit the chances of error in the determination of $(\psi^k)^T$.

This basis suffers moreover from the defaults of oblique multiscale bases mentioned in Remark 3.6.2, and so its use – or that of its more intricate orthogonalized version – should be considered only in special cases, for example, when scale-by-scale resolution with the local basis (3.24) at each scale fails.

3.6.5 Summary for Multiscale Parameterization

Multiscale parameterization has the desirable property that a scale-by-scale resolution will converge to or near to the global minimum for "nicely nonlinear" problems. However, a larger and larger number of d.o.f. is added each time the introduction of a new scale is required: as long as the fit to the data resulting from the optimization at scale k is above the uncertainty level, one has to grasp to the finer scale $k+1$, and introduce in the optimization space $\mathbb{R}^{n_{\text{opt}}}$ *all* the d.o.f. of $\boldsymbol{E}^{k+1}$, which are not in $\boldsymbol{E}^k$, that is, those of $\boldsymbol{W}^{k+1}$. It is then likely that part of them will have little or no influence the output of the model, which shows that the multiscale approach can lead to overparameterization.

3.7 Adaptive Parameterization: Refinement Indicators

We consider in this section an *adaptive* approach for the estimation of distributed parameters, which tries to retain the good properties of multiscale parameterization with respect to stationary points, but avoids the pitfall of overparameterization: one starts with a small number of degrees of freedom (often one), and adds (or substitutes) degrees of freedom one at a time, by using *refinement/coarsening* indicators to choose, among a large number of tentative degrees of freedom, one which incurs, at first order, the strongest decrease of the objective function; the algorithm stops when the data are explained up to the noise level.

Adaptive parameterization is one way of regularizing the inverse problem: it adds the information that one searches for the simpler (in the sense of number of degrees of freedom) parameter function in the chosen class.

Moreover, if the unknown parameter is a function and the tentative degrees of freedom are chosen at each step at a scale finer than the current one, one can expect that the algorithm will retain, when the problem is "nicely nonlinear," the good properties of scale-by-scale optimization with respect to stationary points (Sect. 3.6.3).

3.7.1 Definition of Refinement Indicators

We define in this section refinement indicators independently of the nature of the unknown parameter (vector or function) and of the objective function (provided it is derivable). We make the assumption that the choice of a "good" parameterization depends primarily on the forward map φ to be inverted, and we neglect the constraints in the construction of our indicators. We use the notations (see Sect. 3.3)

$$\left\{\begin{array}{lll} E = \mathbb{R}^{n_{\text{sim}}} & : & \text{space of simulation parameter,} \\ J : \mathbb{R}^{n_{\text{sim}}} \rightsquigarrow \mathbb{R}^{+} & : & \text{a derivable objective function,} \\ \boldsymbol{E} \subset E & : & \text{the current optimization space,} \\ x \in \boldsymbol{E} & : & \text{current optimization parameter.} \end{array}\right. \tag{3.29}$$

The space E and its subspace $\boldsymbol{E}$ are equipped with the Euclidean scalar product in $\mathbb{R}^{n_{\text{sim}}}$, which approximates, up to a multiplicative constant, the L^2-scalar product when x is a function (see Sect. 3.1.1). The current solution and value of the problem are (we ignore the constraints...)

$$\hat{x} = \arg \min_{x \in \boldsymbol{E}} J(x), \qquad \widehat{J} = J(\hat{x}). \tag{3.30}$$

When the minimum value $\widehat{J}$ is not satisfying (e.g., if it is above the noise level in the case of least squares), the question arises of which degree of freedom to add to the *current optimization space* $\boldsymbol{E}$ such that it produces a significant decrease of the objective function J.

So let $\boldsymbol{T} \subset E$, $\boldsymbol{T} \cap \boldsymbol{E} = \emptyset$ be a set of tentative new basis vectors ϵ available at the current parameterization step. We shall refer to $\boldsymbol{T}$ as the set of *tentative degrees of freedom*. It can be very large, as its vectors need not to be linearly independent. We shall suppose that its elements have been normalized:

$$\forall \epsilon \in \boldsymbol{T}, \ \|\epsilon\|_{\boldsymbol{T}} = 1 \tag{3.31}$$

for some norm $\| \cdot \|_{\boldsymbol{T}}$. This norm is not necessarily the Euclidean norm on $E = I\!R^{n_{sim}}$: for example, when the unknown parameter is a real valued function and the vector $x \in E$ is made of the values of the function over the simulation mesh $\mathcal{T}_h$ (Sect. 3.7.2 below), one often uses the ℓ^∞ norm:

$$\|\epsilon\|_{\boldsymbol{T}} = \|\epsilon\|_\infty = \max_{i=1\ldots n_{sim}} |\epsilon_i|. \tag{3.32}$$

To select one – or a few – degree(s) of freedom ϵ in $\boldsymbol{T}$, one associates to any tentative ϵ a *refinement indicator* λ in such a way that large $|\lambda|$'s are associated to ϵ's, which have a strong potential for the decrease of J:

1. *Nonlinear indicators:* One defines

$$\lambda^{\mathrm{NL}} \stackrel{\mathrm{def}}{=} \Delta J = \widehat{J} - \widetilde{J}, \tag{3.33}$$

where the $\widetilde{J}$ is the new value of the problem, computed with the full nonlinear model:

$$\widetilde{J} = J(\tilde{x} + \tilde{y}\epsilon), \qquad (\tilde{x}, \tilde{y}) = \arg \min_{x \in \boldsymbol{E},\ y \in I\!R} J(x + y\epsilon). \tag{3.34}$$

This is the most precise indicator, as it ensures by definition that the ϵ associated to the largest λ^{NL} produces the strongest decrease of the objective function!

But it is also the most computationally intensive one, as its computation requires the solution of the full nonlinear optimization problem. Hence it is impossible to use $\lambda^{\mathrm{NL}} = \Delta J$ as an indicator to choose ϵ among a large number of degrees of freedom.

2. *Gauss–Newton indicators:* In the case of nonlinear least squares problems, where $J(x) = \frac{1}{2}\|\varphi(x) - z\|^2$, one can define

$$\lambda^{\mathrm{GN}} \stackrel{\mathrm{def}}{=} \Delta J^{\mathrm{GN}} = \widehat{J} - \widetilde{J}^{\mathrm{GN}}, \tag{3.35}$$

where the $\widetilde{J}^{\mathrm{GN}}$ is the new value of the problem, computed with the Gauss–Newton approximation to the forward map φ:

$$\widetilde{J}^{\mathrm{GN}} = J^{\mathrm{GN}}(\delta\tilde{x}^{\mathrm{GN}}), \quad \delta\tilde{x}^{\mathrm{GN}} = \arg \min_{\delta\tilde{x} \in \mathrm{span}\{\boldsymbol{E}, \epsilon\}} J^{\mathrm{GN}}(\delta\tilde{x}), \tag{3.36}$$

where

$$J^{\mathrm{GN}}(\delta x) = \frac{1}{2}\|\varphi'(\hat{x})\delta x - \Delta z\|^2 \ \text{ with } \ \Delta z = z - \varphi(\hat{x}). \tag{3.37}$$

The evaluation of λ^{GN} requires now one resolution of a linear least squares problem for each tentative degree of freedom ϵ. This approach can be used (see, e.g., [42, 60]) when the size and computational cost of the problem makes its resolution possible by a Gauss–Newton optimization algorithm (one evaluation of the refinement indicator is then computationally equivalent to one Gauss–Newton iteration). But the number of tested degrees of freedom is still limited in this approach.

3. *First order indicators:* We return now to the general situation (3.29). The optimal objective function $\tau \in I\!R \rightsquigarrow J^*_\tau \geq 0$ in the direction $\epsilon \in \boldsymbol{T}$ is defined by

$$J^*_\tau = J(x^*_\tau + \tau\epsilon), \qquad x^*_\tau = \arg\min_{x \in \boldsymbol{E}} J(x + \tau\epsilon). \tag{3.38}$$

Comparison with (3.34) shows that

$$\begin{cases} x^*_0 &= \hat{x}, \\ J^*_0 &= \widehat{J}, \end{cases} \qquad \begin{cases} x^*_{\tilde{y}} &= \tilde{x}, \\ J^*_{\tilde{y}} &= \widetilde{J}, \end{cases}$$

and (3.33) becomes

$$\lambda^{\mathrm{NL}} = \Delta J = J^*_{\tilde{y}} - J^*_0 = \frac{\mathrm{d}J^*_\tau}{\mathrm{d}\tau}|_{\tau=0}\, \tilde{y} + \ldots \tag{3.39}$$

In absence of information on $\tilde{y}$, one can take a chance and chose ϵ according to the modulus of $\mathrm{d}J^*_\tau/\mathrm{d}\tau$ in (3.39). This choice is comforted by the remark that, because of (3.31), perturbing $\hat{x}$ of a given amount $\|y\,\epsilon\|$ in the direction ϵ will produce a large decrease, at first order, of the optimal objective function for those ϵ's that exhibit $\mathrm{d}J^*_\tau/\mathrm{d}\tau$'s with large modulus. Hence one is led to the ...

Definition 3.7.1 *The* first order refinement indicator *associated to the tentative degree of freedom* $\epsilon \in \boldsymbol{T}$ *at the current optimal parameter* $\hat{x}$ *is*

$$\lambda = \frac{\mathrm{d}J_\tau^*}{\mathrm{d}\tau}|_{\tau=0}. \tag{3.40}$$

It is given by

$$\lambda = \langle \nabla J(\hat{x}), \epsilon \rangle_E = D_\epsilon J(\hat{x}), \tag{3.41}$$

where $\nabla J(\hat{x})$ *denotes the gradient of the objective function with respect to* simulation parameters $x \in E$.

Proof. Derivation of (3.38) with respect to τ at $\tau = 0$ gives, when x_τ^* is a derivable function of τ,

$$\lambda = \left\langle \nabla J(x_0^*), \frac{\mathrm{d}x_\tau^*}{\mathrm{d}\tau} \right\rangle_E + \langle \nabla J(x_0^*), \epsilon \rangle_E. \tag{3.42}$$

But $\mathrm{d}x_\tau^*/\mathrm{d}\tau \in \boldsymbol{E}$ and $\langle \nabla J(\hat{x}), \delta x \rangle_E = 0 \ \forall \delta x \in \boldsymbol{E}$ (c.f. the definition (3.30) of $\hat{x}$), so that the first term vanishes in the right-hand side of (3.42) and (3.41) is proved. ∎

Hence the evaluation of λ for a tentative degree of freedom $\epsilon \in \boldsymbol{T}$ requires only the scalar product of ϵ, in the simulation parameter space $I\!R^{n_{\mathrm{sim}}}$, with the known vector $\nabla J(\hat{x})$! This makes it easy to test a very large number of tentative degrees of freedom before making up one's mind for the choice of a new one.

Remark 3.7.2 *One could argue that, because of the potentially large dimension* n_{sim} *of* x, *the gradient* $\nabla J(\hat{x})$ *can be computationally unaffordable. However, when the implementation of the inversion is based, according to the recommendations of Sect. 3.8.1 below, on the adjoint state approach of Sect. 2.3, the gradient* $\nabla J(\hat{x})$ *with respect to simulation parameters is available as a byproduct of the computation of* $\hat{x}$ *(search for* $\nabla_{x_{\mathrm{sim}}} J$ *in the lower right corner of Fig. 3.8).* ∎

Remark 3.7.3 *The first order refinement indicators* λ *were originally defined as Lagrange multipliers in the context of estimation of piecewise* *constant parameter functions [23, 10]. We have used a more direct – and hopefully simpler – presentation above, but the two definitions coincide: (3.38) can be seen as a constrained optimization problem:*

$$(x_\tau, y_\tau) = \arg \min_{\substack{x \in \mathbf{E}\,,\; y \in I\!R \\ y-\tau=0}} J(x + y\,\epsilon),$$

whose Lagrangian is

$$\mathcal{L}_\tau(x, y, \lambda) = J(x + y\,\epsilon) - \lambda(y - \tau).$$

The necessary Lagrange optimality conditions are then, for a given $\tau \in I\!R$,

$$\begin{cases} \dfrac{\partial \mathcal{L}}{\partial x}((x,y,\lambda)\delta x &= \langle \nabla J(x + y\,\epsilon), \delta x\rangle_E &= 0 \quad \forall \delta x \in \mathbf{E}, \\[2ex] \dfrac{\partial \mathcal{L}}{\partial y}((x,y,\lambda) &= \langle \nabla J(x + y\,\epsilon), \epsilon\rangle_E - \lambda &= 0, \\[2ex] \dfrac{\partial \mathcal{L}}{\partial \lambda}((x,y,\lambda) &= y - \tau &= 0. \end{cases}$$

For $\tau = 0$*, the second equation gives (3.41) and property (3.40), which we use now as definition, is the well-known property that a Lagrange multiplier is the derivative of the optimal objective function with respect to the right-hand side of the corresponding constraint.* ■

We return now to the case of nonlinear least-squares problems, where

$$J(x) = \frac{1}{2}\|\varphi(x) - z\|_F^2,$$

and investigate the relation between the first order indicator λ and the Gauss–Newton indicator λ^{GN}. We use notations (3.35) through (3.37), and denote by

$$\begin{array}{lll} F & \text{the data or observation space} & I\!R^q, \\ \Phi & \text{the } q \times n_{sim} \text{ matrix associated to} & \varphi'(\hat{x}), \end{array} \qquad (3.43)$$

when $E = I\!R^{n_{\mathrm{sim}}}$ and $F = I\!R^q$ are equipped with the Euclidean scalar products. Because the principle of adaptive parameterization is to stop adding degrees of freedom before overparameterization is attained, it is not a restriction to suppose that:

$$\Phi \text{ is injective over span}\{\boldsymbol{E}, \epsilon\} \tag{3.44}$$

By definition one has

$$\begin{aligned}
\lambda^{\text{GN}} &= \widehat{J} - \widetilde{J}^{\text{GN}}, \\
&= \frac{1}{2}\|\Delta z\|_F^2 - \frac{1}{2}\|\Phi\, \delta\tilde{x}^{\text{GN}} - \Delta z\|_F^2, \\
&= \langle \Phi\, \delta\tilde{x}^{\text{GN}}, \Delta z\rangle_F - \frac{1}{2}\|\Phi\, \delta\tilde{x}^{\text{GN}}\|_F^2.
\end{aligned} \tag{3.45}$$

The Euler equation for the Gauss–Newton optimization problem (3.36) is

$$\begin{cases} \text{find } \delta x^{\text{GN}} \in \text{span}\{\boldsymbol{E}, \epsilon\} \subset I\!R^{n_{sim}} \text{ such that} \\ \langle \Phi\, \delta\tilde{x}^{\text{GN}} - \Delta z, \Phi\, \delta x\rangle_F = 0 \quad \forall \delta x \in \text{span}\{\boldsymbol{E}, \epsilon\}. \end{cases} \tag{3.46}$$

Choosing $\delta x = \delta\tilde{x}^{\text{GN}}$ in (3.46) and combining with (3.45) gives first

$$\lambda^{\text{GN}} = \frac{1}{2}\|\Phi\, \delta\tilde{x}^{\text{GN}}\|_F^2. \tag{3.47}$$

Subtracting then from (3.46), the Euler equation for $\hat{x}$

$$\langle -\Delta z, \Phi\, \delta x\rangle_F = 0 \quad \forall \delta x \in \boldsymbol{E}$$

and the Definition (3.41) of λ

$$\langle -\Delta z, \Phi\, \epsilon\rangle_F = \lambda$$

gives

$$\begin{aligned}
\langle \Phi\, \delta\tilde{x}^{\text{GN}}, \Phi\, \delta x\rangle_F &= 0 \qquad \forall \delta x \in \boldsymbol{E}, \\
\langle \Phi\, \delta\tilde{x}^{\text{GN}}, \Phi\, \epsilon\rangle_F &= -\lambda.
\end{aligned}$$

Hence we see that

$$\delta\tilde{x}^{\text{GN}} = -\lambda\, \eta(\epsilon),$$

where $\eta(\epsilon) \in \text{span}\{\boldsymbol{E}, \epsilon\}$ is the unique solution (which exists because of hypothesis (3.44)) of

$$\langle \Phi\, \eta, \Phi\, \delta x\rangle_F = 0 \qquad \forall \delta x \in \boldsymbol{E}, \tag{3.48}$$

$$\langle \Phi\, \eta, \Phi\, \epsilon\rangle_F = 1. \tag{3.49}$$

Combining this result with (3.47) gives

Proposition 3.7.4 *Let notation (3.43) and hypothesis (3.44) hold. Then the Gauss–Newton and first order indicators are related by*

$$\lambda^{\mathrm{GN}} = \frac{\lambda^2}{2} \|\Phi\, \eta(\epsilon)\|_F^2, \tag{3.50}$$

where $\eta(\epsilon)$ is defined by (3.48) and (3.49).

This confirms that the first order indicator carries one part of the information on the variation of λ^{GN} with the tentative degree of freedom ϵ.

It is possible to explicit the calculations needed for the evaluation of the coefficient $\|\Phi\, \eta(\epsilon)\|_F^2$. We return for this to the optimization variables. The dimension of the current optimization space $\boldsymbol{E}$ is n_{opt}, and let

$$e_1 \dots e_{n_{\mathrm{opt}}} \in \boldsymbol{E} \subset E = I\!R^{n_{\mathrm{sim}}}$$

be the current optimization basis, and $\Psi(\epsilon)$ be the $n_{\mathrm{sim}} \times (n_{\mathrm{opt}} + 1)$ parameterization matrix (3.16) for the tentative optimization space associated to ϵ

$$\Psi(\epsilon) = [e_1 \dots e_{n_{opt}}\ \epsilon]. \tag{3.51}$$

Then the vectors $\eta \in \mathrm{span}\{\boldsymbol{E}, \epsilon\}$, $\delta x \in \boldsymbol{E}$ and $\epsilon \in \boldsymbol{T}$, which appear in (3.48) and (3.49), are of the form

$$\begin{array}{rcll} \eta & = & \Psi(\epsilon)\, \eta_{\mathrm{opt}} & \text{for some } \eta_{\mathrm{opt}} \in I\!R^{n_{\mathrm{opt}}+1}, \\ \delta x & = & \Psi(\epsilon)\, (\delta x_{\mathrm{opt}}, 0) & \text{for some } \delta x_{\mathrm{opt}} \in I\!R^{n_{\mathrm{opt}}}, \\ \epsilon & = & \Psi(\epsilon)\, (\underbrace{0 \dots 0}_{n_{\mathrm{opt}} \text{ times}}, 1). & \end{array}$$

Hence (3.48) and (3.49) become

$$\begin{array}{rcll} \langle \Phi\Psi(\epsilon)\, \eta_{\mathrm{opt}}, \Phi\Psi(\epsilon)\, (\delta x_{\mathrm{opt}}\ 0) \rangle_F & = & 0 & \forall \delta x_{\mathrm{opt}} \in I\!R^{n_{\mathrm{opt}}}, \\ \langle \Phi\Psi(\epsilon)\, \eta_{\mathrm{opt}}, \Phi\Psi(\epsilon)\, (\underbrace{0 \dots 0}_{n_{\mathrm{opt}} \text{ times}}\ 1) \rangle_F & = & 1. & \end{array}$$

This rewrites

$$\begin{array}{l} \langle \Phi\Psi(\epsilon)\, \eta_{\mathrm{opt}}, \Phi\Psi(\epsilon)\, \delta x_{\mathrm{opt}} \rangle_F \quad = \\ \qquad \langle (\underbrace{0 \dots 0}_{n_{opt} \text{ times}}\ 1), \delta x_{\mathrm{opt}} \rangle_{I\!R^{n_{\mathrm{opt}}+1}} \quad \forall \delta x_{\mathrm{opt}} \in I\!R^{n_{\mathrm{opt}}+1}, \end{array}$$

that is,

$$\Psi(\epsilon)^{\mathrm{T}}\Phi^{\mathrm{T}}\Phi\,\Psi(\epsilon)\,\eta_{\mathrm{opt}} = (\,\underbrace{0\ldots0}_{n_{\mathrm{opt}}\ \mathrm{times}}\ 1).$$

Solving for η_{opt} and substituting in (3.50) gives an alternative expression of λ^{GN}:

$$\lambda^{\mathrm{GN}} = \frac{\lambda^2}{2}\|(\,\underbrace{0\ldots0}_{n_{\mathrm{opt}}\ \mathrm{times}}\ 1)\|^2_{(\Psi(\epsilon)^{\mathrm{T}}\Phi^{\mathrm{T}}\Phi\,\Psi(\epsilon))^{-1}}. \tag{3.52}$$

This formula is derived in [11] starting from the Lagrange multipliers definition of first order indicators. It will be useful in Sect. 3.7.3 when applying refinement indicators to segmentation of black and white images.

3.7.2 Multiscale Refinement Indicators

We consider in this section the case where the unknown parameter is a *piecewise constant real valued function* $a : \xi \in \Omega \rightsquigarrow a(\xi) \in I\!R$ over some domain $\Omega \subset I\!R^2$. First order indicators have been first introduced [23, 10], and used [69, 35] in this context, and some convergence results have been obtained in [9]. But multiscale refinement indicators are not limited to this case:

- They can be adapted to the estimation of smooth functions (see [33] for the determination of a continuously differentiable function of two variables)
- The method generalizes naturally to domains of higher dimension
- And results for the case of *vector valued parameters* can be found in [44, 11]

Let $\mathcal{T}_h$ be the simulation mesh covering Ω, made of n_{sim} cells K. We approximate the function $a : \Omega \rightsquigarrow I\!R$ by a function a_h which takes a constant value a_K over each cell K of $\mathcal{T}_h$, and choose as simulation space $E = I\!R^{n_{\mathrm{sim}}}$, and as *simulation parameter*:

$$a_{\mathrm{sim}} = (a_{\mathrm{sim},K}, K \in \mathcal{T}_h) \in I\!R^{n_{\mathrm{sim}}} \quad \text{with} \quad a_{\mathrm{sim},K} = a_K. \tag{3.53}$$

For simplicity, we have chosen not to adimensionalize a, as all parameters a_K correspond to the same physical quantity.

Because of the large size of the discretization mesh $\mathcal{T}_h$, the data are usually not sufficient to estimate the value of the unknown parameter in each cell K.

We consider here the case where the additional information that is added to regularize the problem is that a is *constant over a number* n_{opt} *of subdomains (much) smaller than the number* n_{sim} *of elements* K (this is "regularization by parameterization" as described in Sect. 1.3.4, see also [38]). We call such a partition $\mathcal{T}$ of $\mathcal{T}_h$ into n_{opt} subsets $\mathcal{T}_j$ *zonation*, each one made of cells K of $\mathcal{T}_h$ (with the notations of Sect. 3.6.2, $\mathcal{T}_h$ is a sub-mesh of $\mathcal{T}$), and we associate to a function $a : \Omega \rightsquigarrow I\!R$, which takes a constant value a_j on each zone $\mathcal{T}_j$ the vector $a_{\text{opt}} \in I\!R^{n_{\text{opt}}}$ of its values on each zone *calibrated* according to (3.1) – once again we do not adimensionalize for simplicity:

$$a_{\text{opt}} = (a_{\text{opt},j}, j = 1 \dots n_{\text{opt}}) \in I\!R^{n_{\text{opt}}} \quad \text{with} \quad a_{\text{opt,j}} = \left(\frac{|\mathcal{T}_j|}{|\Omega|}\right)^{\frac{1}{2}} a_j. \tag{3.54}$$

The *parameterization map* $\Psi : a_{\text{opt}} \rightsquigarrow a_{\text{sim}} \in I\!R^{n_{n_{\text{sim}}}}$ defined in (1.16) is hence the $n_{\text{sim}} \times n_{\text{opt}}$ matrix:

$$\Psi = [e_1^{\mathcal{T}} \; \dots \; e_{n_{\text{opt}}}^{\mathcal{T}}], \tag{3.55}$$

where the vectors $e_j^{\mathcal{T}} \in I\!R^{n_{\text{sim}}}$, $j = 1 \dots n_{\text{opt}}$, are defined by (compare with (3.19))

$$e_{j,K}^{\mathcal{T}} = \begin{cases} 0 & \text{if } K \notin \mathcal{T}_j, \\ \dfrac{|\Omega|^{\frac{1}{2}}}{|\mathcal{T}_j|^{\frac{1}{2}}} a_{opt,j} & \text{if } K \in \mathcal{T}_j, \end{cases} \tag{3.56}$$

and the *current optimization space* (see (3.29)) is $\boldsymbol{E} = \text{span}\{e_1^{\mathcal{T}} \dots e_{n_{\text{opt}}}^{\mathcal{T}}\}$.

With this choice of optimization parameter, the Euclidean scalar product on $\boldsymbol{E} \subset I\!R^{n_{\text{opt}}}$ is proportional to the L^2-scalar product of functions over Ω, which will preserve the conditioning of the original problem, and the Euclidean norm $\|a_{\text{opt}}\|$ is a mean value of the function a over Ω (Sect. 3.1.1).

The "easy" – or at least more classic – part of the problem is the minimization of the NLS objective function $a_{opt} \rightsquigarrow J(\Psi\, a_{\text{opt}})$ over the admissible parameter set for a *given zonation* $\mathcal{T}$; but the difficult part is that the zonation $\mathcal{T}$ itself is not known a priori! Hence solving the *regularized* problem amounts to *find* an *"optimal" zonation* $\widehat{\mathcal{T}}$ *of* Ω such that:

1. The data are explained up to the noise level (i.e., the minimum of $J \circ \widehat{\Psi}$ is small enough)

2. The number of zones is small enough to avoid overparameterization (i.e., the Jacobian of $\varphi \circ \widehat{\Psi}$ is injective)

The determination of the zonation is a difficult geometric problem. Various approaches to its solution have been proposed: displacement of the boundaries between zones using geometric gradient methods, level set methods, etc. We develop below the multiscale refinement indicator approach, which is particularly well suited for "nicely nonlinear problems" of Definition 3.6.1, (e.g., the estimation of the diffusion coefficient or the source term in Sects. 1.4–1.6 of Chap. 1), where a coarse-to-fine scale-by-scale resolution has been shown in Sect. 3.6.3 to relieve the stationary point problem. The platform Ref-indic for the use of these indicators is available at `http://refinement.inria.fr/ref-indic/`.

The multiscale refinement indicators proceed as follows: one starts with $\mathcal{T}^0$ made of one or two zones, and increases the number of zones one at a time, by splitting one zone of the current zonation $\mathcal{T}^k$ into two subzones, thus producing a new zonation $\mathcal{T}^{k+1}$. This ensures at each step that the new degree of freedom is (locally) at a finer scale than the current ones, hence the name "multiscale" given to these indicators. The zone to be split and the way it is split are chosen according to refinement indicators. By construction of the refinement indicators, this ensures that each new optimization problem is likely to produce the best decrease of the objective function among all tested degrees of freedom.

The scale-by-scale optimization is stopped, as described in Sect. 3.6.2, at the first k for which the data are explained up to the noise level. But the parsimony with which the degrees of freedom are introduced ensures that the Jacobian of $\varphi \circ \Psi$ has a full rank at all successive minimizers $\hat{a}^k_{\text{opt}}$, and hence overparameterization is avoided.

Let $\mathcal{T} = \mathcal{T}^k$ denote the current zonation, and Ψ and $\boldsymbol{E}$ denote the corresponding current parameterization matrix and optimization space. We describe now the construction of $\mathcal{T}^{k+1}$. Cutting one domain $\mathcal{T}_j$ of $\mathcal{T}$ into two subdomains $\mathcal{T}_{j,+}$ and $\mathcal{T}_{j,-}$ amounts to add to the basis $e_1^{\mathcal{T}} \dots e_{n_{\text{opt}}}^{\mathcal{T}}$ of $\boldsymbol{E}$ defined in (3.56) the vector $\epsilon = (\epsilon_K \;,\; K \in \mathcal{T}_h)$ defined by

$$\epsilon = (\epsilon_K \;,\; K \in \mathcal{T}_h) \in I\!R^{n_{\text{sim}}} \quad \text{with} \quad \epsilon_K = \begin{cases} 0 & \text{if } K \notin \mathcal{T}_j, \\ +1 & \text{if } K \in \mathcal{T}_{j,+}, \\ -1 & \text{if } K \in \mathcal{T}_{j,-}. \end{cases} \tag{3.57}$$

The (first order) refinement indicator (3.41) associated to this cut is

$$\lambda = \langle \nabla_{x_{\text{sim}}} J, \epsilon \rangle_{I\!R^{n_{\text{sim}}}} = \sum_{K \in \mathcal{T}_{j=+}} \left(\nabla_{x_{\text{sim}}} J\right)_K - \sum_{K \in \mathcal{T}_{j=-}} \left(\nabla_{x_{\text{sim}}} J\right)_K, \tag{3.58}$$

where $\nabla_{x_{\text{sim}}} J$ is evaluated at the point $\Psi\hat{a}_{\text{opt}}^{\mathcal{T}}$ of $I\!R^{n_{\text{sim}}}$. So we see that, once $\nabla_{x_{\text{sim}}} J$ has been made available as by-product of the minimization over the current zonation, the first order indicator associated to any cut of any domain $\mathcal{T}_j$ is obtained by a simple summation of the components of $\nabla_{x_{\text{sim}}} J$ over the new subdomains. This makes it possible to test a very large set $\boldsymbol{T}$ of tentative refinements of the current zonation $\mathcal{T} = \mathcal{T}^k$. Once the new zonation $\mathcal{T}^{k+1}$ has been chosen, the optimization is performed with respect to the local (at the zonation level) variables $a_{\text{opt},j}, j = 1 \dots n_{\text{opt}}^{k+1}$ as defined in (3.54).

For example, if the simulation mesh $\mathcal{T}_h$ is made of rectangles (see Fig. 3.6), one can choose, for the set $\boldsymbol{T}$ of tentative degrees of freedom, the collection of the cuts dividing each current zone Z_j by a vertical or horizontal line located on one edge of the mesh, and by a "checkerboard cut" centered at each node of the zone (see Remark 3.7.9 below). Using these cuts will lead to zones

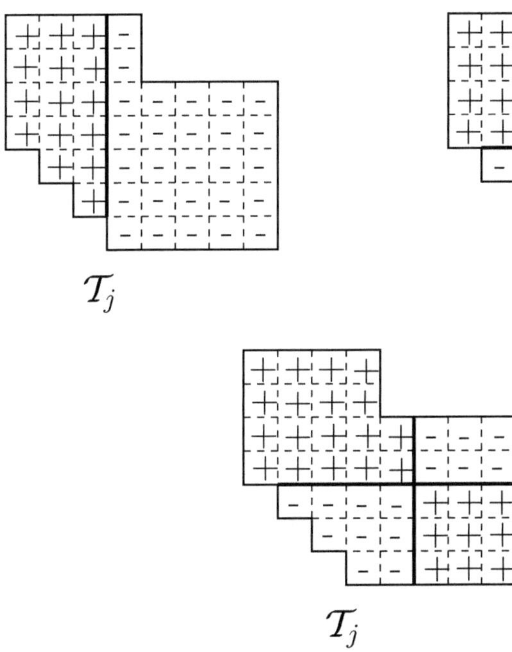

Figure 3.6: Examples of vertical, horizontal, and checkerboard cuts for one zone $\mathcal{T}_j$ of a mesh $\mathcal{T}_h$ made of rectangles. The vector ϵ takes value $+1$ in cells with a plus sign, and -1 in cells with a minus

that have "simple" shapes, which is often an implicit constraint. We refer to [10] for more details and numerical results. But the above list of cuts is not exhaustive, and one can test any cut suggested to the user by its experience or intuition.

If the complexity of the zones is not a problem, one can use a set $\boldsymbol{T}$ containing only one cut in each zone $\mathcal{T}_j$, namely the one that produces the *largest refinement indicator* $\lambda_{j,\max}$:

$$\lambda_{j,\max} = \sum_{K \in \mathcal{T}_j} |(\nabla_{x_{\text{sim}}} J)_K|. \tag{3.59}$$

It corresponds to the vector ϵ given by

$$\epsilon_K = \begin{cases} 0 & \text{if } K \notin T_j, \\ \text{sign}(\nabla_{x_{\text{sim}}} J)_K & \text{if } K \in \mathcal{T}_j. \end{cases} \tag{3.60}$$

This largest refinement indicator approach has been used in [23] for an oil field modeling problem, and is applied in Sect. 3.7.3 below to the segmentation of black and white images.

Notice that the vectors ϵ associated to these cuts by (3.57) and (3.60) satisfy the normalization (3.32) for $\boldsymbol{T}$.

Remark 3.7.5 *The refinement indicators have been defined in the context of unconstrained optimization. In this case, the optimum $\hat{a}^{\mathcal{T}}_{\text{opt}}$ at the current zonation $\mathcal{T}$ satisfies*

$$\partial J/\partial x_{\text{opt},j} = \langle \nabla_{x_{\text{sim}}} J,\, e_j^{\mathcal{T}} \rangle_{I\!R^{n_{\text{sim}}}} = 0 \quad \forall j = 1 \dots n_{\text{opt}}, \tag{3.61}$$

so that formula (3.58) is equivalent to

$$\begin{aligned} \lambda &= 2 \textstyle\sum_{K \in \mathcal{T}_{j=+}} \left(\nabla_{x_{\text{sim}}} J\right)_K \\ &= -2 \textstyle\sum_{K \in \mathcal{T}_{j=-}} \left(\nabla_{x_{\text{sim}}} J\right)_K. \end{aligned} \tag{3.62}$$

In practical applications, however, constraints are taken into account for the determination of $\hat{a}^{\mathcal{T}}_{\text{opt}}$, so that (3.61) does not necessarily hold for all zones (unless all constraints are taken into account via regularization, see Sect. 3.8.1 below). It is hence recommended to compute the refinement indicators by formula (3.58) rather than by (3.62). ■

3.7.3 Application to Image Segmentation

Let Ω be a rectangle covered by a large uniform mesh $\mathcal{T}_h$ made of $n_{\rm sim}$ rectangular (usually square) elements K called pixels, and $z = (z_K\,,\ K \in \mathcal{T}_h)$ be a black and white image on the domain Ω, where $z_K \in [0, 256]$, for example, represents the gray level of pixel K. The segmentation of this image consists in finding a *zonation* $\mathcal{T}$ of $\mathcal{T}_h$, that is, a partition of $\mathcal{T}_h$ into a *small number* $n_{\rm opt}$ of subsets $\mathcal{T}_j$, and an *image* $\hat{a}$ with a constant gray level in each zone $\mathcal{T}_j\,,\ j = 1 \dots n_{\rm opt}$, such that the number $n_{\rm opt}$ of zones of $\mathcal{T}$ is small and $\hat{a}$ is close to the given image z. This fits clearly in the framework of multiscale refinement indicators of Sect. 3.7.2, with the particularly simple forward map:

$$\varphi:\ a \in C = [0, 256]^{n_{\rm opt}} \subset E = I\!R^{n_{\rm sim}} \rightsquigarrow a \in F = I\!R^{n_{\rm sim}}. \tag{3.63}$$

Because of the linearity of the forward map, the nonlinear refinement indicator $\lambda^{\rm NL}$ coincides here with the Gauss–Newton indicator $\lambda^{\rm GN}$. The question that arises then naturally is: for such a simple inverse problem, does the largest first-order indicator λ also hint at the largest decrease of the objective function? To answer this question, we explicit the relation (3.52) between $\lambda^{\rm NL} = \lambda^{\rm GN}$ and λ in the case of image segmentation. The matrix Φ is here the $n_{\rm sim} \times n_{\rm sim}$ identity matrix, and so we are left to determine $\Psi(\epsilon)^T\, \Psi(\epsilon)$.

The resolution of the minimization problem (2.5) for a given zonation is straightforward when $\varphi = I_d$, as it amounts to affect to each zone the mean gray level of the zone in the image! So there is no need here to calibrate the optimization parameters to maintain the conditioning of the optimization problem, and we replace the Definition 3.54 of optimization parameters simply by

$$a_{\rm opt} = (a_{{\rm opt},j}, j = 1 \dots n_{\rm opt}) \in I\!R^{n_{\rm opt}} \quad \text{with} \quad a_{{\rm opt},j} = a_j. \tag{3.64}$$

Formula (3.56) for the definition of the basis $e_j^{\mathcal{T}} \in I\!R^{n_{sim}}$, $j = 1 \dots n_{\rm opt}$, of $\boldsymbol{E}$ becomes now

$$e_{j,K}^{\mathcal{T}} = \begin{cases} 0 & \text{if } K \notin \mathcal{T}_j, \\ 1 & \text{if } K \in \mathcal{T}_j. \end{cases} \tag{3.65}$$

The parameterization matrix $\Psi(\epsilon)$ associated to the tentative splitting of the j-zone $\mathcal{T}_j$ into two subzones $\mathcal{T}_{j,+}$ and $\mathcal{T}_{j,-}$ is given by (3.51), where $e_j^{\mathcal{T}}$ and ϵ are given by (3.65) and (3.57). Hence, the $(n_{\rm opt} + 1) \times (n_{\rm opt} + 1)$ matrix

$\Psi(\epsilon)^T \Psi(\epsilon)$ is (only nonzero elements are represented):

$$\Psi(\epsilon)^{\mathrm{T}} \Psi(\epsilon) = \begin{bmatrix} p_1 & & & & & \\ & \ddots & & & & \\ & & p_j & & & p_j^+ - p_j^- \\ & & & \ddots & & \\ & & & & p_{n_{\mathrm{opt}}} & \\ & & p_j^+ - p_j^- & & & p_j \end{bmatrix}, \tag{3.66}$$

where $p_1 \dots p_{n_{\mathrm{opt}}}$ are the numbers of pixels in each zone of the current zonation $\mathcal{T}$, and $p_{j,+}$ (respectively, $p_{j,-}$) is the number of pixels in the subzone $\mathcal{T}_{j,+}$ (respectively, $\mathcal{T}_{j,-}$) of the tentative cut. It is now a simple calculation to deduce from (3.52) and (3.66) that

$$\lambda_j^{\mathrm{NL}} = \lambda_j^{\mathrm{GN}} = \frac{\lambda_j^2}{8} \frac{p_j}{p_{j,+} p_{j,-}}. \tag{3.67}$$

This shows that the largest first order indicator λ_j does not always correspond to the largest decrease of the objective function among all tentative refinements $\boldsymbol{T}$, even when the forward map is the identity. But for large values of α, the function $\tau \rightsquigarrow \alpha/(\tau(\alpha - \tau))$ is almost flat near $\alpha/2$, so there tend to be a good correlation between λ^2 and the actual decrease of the objective function, at least during the first iterations of the algorithm.

It is hence natural, when the complexity of the zonation is not a problem, which is usually the case in image segmentation, to search at each iteration for the new degree of freedom in a tentative set $\boldsymbol{T}$ containing only, in each zone $\mathcal{T}_j$ of the current parameterization, the degree of freedom ϵ given by (3.59), which corresponds to the *largest refinement indicator* $\lambda_{j,\max}$ defined by (3.60).

This approach is developed in [11] for the segmentation of B and W images, together with a generalization to segmentation of color images. The platform Ref-image for the application of refinement indicators to image segmentation is available at `http://refinement.inria.fr/ref-image/`.

3.7.4 Coarsening Indicators

We return to the general context of the estimation of piecewise constant parameters. Suppose that the refinement indicators of Sect. 3.7.2 have suggested to split one current zone $\mathcal{T}_j$ into two subzones $\mathcal{T}_{j,+}$ and $\mathcal{T}_{j,-}$. Before deciding to add this new degree of freedom, one can ask whether *aggregating one subzone*, say $\mathcal{T}_{j,+}$, to one adjacent neighboring zone, say $\mathcal{T}_\ell$, would not be already beneficial to the objective function, thus leaving unchanged the number of degrees of freedom.

Coarsening indicators that evaluate the interest of such an aggregation have been defined in [10] via Lagrange multipliers. We give here a more direct definition:

Definition 3.7.6 *The* coarsening indicator *for the aggregation of the subzone $\mathcal{T}_{j,+}$ with one adjacent zone $\mathcal{T}_\ell$ is (see Fig. 3.7)*

$$\Delta J^*_{j+,\ell} = -\mu_+(a_{\text{opt},\ell} - a_{\text{opt},j}),$$

where μ_+ is the first order indicator associated by Definition 3.7.1 to the direction

$$\eta = (\eta_K, K \in \mathcal{T}_h), \text{ where } \eta_K = \begin{cases} 0 & \text{if } K \notin T_{j,+}, \\ +1 & \text{if } K \in \mathcal{T}_{j,+}. \end{cases}$$

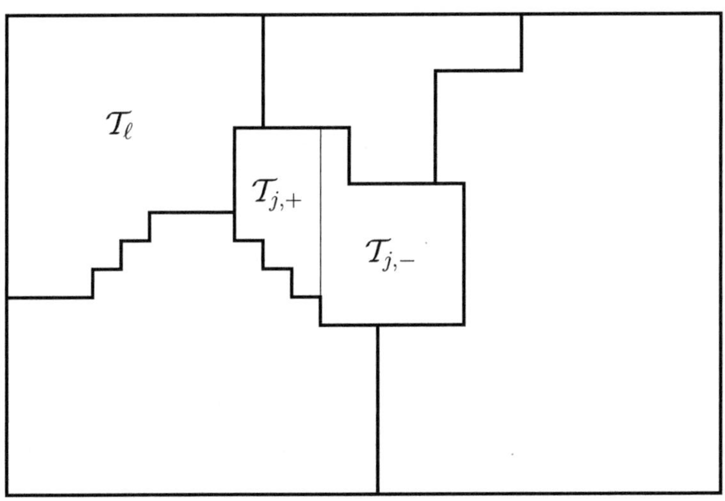

Figure 3.7: Coarsening indicators: the current zonation $\mathcal{T}$ (*thick lines*, 5 zones), the tentative division of zone $\mathcal{T}_j$ proposed by the refinement indicators (*thin line*), and one zone $\mathcal{T}_\ell$ (adjacent to $\mathcal{T}_{j,+}$) to be tested for aggregation with $\mathcal{T}_{j,+}$

It is given by

$$\mu_+ = \sum_{K \in \mathcal{T}_{j,+}} \left(\nabla_{x_{\rm sim}} J\right)_K \tag{3.68}$$

According to Definition 3.7.1, the coarsening indicator $\Delta J^*_{j+,\ell}$ gives the first order *decrease* of the objective function incurred by changing, on $\mathcal{T}_{j,+}$, the parameter from its current value $a_{{\rm opt},j}$ to the value $a_{{\rm opt},\ell}$ in the adjacent zone $\mathcal{T}_\ell$.

$\Delta J^*_{j+,\ell}$ can be positive or negative, but if, for some adjacent zone $\mathcal{T}_\ell$, one finds that $\Delta J^*_{j+,\ell}/J^*$ is larger than some a-priori given percentage, one can consider aggregating the zones $\mathcal{T}_{j,+}$ and $\mathcal{T}_\ell$, (rather than splitting the zone $\mathcal{T}_j$ into two). In this case, the new zonation $\mathcal{T}^{k+1}$ will have the same number of zones than $\mathcal{T}^k$, but will nevertheless produce a better fit to the data.

Remark 3.7.7 *For the aggregation of $\mathcal{T}_{j,-}$ one has to compute (see (3.68))*

$$\mu_- = \sum_{K \in \mathcal{T}_{j,-}} \left(\nabla_{x_{\rm sim}} J\right)_K.$$

Hence μ_- and μ_+ are linked to refinement indicators by

$$\lambda = \mu_+ - \mu_-.$$

Also, $\hat{x}^{\mathcal{T}}_{\rm opt}$ minimizes J over the zonation $\mathcal{T}$, so that

$$\mu_+ + \mu_- = \sum_{K \in \mathcal{T}_j} \left(\nabla_{x_{\rm sim}} J\right)_K = \frac{\partial J}{\partial x_{{\rm opt},j}}(\hat{a}^{\mathcal{T}}_{\rm opt}) \simeq 0,$$

which shows that μ_+ and μ_- are likely to be of opposite signs. ■

3.7.5 A Refinement/Coarsening Indicators Algorithm

We describe here an example of an adaptive optimization algorithm that uses the refinement and coarsening indicators of Sects. 3.7.2 and 3.7.4 to estimate a function a defined over a domain Ω covered by a simulation mesh $\mathcal{T}_h$.

This algorithm uses a conservative strategy for the choice of the refinement: because the refinement indicators are only first order ones, the indicator(s) with maximum absolute value do(es) not always produce the larger decrease of the objective function. Hence the algorithm takes a better chance by computing the actual decrease of the objective function for a set of indicators with large absolute values before taking a decision. The situation is different for the coarsening indicators $\Delta J^*_{j\pm,\ell}$, which give directly an estimation of ΔJ^* for a given aggregation: the algorithm will decide to aggregate or not at the sole view of the coarsening indicators.

Of course, many variants of this algorithm are possible: for example, one could postpone the decision to aggregate after the actual decrease of J^* has been computed for a family of potentially interesting aggregations, which would require the resolution of a few more minimizations problems at each step. We describe now the algorithm.

As a preliminary step, one has to choose the family $\boldsymbol{T}$ of tentative refinements of the current zonation $\mathcal{T}$, which will be explored by the algorithm to define the next zonation (e.g., the set of cuts of Fig. 3.6 for each zone $\mathcal{T}_j$ of $\mathcal{T}$).

Once this is done, the algorithm can go as follows:

1. Choose an initial zonation $\mathcal{T}$.
2. **Do** until data are satisfactorily fitted:
3. Estimate a on the current zonation $\mathcal{T}$ by minimizing J with respect to $a^{\mathcal{T}}_{\text{opt}}$.
4. Compute the refinement indicators λ for all tentative refinements $\boldsymbol{T}$:

   For every zone $\mathcal{T}_j$ of $\mathcal{T}$ and for every cut of $\mathcal{T}_j$ **do**
   compute the corresponding refinement indicator λ
   Enddo

5. Compute $|\lambda|_{\max}$ the largest absolute value of all computed refinement indicators. Select a subset $\boldsymbol{T}_{80\%}$ of cuts corresponding to refinement indicators, which are larger than 80% of $|\lambda|_{\max}$ (this percentage can be adjusted)
6. Minimize J successively over the zonations associated to the cuts of $\boldsymbol{T}_{80\%}$, and let $(\Delta J^*_{\text{cuts}})_{\max}$ denote the best *decrease* obtained.
7. Compute the coarsening indicators for all selected cuts of $\boldsymbol{T}_{80\%}$:

   For every cut of $\boldsymbol{T}_{80\%}$,
   every subzone $\mathcal{T}_{j,\pm}$,
   every adjacent zone $\mathcal{T}_\ell$ **do**
   compute the coarsening indicator $\Delta J^*_{j\pm,\ell}$
   Enddo

8. **If** $(\Delta J^*_{j\pm,\ell})_{\max}$ is larger than 50% of $(\Delta J^*_{\text{cuts}})_{\max}$ (this percentage can be adjusted)

   then aggregate $\mathcal{T}_{j,\pm}$ and $\mathcal{T}_\ell$
   else refine according to the best cut found at step 6.
   Endif

9. Update the current zonation according to the retained cut or aggregation.

   Enddo

Remark 3.7.8 *The above procedure is by nature interactive: after completion of step 7, one can ask the algorithm to display the selected cuts and aggregations, and the user can mark those that seem of particular interest to him, or even figure out a new refinement pattern that incorporates his or her insight of the problem. The minimizations of step 8 can then be limited to the marked items.* ■

Remark 3.7.9 *If one wants the parameter to be constant on zones $\mathcal{T}_j$ made of a single connected component (i.e., made of "one piece"), it is necessary to add a step 5 as*

If *some cuts of $\boldsymbol{T}_{80\%}$ generate subdomains with more than one connected component* **then**

compute the refinement indicators corresponding to the subcuts associated to each connected component (this will be the case each time a checker board cut is selected!).
Update the set $\boldsymbol{T}_{80\%}$ of selected cuts according to the 80% rule.

Endif ■

3.8 Implementation of the Inversion

As we have seen in Sect. 3.3, it is important for the inversion code to be flexible with the choice of optimization parameters. The organization proposed in this section takes this into account.

3.8.1 Constraints and Optimization Parameters

Constrained optimization is a delicate and difficult matter as soon as constraints other than box constraints are imposed. It is hence often unpractical to require that the optimization routine takes in charge the constraints on x_{opt}, which would ensure that $x_{\mathrm{sim}} \in C$. An unconstrained optimization routine is hence often used. We consider below the case where the optimization routine requires only as input the parameter x, the values $J(x)$ of the objective function, and $\nabla J(x)$ of its gradient. But the results can be easily adapted to the case where the Jacobian $\varphi'(x)$ is also required (e.g., in Confidence Regions methods). We refer to [13, 68] for a presentation and analysis of optimization algorithms.

The input x_{sim} to the direct+adjoint routine is then computed from the output x_{opt} of the optimizer in two steps (left part of fig. 3.8):

- *Return to simulation parameters:* A provisory simulation vector $\tilde{x}_{\mathrm{sim}}$ is first computed from x_{opt} using the chosen parameterization formula ψ:

$$\tilde{x}_{\mathrm{sim}} = \psi(x_{\mathrm{opt}}). \tag{3.69}$$

- *Take care of the "explicit" constraints:* Because of the unconstrained optimizer used, the vector $\tilde{x}_{\mathrm{sim}}$ does not necessarily belong to the admissible parameter set C. Let us denote by $c_\ell, \ell \in L$ the constraints defining C:

$$C = \{x \in I\!R^{n_{\mathrm{sim}}} \mid c_\ell(x) \le 0 \ \forall \ell \in L\}.$$

  Some of these constraints – say $\ell \in L_{\mathrm{explicit}}$ – correspond to an explicit projection operator P: they are simply taken into account by projecting $\tilde{x}_{\mathrm{sim}}$ on $C_{\mathrm{explicit}} = \{x \in I\!R^n \mid c_\ell(x) \le 0 \ \forall \ell \in L_{\mathrm{explicit}}\}$ before using it as input to the direct+adjoint routine:

$$x_{\mathrm{sim}} = P(\tilde{x}_{\mathrm{sim}}). \tag{3.70}$$

  An important practical case is that of the box constraints:

$$x_{j,\min} \le x_j \le x_{j,\max}, \qquad j = 1 \dots N,$$

  which correspond to the very simple projection operator

$$\big(P(x)\big)_j = \begin{cases} x_{j,\min} & \text{if } \ x_j \le x_{j,\min}, \\ x_j & \text{if } \ x_{j,\min} \le x_j \le x_{j,\max} \\ x_{j,\max} & \text{if } \ x_j \ge x_{j,\max}. \end{cases} \qquad j = 1 \dots N,$$

  The direct+adjoint routine will work properly only when supplied with simulation parameters that satisfy some sign and/or range conditions, and it is a safe decision to impose at least these constraints through the projection operator P.

The remaining constraints – say $\ell \in L_{\text{implicit}}$ – correspond to an implicit projection operator, which cannot be evaluated by simple calculations. For example, the projection operator associated to linear constraints (other than box constraints!) would require the resolution of a quadratic programming problem for each evaluation of $P(x)$, which is something one usually cannot afford. These constraints can be taken care of in a soft way by adding to the objective function a *penalization term* J_η^{pen}(see center part of Fig. 3.8):

$$J_\eta^{\text{pen}}(x) = \frac{1}{\eta^2} \sum_{\ell \in L_{\text{implicit}}} (c_\ell(x)^+)^2, \qquad \eta > 0, \tag{3.71}$$

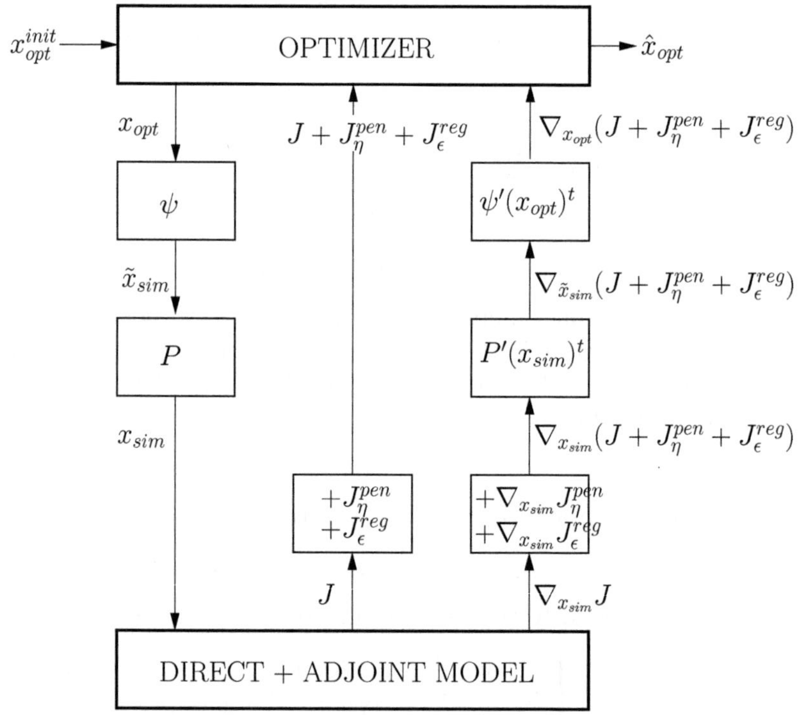

Figure 3.8: Organization of a nonlinear inversion code. *Bottom:* direct and adjoint simulation code, *top:* optimization code, ψ: parameterization routine, P: projection on box constraints, $\psi'(x_{\text{opt}})^t$: adjoint parameterization routine, $P'(x_{\text{sim}})^t$: adjoint projection routine

where

$$c_\ell(x)^+ = \begin{cases} c_\ell(x) & \text{if} \quad c_\ell(x) \geq 0, \\ 0 & \text{if} \quad c_\ell(x) \leq 0. \end{cases}$$

Using a small penalization parameter η will produce a small violation of the constraints, and also a poorly conditioned optimization problem, in practice experimentation is needed to choose η.

Introduction of additional a-priori information via LMT-regularization is easily taken into account by addition of a *regularizing functional* J_ϵ^{reg} (see center part of Fig. 3.8). For example, when the information is that the parameter is "not too far" from some a-priori value x_0, J_ϵ^{reg} has the form (see Sect. 5.1 in Chap. 5, and (1.25) in Chap. 1)

$$J_\epsilon^{\text{reg}} = \frac{\epsilon^2}{2}\|x - x_0\|_E^2.$$

Remark 3.8.1 *It can happen that the regularizing functional depends on the parameter x not only explicitly, but also through the state vector y_x, as, for example, in the adapted regularization considered in Sect. 5.4.3, or in the state-space regularization approach of Chap. 5. For a functional $J_\epsilon^{\text{reg}}(x, y_x)$ like this, the gradient has to be computed together with that of J by the adjoint state technique: this leads to add $\nabla_y J_\epsilon^{\text{reg}}(x, y_x)$ in the right-hand side of the adjoint equation, and a term $\nabla_x J_\epsilon^{\text{reg}}(x, y_x)$ in the formula that give the gradient with respect to x.* ■

3.8.2 Gradient with Respect to Optimization Parameters

As we can see in the right part of Fig. 3.8, the gradient with respect to optimization parameters is obtained from the gradient with respect to simulation parameters in two steps:

- Sum up the gradients of J, J_η^{pen}, and J_ϵ^{reg} with respect to x_{sim}
- Apply, in reverse order, the transpose of the derivatives of the transformations that had been applied to the parameter in the left part of Fig. 3.8.

We detail in this section the implementation of the subroutine that computes $\psi'(x_{\text{opt}})^{\text{T}}\, g_{\text{sim}}$ for any given vector $g_{\text{sim}} \in I\!R^{n_{\text{sim}}}$, where $\psi : x_{\text{opt}} \rightsquigarrow x_{\text{sim}}$ is the parameter change in the left part of Fig. 3.8.

The first thing is to choose a *convenient "direct" sequence of calculations* for the computation of $x_{\rm sim}$ once $x_{\rm opt}$ is given (block ψ in the left part of the diagram). For example, if the same quantity is used at different points of the calculations, it can be computationally efficient to make it an *intermediate variable*, which will be evaluated only once, and used later as many times as needed.

The second thing is to determine a *convenient "adjoint" sequence of calculations* to compute the result $g_{\rm opt} \in I\!R^{n_{\rm opt}}$ of the action of the block $\psi'(x_{\rm opt})^{\rm T}$ (in the right part of the diagram) on any vector $g_{\rm sim} \in I\!R^{n_{\rm sim}}$:

$$g_{\rm opt} = \psi'(x_{\rm opt})^T g_{\rm sim}, \tag{3.72}$$

(the adjoint sequence of calculations will be applied to $g_{\rm sim} = \nabla_{\tilde{x}_{\rm sim}}(J + J_\eta^{\rm pen} + J_\epsilon^{\rm reg})$ to produce $g_{\rm opt} = \nabla_{x_{\rm opt}}(J + J_\eta^{\rm pen} + J_\epsilon^{\rm reg})$). We shall distinguish two cases:

- The *direct sequence of calculations* required for the evaluation of $x_{\rm sim}$ from $x_{\rm opt}$ *does not involve intermediate variables*. In this case, the adjoint sequence of calculations can be determined simply by application of the definition of transposition (as in **Example 7** below)

- The *direct sequence of calculations* required for the evaluation of $x_{\rm sim}$ from $x_{\rm opt}$ *does involve intermediate variables*. It is then necessary to use the adjoint state technique of Chap. 2 to determine the adjoint sequence of calculations (**Example 8** below)

Example 7: Continuous Piecewise Linear Parameterization on a Coarse Mesh

We consider here the case where the unknown parameter x is a function $a : \xi \in \Omega \subset I\!R^2 \leadsto I\!R$, and where the domain Ω is covered by two triangular meshes:

- A fine *simulation mesh* $\mathcal{T}_h$ made of triangles K. We shall denote by $\partial\mathcal{T}_h$ the set of nodes M of $\mathcal{T}_h$

- A coarse *optimization mesh* $\mathcal{T}_{\rm opt}$ made of triangles L. Similarly, we shall denote by $\partial\mathcal{T}_{\rm opt}$ the set of nodes P of $\mathcal{T}_{\rm opt}$

According to Sect. 3.1.1, the simulation and optimization parameters are defined, for continuous piecewise linear approximations, by

$$a_{\mathrm{sim},M} = \left(\frac{\alpha_{\mathrm{sim},M}}{|\Omega|}\right)^{\frac{1}{2}} \frac{a_M}{a_{\mathrm{ref}}}, \qquad \forall M \in \partial\mathcal{T}_h,$$

$$a_{\mathrm{opt},P} = \left(\frac{\alpha_{\mathrm{opt},P}}{|\Omega|}\right)^{\frac{1}{2}} \frac{a_P}{a_{\mathrm{ref}}}, \qquad \forall P \in \partial\mathcal{T}_{opt},$$

where $\alpha_{\mathrm{sim},M}$ and $\alpha_{\mathrm{opt},M}$ are defined, with obvious adaptation, by (2.61), and a_M and a_P are the value of the parameter a at nodes M and P. The simplest parameterization map $\psi : a_{\mathrm{opt}} \rightsquigarrow a_{\mathrm{sim}}$ is obtained by computing a_M by interpolation between the values a_P on the mesh $\mathcal{T}_{\mathrm{opt}}$:

$$a_M = \sum_{P\in\partial\mathcal{T}_{\mathrm{opt}}(M)} \zeta_{M,P}\, a_P, \qquad \forall M \in \partial\mathcal{T}_h,$$

where

$$\partial\mathcal{T}_{\mathrm{opt}}(M) = \{\text{vertices of the triangle } L \text{ of } \mathcal{T}_{\mathrm{opt}} \text{ containing } M\},$$

and where $(\zeta_{M,P},\ P \in \partial\mathcal{T}_{\mathrm{opt}}(M))$ are the barycentric coordinates of M in the triangle L of $\mathcal{T}_{\mathrm{opt}}$ containing M. They are given, for every $M \in \partial\mathcal{T}_h$, by

$$M = \sum_{P\in\partial\mathcal{T}_{\mathrm{opt}}(M)} \zeta_{M,P},\, P \qquad 1 = \sum_{P\in\partial\mathcal{T}_{\mathrm{opt}}(M)} \zeta_{M,P}.$$

The parameterization map ψ is then

$$a_{\mathrm{sim},M} = \sum_{P\in\partial\mathcal{T}_{\mathrm{opt}}(M)} \psi_{M,P}\, a_{\mathrm{opt},P}, \quad \forall M \in \partial\mathcal{T}_h, \tag{3.73}$$

where the nonzero coefficients of ψ are given by

$$\psi_{M,P} = \left(\frac{\alpha_{\mathrm{sim},M}}{\alpha_{\mathrm{opt},P}}\right)^{\frac{1}{2}} \zeta_{M,P} \quad \forall M \in \mathcal{T}_h \text{ and } \forall P \in \partial\mathcal{T}_{\mathrm{opt}}(M).$$

The *direct sequence of calculation* for this parameterization is given by (3.73): it does not involve any intermediate variable (the coefficients $\psi_{M,P}$ are computed once for all).

So we shall simply use the definition of transposition to determine the *adjoint sequence of calculation*: given $g_{\text{sim}} \in I\!R^{n_{\text{sim}}}$, the vector $g_{\text{opt}} = \psi^T g_{\text{sim}}$ satisfies by definition, for any $\delta a_{\text{opt}} \in I\!R^{n_{\text{opt}}}$:

$$\begin{aligned}\langle g_{\text{opt}}, \delta a_{\text{opt}} \rangle_{I\!R^{n_{\text{opt}}}} &= \langle g_{\text{sim}}, \psi\, \delta a_{\text{opt}} \rangle_{I\!R^{n_{\text{sim}}}} \\ &= \sum_{M \in \partial\mathcal{T}_h} g_{\text{sim},M} \Big(\sum_{P \in \partial\mathcal{T}_{\text{opt}}(M)} \psi_{M,P}\, \delta a_{\text{opt},P} \Big) \\ &= \sum_{P \in \partial\mathcal{T}_{\text{opt}}} \underbrace{\Big(\sum_{M \in \partial\mathcal{T}_h(P)} g_{\text{sim},M}\, \psi_{M,P} \Big)}_{g_{\text{opt},P}} \delta a_{\text{opt},P},\end{aligned}$$

where

$$\partial\mathcal{T}_h(P) = \{\text{nodes of } \mathcal{T}_h \text{ in the elements } L \text{ of } \mathcal{T}_{opt} \text{ surrounding } P\}.$$

Hence the sought after *adjoint sequence of calculation* is

$$g_{\text{opt},P} = \sum_{M \in \partial\mathcal{T}_h(P)} g_{\text{sim},M}\, \psi_{M,P}.$$

Example 8: Polar Coordinates

We compute (3.72) by the step-by-step approach of Sect. 2.4.

Step 0: Forward Map and Objective Function: The forward map is here

$$\psi \ : \ \theta = (\theta_1 \cdots \theta_{n-1}) \rightsquigarrow x = (x_1 \cdots x_n),$$

and it is conveniently determined by the sequence of calculation:

$$\begin{aligned} x_1^0 &= 1, && && (3.74)\\ u^k &= \cos\theta_k && \forall k = 1 \cdots n-1, && (3.75)\\ x_j^k &= x_j^{k-1} u^k && \forall k = 1 \cdots n-1 \quad \forall j = 1 \cdots k, && (3.76)\\ x_{k+1}^k &= \sin\theta_k && \forall k = 1 \cdots n-1, && (3.77)\end{aligned}$$

followed by

$$x_j = x_j^{n-1} \qquad \forall j = 1 \cdots n.$$

To compute $g_\theta = \psi'(\theta)^{\mathrm{T}} g_x$ for a given $g_x \in I\!R^n$, one defines the objective function $\theta \rightsquigarrow G(\theta, \psi(\theta))$ by (c.f. (2.13)):

$$G(\theta, x) = \langle x, g_x \rangle_{I\!R^n} = \sum_{j=1}^{n} x_j\, g_{x,j}.$$

Step 1: State-Space Decomposition:

$$\begin{aligned}
\text{parameter} &: (\theta_1 \cdots \theta_n),\\
\text{state vector} &: y = (u^k, x_j^k, k = 1 \cdots n-1, j = 1 \cdots k+1),\\
\text{state equation} &: (3.75), (3.76), and (3.77),\\
\text{observation operator} &: M : y \rightsquigarrow x = x^{n-1} \in I\!R^n,
\end{aligned}$$

Step 2: Lagrangian Formula (2.25) gives here

$$\begin{aligned}
L(\theta, y, \Lambda) &= G(\theta, M(y)) - \langle e(\theta, y), \Lambda > \\
&= \sum_{j=1}^{n} x_j^{n-1}\, g_{x,j} - \sum_{k=1}^{n-1} (u^k - \cos\theta_k)\, \lambda^k \\
&\quad - \sum_{k=1}^{n-1} \sum_{j=1}^{k} (x_j^k - x_j^{k-1} u^k)\, \mu_j^k \\
&\quad - \sum_{k=1}^{n-1} (x_{k+1}^k - \sin\theta_k)\, \mu_{k+1}^k,
\end{aligned}$$

where the Lagrange multiplier vector is

$$\Lambda = (\lambda, \mu) = (\lambda^k, \mu_j^k, k = 1 \cdots n-1, j = 1 \cdots k+1).$$

As expected, the Lagrangian is an explicit function of all its arguments.

Step 3: Adjoint Equation

$$\begin{aligned}
\frac{\partial L}{\partial y}\delta y &= \sum_{j=1}^{n} \delta x_j^{n-1}\, g_{x,j} - \sum_{k=1}^{n-1} \delta u^k\, \lambda^k \\
&\quad - \sum_{k=1}^{n-1} \sum_{j=1}^{k} (\delta x_j^k - \delta x_j^{k-1} u^k - x_j^{k-1} \delta u^k)\, \mu_j^k \\
&\quad - \sum_{k=1}^{n-1} \delta x_{k+1}^k\, \mu_{k+1}^k.
\end{aligned}$$

Factorizing $\delta y = (\delta u^k, \delta x_j^k, k = 1 \cdots n-1, j = 1 \cdots k+1)$ gives

$$\begin{aligned}\frac{\partial L}{\partial y}\delta y &= -\sum_{k=1}^{n-1} \delta u^k (\lambda^k - \sum_{j=1}^{k} x_j^{k-1} \mu_j^k) \\ &+ \sum_{j=1}^{n} \delta x_j^{n-1}\, g_{x,j} - \sum_{k=1}^{n-1}\sum_{j=1}^{k+1} \delta x_j^k\, \mu_j^k + \sum_{k=1}^{n-1}\sum_{j=1}^{k} \delta x_j^{k-1} u^k\, \mu_j^k.\end{aligned}$$

Making the index shift $k \rightsquigarrow k-1$ in the last summation gives, as $\delta x_1^0 = 0$,

$$\begin{aligned}\frac{\partial L}{\partial y}\delta y &= -\sum_{k=1}^{n-1} \delta u^k (\lambda^k - \sum_{j=1}^{k} x_j^{k-1} \mu_j^k) \\ &+ \sum_{j=1}^{n} \delta x_j^{n-1} (g_{x,j} - \mu_j^{n-1}) \\ &- \sum_{k=1}^{n-2}\sum_{j=1}^{k+1} \delta x_j^k (\mu_j^k - u^{k+1}\, \mu_j^{k+1}),\end{aligned} \tag{3.78}$$

which is of the form (2.26). Following (2.27), the *adjoint state* Λ_θ is the solution obtained by equating to zero the coefficients of $\delta x_j^{n-1}, \delta x_j^k$, and δu^k in (3.78):

$$\begin{aligned}\mu_j^{n-1} &= g_{x,j} & \forall j &= 1 \cdots n, \\ \mu_j^k &= u^{k+1}\, \mu_j^{k+1}) & \forall k &= n-2 \cdots 1\ ,\ \forall j = 1 \cdots k+1, \\ \lambda^k &= \sum_{j=1}^{k} x_j^{k-1} \mu_j^k & \forall k &= n-1 \cdots 1.\end{aligned}$$

Step 4: Gradient Equation According to (2.28), $g_\theta = \psi'(\theta)^T g_x$ is given by

$$\begin{aligned}g_{\theta,k} &= \frac{\partial L}{\partial \theta_k}(\theta, y_\theta, \Lambda_\theta) & \forall k &= 1 \cdots n-1, \\ &= -\underbrace{\sin\theta_k}_{x_{k+1}^k} \lambda^k + \underbrace{\cos\theta_k}_{u^k} \mu_{k+1}^k & \forall k &= 1 \cdots n-1, \\ g_{\theta,k} &= -x_{k+1}^k \lambda^k + u^k \mu_{k+1}^k & \forall k &= 1 \cdots n-1.\end{aligned}$$

3.9 Maximum Projected Curvature: A Descent Step for Nonlinear Least Squares

We have seen in Sect. 3.8.1 how to transform the original constrained nonlinear least squares problem into the formally unconstrained problem:

$$\hat{x}_{\rm opt} \text{ minimizes } J(x_{\rm opt}) \stackrel{\rm def}{=} \frac{1}{2}\|F(x_{\rm opt})\|^2 \text{ over } I\!R^{n_{\rm opt}}, \tag{3.79}$$

where (see (3.69) and (3.70) or left part of Fig. 3.8):

$$F(x_{\rm opt}) \stackrel{\rm def}{=} \varphi\big(P(\Psi(x_{\rm opt}))\big) - z. \tag{3.80}$$

For the rest of this section, we shall drop the index *opt*, so that x will denote the optimization parameter.

We have seen that the determination of the gradient ∇J of the objective function by adjoint state was required both in the preliminary study of the problem (for the determination of the number of independant parameters that can be retrieved from the data, in Sect. 3.2), and in adaptive parameterization algorithms (for the determination of first order refinement indicators, Sect. 3.7, Remark 3.7.2).

It is hence natural to use gradient-based descent optimization algorithms for the solution of (3.79), and it becomes the only possible choice when the number of optimization parameters is large enough to make the resolution of the linearized problem – and hence confidence regions – unaffordable [78, 79]. But the efficiency of descent methods is hampered by the back-tracking – and the many function evaluations – required at each iteration to obtain an *acceptable descent step*. We describe in this section the *Maximum Projected Curvature* (MPC) step [22, 2, 3], which takes advantage of the nonlinear least squares structure of the problem: it uses the first and second derivatives of the forward map F in the descent direction to determine a descent step that is always acceptable.

3.9.1 Descent Algorithms

We recall first the organization of a descent algorithm for the resolution of problem (3.79). Given the current estimate x_k, the algorithm

1. Determines a *descent direction* y_k, such that

$$\langle y_k \,, \nabla J(x_k)\rangle \; < 0. \tag{3.81}$$

It is the way the descent direction y_k is chosen, which gives its name to the optimization algorithm class: Steepest Descent, Quasi-Newton, Gauss–Newton, Levenberg–Marquardt, etc.

2. Chooses a *search curve* $g_k : I\!R^+ \to I\!R^n$ *in the parameter space, such that*

$$g_k(0) = x_k, \quad g'(0) = y_k. \tag{3.82}$$

In practice, all descent algorithms perform a straight line *search*, along the half-line originating at x_k pointing in the direction y_k:

$$g_k(\alpha) = x_k + \alpha\, y_k \qquad \alpha \geq 0. \tag{3.83}$$

The reason behind this choice is mostly its simplicity. But we shall also consider other search curves, in particular, if one can afford to compute the Jacobian $F'(x_k)$, the more intrinsic *geodesic search curves* (Definition 3.9.1 below).

3. Computes an *acceptable descent step* α_k, such that
• α_k decreases sufficiently the objective function:

$$J(g_k(\alpha_k)) \;\leq\; J(x_k) + \omega\, \alpha_k \langle y_k\, , \nabla J(x_k)\rangle \qquad \text{(Armijo condition)}, \tag{3.84}$$

where $0 < \omega \leq 1/2$ is a given number (the same through all iterations of course),
• α_k is not too small:

$$\begin{cases} \langle g'(\alpha_k), \nabla J(g_k(\alpha_k))\rangle \;\geq\; \omega' \langle y_k, \nabla J(x_k)\rangle & \text{(Wolfe condition)}, \\ \text{or} & \\ J(g_k(\alpha_k)) \geq J(x_k) + \omega'\, \alpha_k \langle y_k\, , \nabla J(x_k)\rangle & \text{(Goldstein condition)}, \end{cases}$$

where $\omega < \omega' < 1$ is also a given number. The determination of α_k is the *linesearch part* of the optimization algorithm. It differs from one implementation to the other inside the same class of algorithm, and is largely responsible for its performance.

An example of acceptable descent step is the *Curry step* $\bar{\alpha}$: one moves on the search curve g in the parameter space until the first stationary point of $J = 1/2\|F\|^2$ is encountered:

$$\bar{\alpha} = \text{Inf}\{\alpha \geq 0 \text{ such that } \frac{\mathrm{d}}{\mathrm{d}\alpha}\|F(g_k(\alpha))\|^2 = 0\}. \tag{3.85}$$

However, the Curry step cannot be used in practice, as its determination would require to many function evaluations.

4. Once α_k is accepted, the $k+1$ iterate is defined by

$$x_{k+1} = g_k(\alpha_k). \tag{3.86}$$

3.9.2 Maximum Projected Curvature (MPC) Step

We define in this subsection the MPC step α_k for given x_k, y_k, and g_k, and so we drop the iteration index k, with the exception of x_k and x_{k+1} for obvious reasons.

Let P be the curve of the data space, which is the image by F of the search curve g of the parameter space:

$$P = F \circ g : \alpha \in I\!R^+ \rightsquigarrow F(g(\alpha)) \in I\!R^q.$$

We make the hypothesis that

$$P \in W^{2,\infty}(]0, \bar{\alpha}[, I\!R^q), \tag{3.87}$$

where $\bar{\alpha}$ is the Curry step defined in (3.85). Hence the two first distributional derivatives of P with respect to α are in $L^\infty(]0, \bar{\alpha}[, I\!R^q)$, and we can define, for almost every $\alpha \in]0, \bar{\alpha}[$, the *residual* r, the *velocity* V, the *acceleration* A, the *arc length* ν, and the *Curry arc length* $\bar{\nu}$ along P by

$$r = \|P\|, \tag{3.88}$$

$$V = \mathrm{d}P/\mathrm{d}\alpha = D_{g'}F(g), \tag{3.89}$$

$$A = \mathrm{d}^2P/\mathrm{d}\alpha^2 = D^2_{g',g'}F(g) + D_{g''}F(g), \tag{3.90}$$

$$\nu(\alpha) = \int_0^\alpha \|V(\alpha)\|\,\mathrm{d}\alpha, \qquad \bar{\nu} = \nu(\bar{\alpha})\ . \tag{3.91}$$

which satisfy

$$V \in \mathcal{C}^0([0, \bar{\alpha}], I\!R^q), \qquad A \in L^\infty(]0, \bar{\alpha}[, I\!R^q).$$

Definition 3.9.1 *Let F be derivable with $F'(x)$ injective for all $x \in I\!R^n$. We shall say that g is a* geodesic search curve *if it satisfies the differential equation*

$$F'(g)^{\mathrm{T}} F'(g)\, g'' + F'(g)^{\mathrm{T}} D^2_{g',g'} F(g) = 0 \quad \forall \alpha \geq 0. \tag{3.92}$$

The acceleration A is hence orthogonal to the hyperplane $\{\delta y \in I\!R^q \mid \delta y = F'(g)\delta x\}$ tangent to the attainable set $F(I\!R^n)$ at $P(g)$, which shows that P is a geodesic of $F(I\!R^n)$. The velocity V and acceleration A satisfy moreover

$$\begin{aligned} \mathrm{d}\|V\|^2/\mathrm{d}\alpha &= 0 \quad \Longrightarrow \quad \nu(\alpha) = \alpha\|V(0)\| \quad \forall \alpha \geq 0, \\ \|A\|^2 &= \|D^2_{g',g'}F(g)\|^2 - \|D_{g''}F(g)\|^2. \end{aligned} \tag{3.93}$$

We shall consider in the sequel only the *straight search curves* (3.83), or the *geodesic search curves* of Definition 3.9.1.

The Definition (3.85) of the Curry step $\bar{\alpha}$ along P implies that

$$V(\alpha) \neq 0 \quad \forall \alpha \in [0, \bar{\alpha}[, \tag{3.94}$$

and Proposition 8.2.1 shows that the first and second derivatives v and a of P with respect to the arc length ν are given, for almost every $\alpha \in]0, \bar{\alpha}[$, by

$$v = V/\|V\|, \qquad \|v\| = 1 \quad \text{with} \quad \langle v, a\rangle = 0, \tag{3.95}$$

$$a = \big(A - \langle A, v\rangle v\big)/\|V\|^2, \quad \|a\|^2 = \big(\|A\|^2 - \langle A, v\rangle^2\big)/\|V\|^4. \tag{3.96}$$

We make now the additional hypothesis that

$$\exists\ 1/R > 0\ :\ \|A(\alpha)\| \leq 1/R\ \|V(\alpha)\|^2 \quad \text{for almost every } \alpha \in]0, \bar{\alpha}[, \tag{3.97}$$

so that the *radius of curvature* $\rho(\alpha)$ along P satisfies (Proposition 8.2.2)

$$1/\rho(\alpha) \stackrel{\text{def}}{=} \|a(\nu(\alpha))\| \leq 1/R \quad \text{for almost every } \alpha \in]0, \bar{\alpha}[. \tag{3.98}$$

The idea behind the MPC step is then to use the lower bound R on the radius of curvature to find a *calculable* "worst case" lower bound $\bar{\alpha}_W$ to the computationally unaffordable Curry step $\bar{\alpha}$.

We give first an intuitive presentation of the MPC step in the case of a two-dimensional data space $I\!R^q = I\!R^2$. Then P is a plane curve which, as one can see in Fig. 3.9, starts at $F(x_k)$ (arc length $\nu = 0$) in the direction $v = v(0)$. After that, the only thing known about P is, according to (3.98), that its

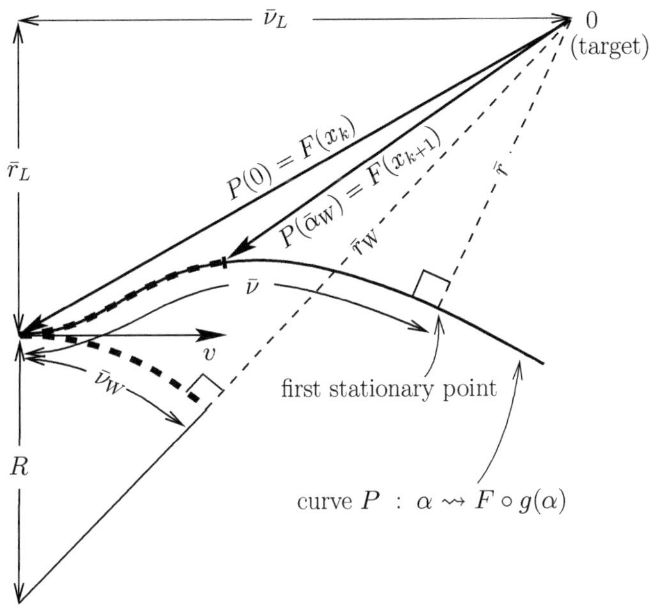

Figure 3.9: Geometrical representation of MPC step and Theorem 3.9.4 in a two-dimensional data space, where projected curvature and curvature coincide. The two *thick dashed lines* have the same arc length

curvature remains smaller than $1/R$. Hence there is no way, as expected, to determine in a computationally efficient way the Curry arc length $\bar{\nu}$!

But as one can see in Fig. 3.9, the curve that leads to the *smallest stationary arc length* $\bar{\nu}_W$ among all curves of curvature smaller than $1/R$ that leave $F(x_k)$ in the direction v is the arc of circle of radius R (lower thick dashed line) that "turns away" from the target 0. This curve is the "worst" (hence the subscript W) from the optimization point of view, as it produces the largest stationary residual $\bar{r}_W$!

The MPC step is then $\bar{\alpha}_W = \nu^{-1}(\bar{\nu}_W)$*: it consists in moving on the search curve* g *in the parameter space up to the point* $\bar{\alpha}_W$ *such that* $x_{k+1} = g(\bar{\alpha}_W)$ *has its image* $F(x_{k+1})$ *at an arc length distance* $\bar{\nu}_W$ *of the origin* $F(x_k)$ *of* P (thick dashed line part of P on Fig. 3.9).

An important property that can be seen on the figure is that the residual $r_{k+1} = \|F(x_{k+1})\|$ at the point obtained by the MPC step is better (smaller) than the worst case stationary residual $\bar{r}_W$. This will ensure for the MPC step a useful "guaranteed decrease property."

We turn now to the rigorous presentation of the MPC step in the general case of a data space of any dimension. The curve P does not necessarily remain anymore in the plane defined by $P(0)$ and $v(0)$, and the curvature that matters for the determination of the worst-case Curry step turns out be the *projected curvature* $1/\rho_{proj}(\alpha)$ of P on the plane defined by the two vectors $P(\alpha)$ and $v(\alpha)$. But this plane is defined only when the two vectors do not point in the same direction, and so we define a set of exceptional points:

$$Z = \{\alpha \in [0, \bar{\alpha}] \text{ such that } P(\alpha) - \langle P(\alpha), v(\alpha) \rangle v(\alpha) = 0\}. \tag{3.99}$$

The measure of Z can be strictly positive, and so one has to give special attention to its points in the definition of projected curvature.

Definition 3.9.2 *The* projected radius of curvature *along P is defined by*

$$1/\rho_{\text{proj}}(\alpha) = |\langle a, n \rangle| \qquad \textit{for a.e. } \alpha \in [0, \bar{\alpha}] \setminus Z, \tag{3.100}$$
$$1/\rho_{\text{proj}}(\alpha) = 0 \qquad \textit{for a.e. } \alpha \in Z, \tag{3.101}$$

where n is a unit vector orthogonal to v in the P, v plane, when it is defined

$$n = (P - \langle P, v \rangle v)/(\|P - \langle P, v \rangle v\|) \tag{3.102}$$

(P, v, a are evaluated at α).

One expects the projected curvature of P to be smaller than its curvature, this is confirmed by the:

Proposition 3.9.3 *Let P satisfy the hypothesis (3.87) and (3.97). Then*

$$1/\rho_{\text{proj}}(\alpha) \le 1/\rho(\alpha) \qquad \textit{for a.e. } \alpha \in [0, \bar{\alpha}] \setminus Z, \tag{3.103}$$
$$1/\rho_{\text{proj}}(\alpha) = 1/\rho(\alpha) = 0 \qquad \textit{for a.e. } \alpha \in Z, \tag{3.104}$$

and the following equality holds:

$$|\langle a, P \rangle| = \|P - \langle P, v \rangle v\| / \rho_{\text{proj}} \quad \textit{for a.e. } \alpha \in]0, \bar{\alpha}[, \tag{3.105}$$

where $P, v, a, \rho_{\text{proj}}$ are evaluated at α.

The proof of the proposition is given in Appendix 1. The next theorem proves the intuitive properties of the MPC step seen on Fig. 3.9:

Theorem 3.9.4 *Let $x_k, y, g, \bar{\alpha}$ be the current estimate, descent direction, search curve, and Curry step at iteration k of a descent optimization algorithm for the resolution of problem (3.79). Let $r(\alpha)$ denote the residual along P as defined in (3.88), and suppose that the curve $P : \alpha \rightsquigarrow F(g(\alpha))$ satisfies (3.87) and (3.97).*

Then one can choose as descent step α_k the MPC step $\bar{\alpha}_W$

$$\alpha_k = \bar{\alpha}_W \quad \textit{defined by} \quad \nu(\bar{\alpha}_W) = \bar{\nu}_W, \tag{3.106}$$

which satisfies

$$\begin{aligned} 0 < \quad & \bar{\alpha}_W \quad \leq \bar{\alpha} & (3.107)\\ r_k \stackrel{\text{def}}{=} r(0) > \quad \bar{r}_W \geq r_{k+1} & \stackrel{\text{def}}{=} r(\bar{\alpha}_W) \quad \geq \bar{r} \stackrel{\text{def}}{=} r(\bar{\alpha}), & (3.108)\end{aligned}$$

where

- *$\bar{\nu}_W$ and $\bar{r}_W$ are the* worst case *stationary arc length and residual:*

$$\begin{aligned} \bar{\nu}_W &= R\tan^{-1}\frac{\bar{\nu}_L}{R+\bar{r}_L}, & (3.109)\\ \bar{r}_W &= ((R+\bar{r}_L)^2+\bar{\nu}_L^2)^{\frac{1}{2}} - R, & (3.110)\end{aligned}$$

- *$\bar{\nu}_L$ and $\bar{r}_L$ are the* linear case *stationary arc length and residual:*

$$\begin{aligned} \bar{\nu}_L &= \left|\left\langle F(x_k), \frac{V}{\|V\|}\right\rangle\right|, & (3.111)\\ \bar{r}_L &= \sqrt{r_k^2 - \bar{\nu}_L^2}, & (3.112)\end{aligned}$$

  where $V = V(0) = D_yF(x_k)$ and $r_k = r(0) = \|F(x_k)\|$.

- *$1/R$ is an upper bound to the* projected curvature *of P over the $[0, \bar{\alpha}]$ interval:*

$$1/\rho_{\text{proj}}(\alpha) \leq 1/R \qquad \textit{for almost every } \alpha \in [0, \bar{\alpha}]. \tag{3.113}$$

The proof of Theorem 3.9.4 is given in Appendix 2. The theorem remains true when the data space $I\!R^q$ is replaced by an infinite dimensional Hilbert space – the proof goes over without change.

Proposition 3.9.5 *Let P satisfy the hypothesis (3.87) and (3.97). Then the MPC step α_k:*

1. *Ensures a* guaranteed decrease *of the objective function:*

$$J(x_k) - J(x_{k+1}) = \frac{1}{2}(r_k^2 - r_{k+1}^2) \geq \frac{1}{2}\bar{\nu}_{\mathrm{W}}\bar{\nu}_{\mathrm{L}} > 0, \tag{3.114}$$

2. *Satisfies the* Armijo condition *(3.84) with the ratio:*

$$\omega = \frac{1}{2}\frac{\overline{\|V\|}(\alpha_k)}{\|V(0)\|} \quad \textit{where} \quad \overline{\|V\|}(\alpha_k) = \frac{1}{\alpha_k}\int_0^{\alpha_k} \|V(\alpha)\|\,\mathrm{d}\alpha. \tag{3.115}$$

   In particular, one can always pretend one has reparameterized the search curve g by the arc length ν along P, as this does not change x_{k+1}! Then $\alpha = \nu$, $V(\alpha) = v(\nu)$, and $\|V(\alpha)\| = \|v(\nu)\| = 1$, so that the Armijo condition is satisfied with $\omega = 1/2$, which corresponds to a quadratical decrease property.

The proof of the proposition is given in Appendix 3. The guaranteed decrease property (3.114) is sharp: for a linear problem, one has $R = +\infty$, so that $\bar{\nu}_{\mathrm{W}} = \bar{\nu}_{\mathrm{L}}$, and equality holds in (3.114).

Remark 3.9.6 ***Choice of the search curve g.*** *One sees from (3.114) that the decrease of the objective function at iteration k is proportional to $\bar{\nu}_{\mathrm{W}}$, which itself is an increasing function, by formula (3.109), of the upper bound $1/R$ on the projected curvature. Let $1/\rho_{\mathrm{straight}}(0)$ and $1/\rho_{\mathrm{geodesic}}(0)$ denote the curvature at $\alpha = 0$ of the curves P_{straight} and P_{geodesic} associated to the straight and geodesic search curves. P_{geodesic} has the smallest curvature at $F(x_k)$ among all curves of $F(I\!R^n)$ passing through $F(x_k)$ in the direction V_k. One of them is P_{straight}, hence,*

$$1/\rho_{\mathrm{geodesic}}(0) \leq 1/\rho_{\mathrm{straight}}(0).$$

This does not imply necessarily that the projected curvatures satisfy the same inequality, as it depends on the relative position of $F(x_k)$, the acceleration $a_{\mathrm{straight}}(0)$ (orthogonal to V_k), and the acceleration $a_{\mathrm{geodesic}}(0)$ (orthogonal to the tangent plane). But over a large number of problems and iterations, one can expect that

$$\textit{in the mean:} \quad 1/\rho_{\mathrm{proj,geodesic}}(0) \leq 1/\rho_{\mathrm{proj,straight}}(0).$$

Similarly, there is no reason for $1/\rho_{\text{proj,geodesic}}(\alpha)$ *to remain smaller than* $1/\rho_{\text{proj,straight}}(\alpha)$ *up to the Curry step* $\bar{\alpha}$*! But one can expect that the upper bounds on the projected curvature satisfy*

$$\textit{in the mean:}\ 1/R_{\text{geodesic}} \le 1/R_{\text{straight}} \implies \bar{\nu}_{\text{W,geodesic}} \ge \bar{\nu}_{\text{W,straight}}.$$

Hence using the geodesic search curve amounts giving oneself a better chance of making a larger descent step! ■

3.9.3 Convergence Properties for the Theoretical MPC Step

We recall first the definition of convergence for a minimization algorithm:

Definition 3.9.7 *A minimization algorithm is convergent if and only if it produces a sequence of iterates that satisfy* $\nabla J(x_k) \to 0$ *when* $k \to +\infty$.

Convergent descent algorithm compute only stationary point – one has to rely on Q-wellposedness results (Chap. 4) to ensure that this stationary point is actually a minimizer.

Proposition 3.9.8 *Define*

$$D = \left\{ x \in I\!R^n \mid J(x) = \frac{1}{2}\|F(x)\|^2 \le J(x_0) \right\}, \tag{3.116}$$

where x_0 *is the initial guess for the resolution of problem (3.79), and suppose that* $F'(x)$ *and* $D^2_{v,v}F(x)$ *exist* $\forall x \in D, \forall y \in I\!R^n$ *and satisfy*

$$\|F'(x)v\| \le M\|v\| \qquad \forall x \in D, \forall y \in I\!R^n, \tag{3.117}$$

$$\|D^2_{v,v}F(x)\| \le 1/R\,\|F'(x)v\|^2 \quad \forall x \in D, \forall y \in I\!R^n, \tag{3.118}$$

for some $M \ge 0$ *and* $1/R \ge 0$. *Then for any current estimate* $x_k \in D$ *and any descent direction* y_k*, the curve* $P_k = F \circ g_k$*, where* g_k *is the associated straight or geodesic search curve, satisfies*

$$P_k \in W^{2,\infty}(]0, \bar{\alpha}_k[, I\!R^q), \tag{3.119}$$

$$\|A_k(\alpha)\| \le 1/R_k\|V_k(\alpha)\|^2 \ \textit{a.e. on}\ ]0, \bar{\alpha}_k[\ \textit{for some}\ R_k \ge R. \tag{3.120}$$

Hence one can apply to the resolution of the nonlinear least squares problem (3.79) the descent algorithm *(3.81), (3.82), and (3.86) with the* MPC step

(3.106) and (3.109) along the straight *search curve (3.83) or the* geodesic *search curve (3.92). The iterates x_k and descent directions y_k satisfy then*

$$\sum_{k \in I\!N} \nabla J(x_k) \cos^2 \theta_k < \infty, \tag{3.121}$$

where $\theta_k \in [0, \pi/2]$ is the angle between $-\nabla J(x_k)$ and the descent direction y_k at iteration k, and the algorithm converges for the following choices of the descent directions:

- $y_k + F_k' F_k = 0,$ *(Steepest descent)*
- $F_k'^{\mathrm{T}} F_k' \, y_k + F_k' F_k = 0,$ *(Gauss-Newton)*
  provided $\exists \beta > 0$ *s.t.* $\|F_k' \, y\| \geq \beta \|y\| \quad \forall y \in I\!R^n \quad \forall k = 0, 1, 2 \ldots$
- $M_k y_k + F_k' F_k = 0,$ *(Quasi-Newton)*
  provided the matrices M_k are uniformly bounded and coercive,
- $F_k'^{\mathrm{T}} F_k' y_k + \lambda_k y_k + F_k' F_k = 0,$ *(Levenberg–Marquardt)*
  provided λ_k is chosen such that $\lambda_k \geq c/(1-c)\|F_k'\|$.

The proof is given in Appendix 4 below. If C is a closed convex subset of D, hypothesis (3.117) and (3.118) show that F, C is a FC problem in the sense of Definition 4.2.1. These hypothesis are satisfied in quite general situations: for the finite dimensional problems considered here, for example, if D is bounded, F smooth, and $F'(x)$ injective for all x of D (see Sect. 4.5).

3.9.4 Implementation of the MPC Step

The exchange of information between the optimization and modeling routine required for the MPC step is as follows:

1. Knowing the current estimate x_k, the modeling routine returns
   - either $\nabla J_k \stackrel{\text{def}}{=} \nabla J(x_k)$ (Steepest Descent or Quasi Newton)
   - or $F_k' \stackrel{\text{def}}{=} F'(x_k)$ (Gauss–Newton or Levenberg–Marquardt)
2. Using these informations, the optimization routine determines a descent direction y_k

3. Knowing y_k, the modeling routine returns:

   - $D^2_{y_k,y_k}F(x_k)$ second directional derivative of the forward model
   - $D_{y_k}F(x_k)$ in the case where only ∇J_k is available in step 1

4. Finally, the optimization routine determines

   - The search curve g_k
   - An MPC descent step α_k along g_k

   and sets $x_{k+1} = g_k(\alpha_k)$

The computation of one or two directional derivatives of the forward model in step 3 represents the additional computational burden (compare with Fig. 3.8) required for the implementation of the MPC step; it is largely compensated by a much smaller amount of back-tracking, as the *computed* MPC step will be directly admissible most of the time (remember that the *theoretical* MPC step defined in the previous section is *always* admissible, Proposition 3.9.5).

We detail in this subsection the calculations involved in step 4. One computes first

$$V_k = D_{y_k}F(x_k) = F'_k\, y_k, \qquad v_k = V_k/\|V_k\|, \tag{3.122}$$

using either $D_{y_k}F(x_k)$, provided in step 3 when only ∇J_k is available, or the Jacobian F'_k when it is available in step 1.

• Determination of g_k

We shall use for g_k a second degree polynomial curve in the parameter space:

$$g_k(\alpha) = x_k + \alpha\, y_k + \frac{\alpha^2}{2} g''_k, \tag{3.123}$$

where g''_k is to be chosen.

1. *Case of a straight search curve.* This choice is the only one for Steepest Descent or Quasi Newton algorithms, as the modeling routine returns only ∇J_k, but not F'_k, so that no information is available on the tangent plane to the attainable set at $F(x_k)$. One can of course use for g_k the linear function (3.83), which corresponds to $g''_k = 0$. But one can also take advantage of the Proposition 3.9.5, which shows that the MPC step satisfies the Armijo condition with $\omega = 1/2$ when the arc length ν

is proportional to α, and use for g_k the second order polynomial (3.123), where g_k'' is chosen such that $\mathrm{d}^2\nu/\mathrm{d}\alpha^2(0) = 0$:

$$g_k'' + \langle v_k, D^2_{y_k,y_k}F(x_k)\rangle y_k = 0.$$

The computational cost is negligible, as $D^2_{y_k,y_k}F(x_k)$ has in any case to be computed for the determination of the MPC step α_k, and this choice increases the chances that the first guess of α_k to be determined below in (3.129) satisfies the Armijo condition.

2. *Case of a geodesic search curve.* This choice is possible only for the Gauss–Newton or the Levenberg–Marquardt algorithms, and in general for all algorithms where the Jacobian $F'(x_k)$ is available. The numerical resolution of the differential equation (3.92) is feasible, but computationally intensive. So we shall use for g_k the second degree polynomial (3.123), where g_k'' is chosen such that the associated curve P_k has a second order contact with the geodesic at $F(x_k) = P_k(0)$:

$$F_k'^{\mathrm{T}}F_k'\, g_k'' + F_k'^{\mathrm{T}}D^2_{y_k,y_k}F(x_k) = 0. \tag{3.124}$$

The additional cost here is the resolution of the linear system (3.124), which turns out to be the same linear system as the Gauss–Newton equation for the descent direction y_k in Proposition 3.9.8, but with a different right-hand side. By construction, this (approximated) geodesic search curve has the same advantages as the straight line search, plus the prospect of producing in the mean larger steps (Remark 3.9.6), and hence a faster decrease of J.

In both cases, the arc length along P_k satisfies $\mathrm{d}^2\nu/\mathrm{d}\alpha^2(0) = 0$, so that the *second order development* of ν at 0 reduces to

$$\nu(\alpha) = \alpha\|V_k\| + \mathrm{o}(\alpha^2). \tag{3.125}$$

- **Determination of α_k**

There are two points in the determination of α_k by Theorem 3.9.4, where an exact calculation unfeasible, and which require an approximation:

1. *Determination of* $1/R_k$. A precise determination of the upper bound $1/R_k$ to the (unknown) projected curvature along P_k up to the (unknown) Curry step $\bar{\alpha}_k$ (condition (3.113)) would require the evaluation

of $1/\rho_{\rm proj}$ at many points along P_k up to $\bar{\alpha}_k$, and would be at least as expensive as the numerical determination of the Curry step $\bar{\alpha}$ itself! But there is one thing available at almost no cost, namely the projected curvature at the origin of P_k. So one can define a tentative R_k by

$$R_k = \kappa_k\, \rho_{{\rm proj},k} \quad \text{with} \quad 0 \leq \kappa_k \leq 1, \tag{3.126}$$

where κ_k is a security factor that accounts for the possible increase of the projected curvature along the path, and $\rho_{{\rm proj},k}$ is given by Proposition 3.9.3:

$$1/\rho_{{\rm proj},k} = |\langle a_k, F_k \rangle| / \| F_k - \langle F_k, v_k \rangle v_k \|,$$

with a_k given by (3.96) and (3.90):

$$\begin{aligned} a_k &= \big(A_k - \langle A_k, v_k\rangle v_k\big)/\|V_k\|^2, & (3.127)\\ A_k &= D^2_{y_k,y_k}F(x_k) + D_{g_k''}F(x_k), & (3.128)\end{aligned}$$

where $D_{g_k''}F(x_k)$ is given by

$$D_{g_k''}F(x_k) = \begin{cases} -\langle v_k, D^2_{y_k,y_k}F(x_k)\rangle V_k & \text{(straight line search)}\\ F_k'\, g_k'' & \text{(geodesic search)}\end{cases}$$

2. *Determination of* α_k. An accurate determination of α_k by (3.106) would require the use of a quadrature formula to evaluate the arc length function $\nu(\alpha)$, and hence a large number of evaluations of F, something one cannot afford. So one replaces $\nu(\alpha)$ by its second order approximation (3.125), and define a tentative MPC step by

$$\alpha_k = \bar{\nu}_{{\rm W},k}/\|V_k\|, \tag{3.129}$$

where $\bar{\nu}_{{\rm W},k}$ is given by (3.109)

$$\bar{\nu}_{{\rm W},k} = R_k \tan^{-1} \frac{\bar{\nu}_{{\rm L},k}}{R_k + \bar{r}_{{\rm L},k}} \quad \text{with} \tag{3.130}$$

$$\bar{\nu}_{{\rm L},k} = |\langle F_k, v_k \rangle|, \qquad \bar{r}_{{\rm L},k} = \sqrt{\|F_k\|^2 - \bar{\nu}^2_{{\rm L},k}}. \tag{3.131}$$

Because of the above approximations, there is no guarantee that, at the difference of the theoretical MPC step, this tentative MPC step satisfies the

Armijo condition (3.84) with $\omega \simeq 1/2$ as stated by Proposition 3.9.5! Hence a tentative α_k will be accepted according to the (less demanding) following Armijo condition, for some $\omega \in]0, 1/2]$ (often $\omega = 10^{-4}$):

$$J(g_k(\alpha_k)) \leq J(x_k) + \omega\, \alpha_k \langle y_k\, , \nabla J(x_k)\rangle. \tag{3.132}$$

- If (3.132) is violated, then R_k is reduced by a factor $\tau \in]0, 1[$:

  $$R_k \leftarrow \tau R_k,$$

  and the calculations (3.129), (3.130), and (3.131) are repeated.

- If (3.132) is satisfied, then α_k is accepted, and x_k is updated:

  $$x_{k+1} = g_k(\alpha_k),$$

with g_k given by (3.123).

3.9.5 Performance of the MPC Step

A comparison of the MPC step (for both straight and geodesic search) with the Armijo, Goldstein, and Wolfe line search is available in [2] for the steepest descent, Gauss–Newton, and Levenberg–Marquardt algorithms, together with a comparison of the best performers with the confidence region and Levenberg Marquardt algorithms of the Matlab library.

The comparison of the algorithms have been conducted on a set of 13 test problems, and for various performance criterions (convergence, number of function evaluations, value of objective function at convergence, etc). We refer to [2] for the presentation of the *performance profiles* used for this comparison, and for a detailed account of the comparisons. We summarize here the overall result of this study:

- Straight vs. geodesic search: The comparison turns to the advantage of the geodesic search whenever it is available (Gauss–Newton and Levenberg–Marquardt algorithms)

- Steepest descent algorithm: The MPC step (with straight line search necessarily) performs better than the classical Armijo and Goldstein stepping algorithms, but worse than the Wolfe stepping (see [13, 2] for the definition of the stepping algorithms)

- Gauss–Newton algorithm: The MPC step with geodesic search performs better than the Armijo, Wolfe, and Goldstein stepping algorithms
- Levenberg–Marquardt algorithms: The MPC step with geodesic search performs better than the Armijo and Goldstein stepping algorithms, but worse than Wolfe stepping
- Best performers: Gauss–Newton with MPC step and geodesic search performs better than the Matlab Levenberg–Marquardt and confidence region algorithms

Appendix 1: Proof of Proposition 3.9.3

Reparameterization of P by Arc Length

The curve $\alpha \rightsquigarrow P(\alpha)$ of the data space satisfies (3.87) and (3.97), and Proposition 8.2.2 part (i) shows that the reparameterization $p : \nu \in [0,]\bar{\nu}] \rightsquigarrow I\!R^q$ of P as a function of the arc lenth ν is a $W^{2,\infty}([0,\bar{\nu}], I\!R^q)$ function: the first and second derivatives v and a of p given by (3.95) and (3.96), the curvature $1/\rho$ defined by (3.98), and the projected curvature $1/\rho_{\text{proj}}$ defined by (3.100) and (3.101) exist almost everywhere on $[0,\bar{\nu}]$, and belong to $L^\infty([0,\bar{\nu}], I\!R^q)$.

The $\alpha \rightsquigarrow \nu$ change of variable is continuously derivable and strictly increasing (see (3.94)), so that zero measure sets of $[0,\bar{\alpha}]$ correspond to zero measure sets of $[0,\bar{\nu}]$. Hence Proposition 3.9.3 will be proved if we show that

$$1/\rho_{\text{proj}}(\nu) \le 1/\rho(\nu) \qquad \text{a.e. on } [0,\bar{\nu}] \setminus Z, \tag{3.133}$$

$$1/\rho_{\text{proj}}(\nu) = 1/\rho(\nu) = 0 \qquad \text{a.e. on } Z, \tag{3.134}$$

where Z is now defined by

$$Z = \{\nu \in [0,\bar{\nu}] \ \text{ such that } p(\nu) - \langle p(\nu), v(\nu)\rangle v(\nu) = 0\}. \tag{3.135}$$

Comparison of Curvature and Projected Curvature

The inequality (3.133) follows immediately from the Definitions (3.98) of ρ and (3.100) and (3.101) of ρ_{proj} for $\alpha \notin Z$. Then (3.134) will be proved if we show that

$$1/\rho(\nu) = \|a(\nu)\| = 0 \qquad \text{for a.e. } \nu \in Z. \tag{3.136}$$

So let $\nu \in Z$ be given. Two cases can happen:

1. ν is an isolated point of Z: There exists $\eta > 0$ such that $]\nu - \eta, \nu + \eta[\cap Z = \{\nu\}$. But the set of isolated points of Z has a zero measure, and so we can ignore these points.

2. ν is *not* an isolated point of Z: $\forall n \in I\!N - \{0\}$, there exists $\nu_n \in Z, \nu_n \neq \nu$ such that $|\nu_n - \nu| \leq 1/k$. Hence $(p = p(\nu), p_n = p(\nu_n)$ etc.):

$$\begin{aligned} 0 &= \big(p_n - \langle p_n, v_n\rangle v_n\big) - \big(p - \langle p, v\rangle v\big) \qquad (3.137)\\ &= p_n - p - \langle p_n - p, v_n\rangle v_n \\ &\qquad -\langle p, v_n - v\rangle v_n \\ &\qquad -\langle p, v\rangle (v_n - v). \end{aligned}$$

Because p and v are derivable almost everywhere on $[0, \bar{\nu}]$, we can suppose that $p'(\nu) = v(\nu)$ and $v'(\nu) = a(\nu)$ exist. Then dividing (3.137) by $\nu_n - \nu$ and passing to the limit gives, as p and v are continuous at ν,

$$\langle p, a\rangle v + \langle p, v\rangle a = 0,$$

which implies, as v and a are orthogonal, that $\langle p, a\rangle v = 0$ and $\langle p, v\rangle a = 0$. By definition of the Curry step $\bar{\nu}$, one has $\langle p, v\rangle < 0$, and hence $a = 0$, and (3.136) is proved.

Appendix 2: Proof of Theorem 3.9.4

We suppose for simplicity that the Curry step satisfies

$$\bar{\alpha} < +\infty,$$

which corresponds to the usual practical situation. But the proof remains true if $\bar{\alpha} = +\infty$, provided one defines $\bar{r}$ by

$$\bar{r} = \lim_{\alpha \to \bar{\alpha}} r(\alpha).$$

As we have seen in Appendix 1, the reparameterization p of P as a function of the arc length ν belongs to $W^{2,\infty}([0, \bar{\nu}], I\!R^q)$.

Reparameterization by the Squared Residual Decrease

Following (3.99), we denote by r the residual along p and by $\bar{r}$ its Curry value:

$$r(\nu) = \|p(\nu)\| \text{ for all } \nu \in [0, \bar{\nu}], \qquad \bar{r} = r(\bar{\nu}).$$

By definition of the Curry step $\bar{\nu}$, the $\nu \rightsquigarrow r$ function is strictly decreasing over the $[0, \bar{\nu}]$ interval. Hence we can use as new parameter along the curve the *decrease* t *of the squared residual*:

$$t = r_0^2 - r(\nu)^2, \quad \text{where} \quad r_0^2 = r(0)^2, \tag{3.138}$$

which satisfies

$$0 \le t \le \bar{t} \stackrel{\text{def}}{=} r_0^2 - \bar{r}^2.$$

Derivation of (3.138) gives

$$\mathrm{d}t/\mathrm{d}\nu = -2\langle p, v\rangle = 2\,|\langle p, v\rangle|, \tag{3.139}$$

and the Curry arc length $\bar{\nu}$ we want to minorate can be written as

$$\bar{\nu} = \frac{1}{2}\int_0^{\bar{t}} \frac{\mathrm{d}t}{|\langle p, v\rangle|}. \tag{3.140}$$

Notice that this integral is singular, as $\langle p, v\rangle \to 0$ when $t \to \bar{t}$.

Stationary Linearized Residual and Projected Curvature

Let now $\bar{r}_{\mathrm{L}}$(where L stands for "linearized") denote the *linearized stationary residual*, that is, the stationary residual along the tangent to p at arc length ν:

$$\bar{r}_{\mathrm{L}}(\nu)^2 = \|p(\nu) - \langle p(\nu), v(\nu)\rangle v(\nu)\|^2 = \|p\|^2 - \langle p, v\rangle^2. \tag{3.141}$$

For a linear problem, p is a straight half line of the data space $I\!R^q$, and so $\bar{r}_{\mathrm{L}}$ is constant. We calculate in this section $|\mathrm{d}\bar{r}_{\mathrm{L}}/\mathrm{d}t|$, which will provide the adequate measure of nonlinearity for our purpose.

Derivation of (3.141) with respect to ν gives, together with (3.139) and (3.105),

$$\left|\frac{\mathrm{d}\bar{r}_{\mathrm{L}}^2}{\mathrm{d}t}\right| = |\langle p, a\rangle| = \frac{\bar{r}_{\mathrm{L}}}{\rho_{\mathrm{proj}}} \quad \text{for a.e. } t \in [0, \bar{t}].$$

The function $t \rightsquigarrow \bar{r}_{\mathrm{L}}$ is derivable a.e. on $[0,\bar{t}] \setminus Z$ (as the square root of the strictly positive, a.e. derivable function $t \rightsquigarrow \bar{r}_{\mathrm{L}}^2$), where (c.f. (3.99) and (3.135))

$$Z = \{t \in]0,\bar{t}[\text{ such that } \bar{r}_{\mathrm{L}}(t) = 0\}.$$

Hence

$$\left|\frac{\mathrm{d}\bar{r}_{\mathrm{L}}}{\mathrm{d}t}\right| = \frac{1}{2\rho_{\mathrm{proj}}} \leq \frac{1}{2R} \quad \text{for a.e. } t \in [0,\bar{t}] \setminus Z. \tag{3.142}$$

We prove now that this property remains true a.e. on Z. Let $t \in Z$ be a point where $\mathrm{d}\bar{r}_{\mathrm{L}}/\mathrm{d}t$ exists. Then $\bar{r}_{\mathrm{L}}(t) = 0$ and $\bar{r}_{\mathrm{L}}(t') \geq 0$ for a.e. $t' \in [0,\bar{t}]$, which implies that $\mathrm{d}\bar{r}_{\mathrm{L}}/\mathrm{d}t = 0$. Hence (3.142) holds necessarily true at any such point, where by definition $1/\rho_{\mathrm{proj}} = 0 \leq 1/R$, and we are left to prove that $\bar{r}_{\mathrm{L}}$ is derivable a.e. on $[0,\bar{t}]$.

We define for that purpose a sequence of functions

$$\eta_k = \max\{\bar{r}_{\mathrm{L}}, 1/k\} \quad k = 1, 2, \ldots,$$

which converges simply to $\bar{r}_{\mathrm{L}}$, and hence in the sense of distributions

$$\eta_k \longrightarrow \bar{r}_{\mathrm{L}} \text{ in } \mathcal{D}'(]0,\bar{t}[) \quad \text{when } k \longrightarrow +\infty. \tag{3.143}$$

This sequence is bounded independently of k in the $L^\infty(0,\bar{t})$ norm:

$$\|\eta_k\|_\infty \leq \max\{\|\bar{r}_{\mathrm{L}}\|_\infty, 1\} \leq \max\{\|p\|_\infty, 1\} \leq \max\{\|p(0)\|, 1\}. \tag{3.144}$$

The functions η_k are derivable a.e. on $[0,\bar{t}]$, as the square root of the a.e. derivable functions $\max\{\bar{r}_L^2, 1/k^2\} \geq 1/k^2 > 0$. Hence,

$$\left|\frac{\mathrm{d}\eta_k}{\mathrm{d}t}(t)\right| = \begin{cases} \left|\dfrac{\mathrm{d}\bar{r}_{\mathrm{L}}}{\mathrm{d}t}(t)\right| = \dfrac{1}{2\rho_{\mathrm{proj}}} \leq \dfrac{1}{2R} & \text{for a.e. } t \text{ such that } \bar{r}_L(t) > \dfrac{1}{k}, \\ 0 & \text{for a.e. } t \text{ such that } \bar{r}_L(t) \leq \dfrac{1}{k}, \end{cases}$$

where we have used (3.142) to evaluate $\mathrm{d}\bar{r}_{\mathrm{L}}/\mathrm{d}t$ when $\bar{r}_{\mathrm{L}} > 1/k$. This implies

$$\left\|\frac{\mathrm{d}\eta_k}{\mathrm{d}t}\right\|_\infty \leq \frac{1}{2R}$$

which, together with (3.144), shows that the sequence η_k, $k = 1, 2 \ldots$, is bounded in, say, $H^1(]0,\bar{t}[)$. Hence there exists a subsequence, still denoted by η_k, and $w \in H^1(]0,\bar{t}[)$ such that

$$\eta_k \rightharpoonup w \text{ weakly in } H^1(]0,\bar{t}[) \subset \mathcal{D}'(]0,\bar{t}[) \quad \text{when } k \longrightarrow +\infty. \tag{3.145}$$

Comparison of (3.143) and (3.145) implies that $\bar{r}_{\rm L} = w \in H^1(]0,\bar{t}\,[)$, which proves the desired result: $\bar{r}_{\rm L}$ is a.e. derivable over $[0,\bar{t}\,]$. Hence

$$\left|\frac{d\bar{r}_{\rm L}}{dt}\right| = \frac{1}{2\rho_{proj}} \leq \frac{1}{2R} \qquad \text{for a.e. } t \in [0,\bar{t}\,], \tag{3.146}$$

and the following majoration holds for the continuous function $\bar{r}_{\rm L}$:

$$\bar{r}_{\rm L}(t) \leq \bar{r}_{\rm L}(0) + \frac{t}{2R} \stackrel{\text{def}}{=} \bar{r}_{\rm L,W}(t) \quad \text{for all } t \in [0,\bar{t}\,], \tag{3.147}$$

where $\bar{r}_{\rm L,W}$ is the worst case stationary linearized residual. As we shall prove in the Sect. 3.9.5, and as one can guess from Fig. 3.9, $\bar{r}_{\rm L,W}$ is actually the stationary linearized residual along the arc of circle $p_{\rm W}$ of radius R (thick lower dashed line on the figure), which turns away from the target 0 in the plane containing $p(0)$ and $v(0)$. But for the time being, one can simply take (3.147) as the definition of $\bar{r}_{\rm L,W}$.

Comparison with the Worst Case

Let $\bar{r}_{\rm L}$ and $\bar{r}_{\rm L,W}$ be defined by (3.141) and (3.147), and define, for $t \geq 0$

$$\begin{aligned} \mu(t) &= t + \bar{r}_{\rm L}(t)^2 \quad = r_0^2 - \langle p(t), v(t)\rangle^2, & (3.148)\\ \mu_{\rm W}(t) &= t + \bar{r}_{\rm L,W}(t)^2, & (3.149)\end{aligned}$$

(the graph of $\mu_{\rm W}$ is a parabola), which satisfy (Fig. 3.10)

$$\begin{aligned} \mu(0) &= \mu_{\rm W}(0) = \bar{r}_{\rm L}(0)^2 < r_0^2, & (3.150)\\ \mu(t) &\leq \mu_{\rm W}(t) \qquad \text{for all } t \geq 0, & (3.151)\end{aligned}$$

where we have used the fact that $\langle v(0), p(0)\rangle < 0$ ($v(0)$ is a descent direction), and the majoration (3.147).

First, the right equality in (3.148) shows that t corresponds to a stationary point of $r^2 = \|p(t)\|^2$, where $\langle p(t), v(t)\rangle = 0$, if and only if $\mu(t) = r_0^2$. Hence the first stationary residual $\bar{r}^2$ and the corresponding parameter $\bar{t}$ are given by

$$\bar{t} = \inf\{t > 0 \mid \mu(t) = r_0^2\}, \qquad \bar{r}^2 = r_0^2 - \bar{t} = \bar{r}_{\rm L}(\bar{t})^2.$$

Next, the worst case function $\mu_{\rm W}$ is monotonous, so we can *define uniquely* $\bar{t}_{\rm W}$ and a worst case stationary residual $\bar{r}_{\rm W}$ by (Fig. 3.10),

$$\mu_{\rm W}(\bar{t}_{\rm W}) = r_0^2, \qquad \bar{r}_{\rm W}^2 = r_0^2 - \bar{t}_{\rm W} = \bar{r}_{\rm L,W}(\bar{t}_{\rm W})^2, \tag{3.152}$$

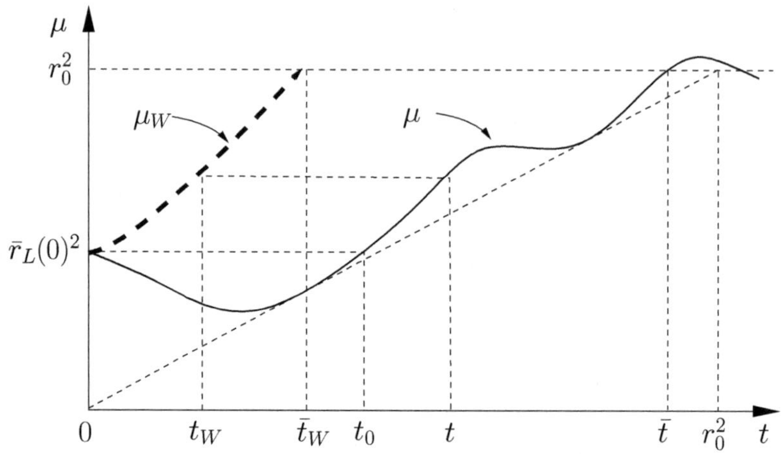

Figure 3.10: Properties of the μ_W and μ functions

which satisfy, using (3.151),

$$\bar{t}_W \leq \bar{t}, \qquad \bar{r}_W \geq \bar{r}. \tag{3.153}$$

Then combining (3.140) with (3.148), one can express $\bar{\nu}$ and *define* $\bar{\nu}_W$ by

$$\bar{\nu} = \frac{1}{2}\int_0^{\bar{t}} \frac{\mathrm{d}t}{(r_0^2 - \mu(t))^{1/2}}, \qquad \bar{\nu}_W = \frac{1}{2}\int_0^{\bar{t}_W} \frac{\mathrm{d}t_W}{(r_0^2 - \mu_W(t_W))^{1/2}}. \tag{3.154}$$

We are now close to our objective, which is to show that $\bar{\nu} \geq \bar{\nu}_W$. We remark first, using (3.151), that for a given t, the first integrand is larger than the second one – this goes in the right direction, but it does not allow to conclude, as the two domains of integrations are not the same! So we make a change of variable in order to obtain integrals defined over the same interval. Define first t_0 by (Fig. 3.10):

$$t_0 = \max\{t \in [0, \bar{t}[\mid \mu(t) = \bar{r}_L(0)^2\}.$$

The range of μ over the $[t_0, \bar{t}]$ interval is then $[\bar{r}_L(0)^2, r_0^2]$, the same as the range of μ_W over $[0, \bar{t}_W]$. Hence, we can define a (nonmonotonous) change of variable $t \rightsquigarrow t_W$ from the $[t_0, \bar{t}]$ interval onto the $[0, \bar{t}_W]$ interval by

$$\mu_W(t_W) = \mu(t) \qquad \forall t \in [t_0, \bar{t}],$$

where t_W is uniquely defined because of the strict monotonicity of μ_W. Because of (3.151), t and t_W satisfy, as in (3.153),

$$t_W \le t, \qquad \bar r_{L,W}(t_W) \ge \bar r_L(t) \qquad \forall t \in [t_0, \bar t]. \tag{3.155}$$

We can now use this change of variable in the integral that gives $\bar\nu_W$:

$$\bar\nu_W = \frac{1}{2}\int_0^{\bar t_W} \frac{dt_W}{(r_0^2 - \mu_W(t_W))^{1/2}} = \frac{1}{2}\int_{t_0}^{\bar t} \frac{\mu'(t)}{\mu'_W(t_W)} \frac{dt}{(r_0^2 - \mu(t))^{1/2}},$$

where $\mu'_W(t_W) > 0$ and $\mu'(t)$ can be positive or negative. But differentiation of (3.148) gives, using (3.155) and (3.146),

$$\begin{aligned}
|\mu'(t)| &= |1 + 2\bar r_L(t)\frac{d\bar r_L}{dt}(t)| \\
&\le 1 + 2\bar r_L(t)|\frac{d\bar r_L}{dt}(t)| \\
&\le 1 + \bar r_{L,W}(t_W)\frac{1}{R} = \mu'_W(t_W).
\end{aligned}$$

Hence $\bar\nu_W$ satisfies

$$\begin{aligned}
\bar\nu_W &\le \frac{1}{2}\int_{t_0}^{\bar t} \frac{|\mu'(t)|}{\mu'_W(t_W)} \frac{dt}{(r_0^2 - \mu(t))^{1/2}} \le \frac{1}{2}\int_{t_0}^{\bar t} \frac{dt}{(r_0^2 - \mu(t))^{1/2}}, \\
&\le \frac{1}{2}\int_0^{\bar t} \frac{dt}{(r_0^2 - \mu(t))^{1/2}} = \bar\nu,
\end{aligned}$$

and (3.107) is proved. We turn now to the proof of (3.108). Let $t(\bar\nu_W)$ be the squared residual decrease along p *at the worst case stationary arc length* $\bar\nu_W$. Formulas (3.154) for the arc length give

$$\bar\nu_W = \frac{1}{2}\int_0^{t(\bar\nu_W)} \frac{dt}{(r_0^2 - \mu(t))^{1/2}} = \frac{1}{2}\int_0^{t_W} \frac{dt}{(r_0^2 - \mu_W(t))^{1/2}}.$$

The first integrand is smaller than the second, so $t(\bar\nu_W)$ has to be larger than t_W, which gives, by definition of t and $\bar r_W$,

$$r_0^2 - r(\bar\nu_W)^2 \ge r_0^2 - \bar r_W^2,$$

which is (3.108).

Determination of the Worst Case

It remains now to prove formula (3.109) and (3.110) for $\bar{\nu}_{\mathrm{W}}$ and $\bar{r}_{\mathrm{W}}$. We calculate first the integral (3.154), which gives $\bar{\nu}_{\mathrm{W}}$. Formula (3.147) for $\bar{r}_{\mathrm{L,W}}$ gives, with the notations $\bar{r}_{\mathrm{L}} = \bar{r}_{\mathrm{L}}(0)$, $\bar{\nu}_{\mathrm{L}}^2 = r_0^2 - r_{\mathrm{L}}(0)^2$:

$$\begin{aligned} r_0^2 - \mu_{\mathrm{W}}(t) &= r_0^2 - \big(t + (\bar{r}_{\mathrm{L}} + \frac{t}{2R})^2\big), \\ &= \bar{\nu}_{\mathrm{L}}^2 + (R + \bar{r}_{\mathrm{L}})^2 - (R + \bar{r}_{\mathrm{L}} + \frac{t}{2R})^2 \\ &= (1 - u^2(t))(\bar{\nu}_{\mathrm{L}}^2 + (R + \bar{r}_{\mathrm{L}})^2), \end{aligned}$$

where u is defined by

$$u(t) = \frac{R + \bar{r}_{\mathrm{L}} + \dfrac{t}{2R}}{\big(\bar{\nu}_{\mathrm{L}}^2 + (R + \bar{r}_{\mathrm{L}})^2\big)^{1/2}}.$$

With this change of variable, formula (3.154) becomes

$$\begin{aligned} \bar{\nu}_W &= R \int_{u(0)}^{1} \frac{du}{(1 - u^2)^{1/2}} \\ &= R\{\frac{\pi}{2} - \sin^{-1} u(0)\} \\ &= R \tan^{-1} \frac{\bar{\nu}_{\mathrm{L}}}{R + \bar{r}_{\mathrm{L}}}, \end{aligned}$$

which is (3.109). We calculate now the worst residual $\bar{r}_{\mathrm{W}}$. The definitions (3.152) of $\bar{t}_{\mathrm{W}}$, $\bar{r}_{\mathrm{W}}$, and (3.147) of $\bar{r}_{\mathrm{L,W}}$ show that

$$\bar{r}_{\mathrm{W}} = \bar{r}_{\mathrm{L,W}}(\bar{t}_{\mathrm{W}}) = \bar{r}_{\mathrm{L}} + \frac{\bar{t}_{\mathrm{W}}}{2R}.$$

Hence the first equation of (3.152) rewrites

$$\begin{aligned} \bar{t}_{\mathrm{W}} + \bar{r}_{\mathrm{W}}^2 &= r_0^2 \\ 2R(\bar{r}_{\mathrm{W}} - \bar{r}_{\mathrm{L}}) + \bar{r}_{\mathrm{W}}^2 &= r_0^2. \end{aligned}$$

The positive root of this second order equation for $\bar{r}_{\mathrm{W}}$ is

$$\bar{r}_{\mathrm{W}} = -R + (R^2 + 2R\bar{r}_{\mathrm{L}} + r_0^2)^{1/2},$$

which is (3.110).

The above expressions for $\bar{\nu}_{\mathrm{W}}$ and $\bar{r}_{\mathrm{W}}$ are those of the first stationary arc length and residual along a circle p_{W} of radius R (thick lower dashed line in Fig. 3.10), which turns away from the target 0 in any plane containing $p(0)$ and $v(0)$ (there can be many if these two vectors are colinear). This proves that the worst case is achieved by the circle(s) p_{W}.

Appendix 3: Proof of Proposition 3.9.5

We prove first the guaranteed decrease property (3.114). One has

$$\begin{aligned} r_k^2 &= r(0)^2 = (\bar{\nu}_{\mathrm{L}}^2 + \bar{r}_{\mathrm{L}}^2), && \text{(by definition)} \\ r_{k+1}^2 &\le \bar{r}_{\mathrm{W}}^2. && \text{(formula (3.108))} \end{aligned}$$

Let $\gamma \in [0, \pi/2[$ be defined by (Fig. 3.9):

$$\sin\gamma = \bar{\nu}_{\mathrm{L}}/((R+\bar{r}_{\mathrm{L}})^2 + \bar{\nu}_{\mathrm{L}}^2)^{\frac{1}{2}}, \tag{3.156}$$

$$\tan\gamma = \bar{\nu}_{\mathrm{L}}/(R+\bar{r}_{\mathrm{L}}). \tag{3.157}$$

Using first the Definition (3.110) of $\bar{r}_{\mathrm{W}}$, then (3.156) and (3.157), and finally (3.109) one finds

$$\begin{aligned} \frac{1}{2}(r_k^2 - r_{k+1}^2) &\ge \frac{1}{2}\Big(\bar{\nu}_{\mathrm{L}}^2 + \bar{r}_{\mathrm{L}}^2 - \big(((R+\bar{r}_{\mathrm{L}})^2 + \bar{\nu}_{\mathrm{L}}^2)^{\frac{1}{2}} - R\big)^2\Big) \\ &= \frac{1}{2}\big(\bar{r}_{\mathrm{L}}^2 - (R+\bar{r}_{\mathrm{L}})^2 - R^2 + 2R\,\bar{\nu}_{\mathrm{L}}/\sin\gamma\big) \\ &= -R\bar{r}_{\mathrm{L}} - R^2 + R\,\bar{\nu}_{\mathrm{L}}/\sin\gamma \\ &= -R(\bar{\nu}_{\mathrm{L}}/\tan\gamma - R) + R\,\bar{\nu}_{\mathrm{L}}/\sin\gamma \\ &= R\,\bar{\nu}_{\mathrm{L}}\frac{1-\cos\gamma}{\sin\gamma} \\ &= \bar{\nu}_{\mathrm{W}}\,\bar{\nu}_{\mathrm{L}}\frac{1-\cos\gamma}{\gamma\sin\gamma} \end{aligned}$$

The function $\gamma \rightsquigarrow (1-\cos\gamma)/(\gamma\sin\gamma)$ increases from $1/2$ to $2/\pi$ when γ goes from 0 to $\pi/2$, which proves (3.114). Then the Definition (3.106) of α_k gives

$$\bar{\nu}_{\mathrm{L}} = \nu(\alpha_k) = \int_0^{\alpha_k} \|V(\alpha)\| \mathrm{d}\alpha = \alpha_k \overline{\|V\|}(\alpha_k),$$

and (3.111) gives, as $\langle F(x_k), V\rangle < 0$,

$$\bar{\nu}_{\mathrm{L}} = -\langle F(x_k), V\rangle/\|V(0)\| = -\langle \nabla J(x_k), y_k\rangle/\|V(0)\|. \tag{3.158}$$

Hence the guaranteed decrease writes

$$\frac{1}{2}\bar{\nu}_{\rm W}\bar{\nu}_{\rm L} = -\frac{1}{2}\frac{\overline{\|V\|}(\alpha_k)}{\|V(0)\|}\alpha_k\langle\nabla J(x_k), y_k\rangle = -\omega\alpha_k\langle\nabla J(x_k), y_k\rangle,$$

which proves with (3.114) that the Armijo condition (3.84) is satisfied for ω given by (3.115), and ends the proof of Proposition 3.9.5.

Appendix 4: Proof of Proposition 3.9.8

Properties (3.119) and (3.120) follow immediately from (3.117) and (3.118) using either (3.90) with $g'' = 0$ when g_k is a straight search curve, or (3.93) when g_k is a geodesic search curve.

Hence the hypothesis of Theorem 3.9.4 are satisfied, and one can use for all k the MPC step $\alpha_k = \bar{\alpha}_W$ computed from R_k by (3.106) and (3.109). The guaranteed decrease property (3.114) writes then at iteration k

$$\begin{aligned} J(x_k) - J(x_{k+1}) &\geq \frac{1}{2}\bar{\nu}_{\rm W,k}\,\bar{\nu}_{\rm L,k} \\ &= \frac{1}{2}\frac{\bar{\nu}_{\rm W,k}}{\bar{\nu}_{\rm L,k}}\,\bar{\nu}_{\rm L,k}^2 \\ &= \frac{1}{2}\frac{R_k}{\bar{\nu}_{\rm L,k}}\tan^{-1}\Big(\frac{\bar{\nu}_{\rm L,k}}{R_k + \bar{r}_{\rm L,k}}\Big)\,\bar{\nu}_{\rm L,k}^2. \end{aligned}$$

The last right-hand side is a decreasing function of $\bar{r}_{\rm L,k}$ and an increasing function of R_k. Hence one obtains using $\bar{r}_{\rm L,k} \leq r_k \leq r_0$ and (3.120)

$$\begin{aligned} J(x_k) - J(x_{k+1}) &\geq \frac{1}{2}\frac{R}{\bar{\nu}_{\rm L,k}}\tan^{-1}\Big(\frac{\bar{\nu}_{\rm L,k}}{R + r_0}\Big)\,\bar{\nu}_{\rm L,k}^2 \\ &\geq \frac{1}{2}\frac{R}{r_0}\tan^{-1}\Big(\frac{r_0}{R + r_0}\Big)\,\bar{\nu}_{\rm L,k}^2, \end{aligned}$$

where we have used $\bar{\nu}_{\rm L,k} \leq r_k \leq r_0$ and the fact that $\dfrac{R}{r}\tan^{-1}\Big(\dfrac{r}{R + r_0}\Big)$ is a decreasing function of r over the $[0, R + r_0]$ interval. We replace now $\bar{\nu}_{\rm L,k}^2$ by its expression (3.158) and use hypothesis (3.117)

$$\begin{aligned} J(x_k) - J(x_{k+1}) &\geq \frac{1}{2}\frac{R}{r_0}\tan^{-1}\frac{r_0}{R + r_0}\,\frac{\langle\nabla J(x_k), y_k\rangle^2}{\|V_k\|^2} \\ &\geq \frac{1}{2M^2}\frac{R}{r_0}\tan^{-1}\frac{r_0}{R + r_0}\,\|\nabla J(x_k)\|^2\cos^2\theta_k \end{aligned}$$

Summing up the last inequalities for $k = 0, \cdots K$ gives

$$J(x_0) \geq J(x_0) - J(x_{K+1}) \geq C(M, R, r_0) \sum_{k=1\cdots K} \|\nabla J(x_k)\|^2 \cos^2 \theta_k,$$

which proves the convergence of the series (3.121). Finally, the conditions associated to each choice of descent direction ensure that $\cos^2 \theta_k \geq c_{min} > 0$. Together with the convergence of the series (3.121), this implies that

$$\|\nabla J(x_k)\|^2 \to 0,$$

which is the definition of convergence of the algorithm. This ends the proof of Proposition 3.9.8.

Chapter 4

Output Least Squares Identifiability and Quadratically Wellposed NLS Problems

We consider in this chapter the nonlinear least squares (NLS) problem (1.10), which we recall here for convenience:

$$\hat{x} \quad \text{minimizes} \quad J(x) = \frac{1}{2}\|\varphi(x) - z\|_F^2 \quad \text{over} \quad C. \tag{4.1}$$

As we have seen in Chap. 1, this *inverse problem* describes the identification of the parameter $x \in C$ from a measurement z of $\varphi(x)$ in F. We suppose that the minimum set of hypothesis (1.12) of Chap. 1 holds:

$$\begin{cases} E & = & \text{Banach space, with norm } \| \ \|_E, \\ C & \subset & E \ \text{ with } C \text{ convex and closed}, \\ F & = & \text{Hilbert space, with norm } \| \ \|_F, \\ z & \in & F \\ \varphi & : & C \rightsquigarrow F \text{ is differentiable along segments of C}, \\ \text{and} & : & \exists\, \alpha_M \geq 0 \ \text{ s.t. } \ \forall x_0, x_1 \in C, \ \forall t \in [0,1], \\ & & \|D_t\,\varphi((1-t)x_0 + t x_1)\|_F \leq \ \alpha_M \|x_1 - x_0\|_E, \end{cases} \tag{4.2}$$

and we recall the definition of stationary points:

G. Chavent, *Nonlinear Least Squares for Inverse Problems: Theoretical Foundations and Step-by-Step Guide for Applications*, Scientific Computation,
DOI 10.1007/978-90-481-2785-6_4,

Definition 4.0.9 *A point* $\tilde{x} \in C$ *is a* stationary point of problem (4.1) *if it satisfies the first-order necessary optimality condition:*

$$\langle \varphi(\tilde{x}) - z, D_{t=0}\,\varphi((1-t)\tilde{x} + tx)\rangle_F \geq 0 \quad \forall x \in C. \tag{4.3}$$

A point $\tilde{x}$ *is a* parasitic stationary point *if* $\tilde{x}$ *satisfies (4.3) but* $\tilde{x}$ *is not a solution of (4.1).*

A local minimum of J *on* C *is a stationary point.*

Our objective is to find additional conditions on C, φ, and z which ensure that (4.1) is both *wellposed* and *optimizable*:

- A wellposed problem in the sense of Hadamard has a unique solution, which depends continuously on the right-hand side of the equation. For the optimization problem (4.1) of interest, uniqueness of the solution $\hat{x}$ can hold only if the data z has a unique projection on $D = \varphi(C)$, which *cannot be true for any* z when D is not convex! So the existence, uniqueness, and stability properties will be required only to hold for data z that stay in some neighborhood ϑ of $D = \varphi(C)$ in F.

- Optimizability means the absence of parasitic stationary points in the quadratic objective function. It is an extremely useful property for nonlinear inverse problems, as it make the least squares formulation numerically constructive, by ensuring that local optimization algorithms will not stop prematurely in a parasitic stationary point.

To ensure constructiveness of the NLS formulation (4.1), these two properties are combined in the Definition 1.3.2 of quadratic (Q)-wellposedness, or, in parameter estimation terms, of output least squares (OLS)-identifiability of x, which we recall here:

Definition 4.0.10 *Let* φ *and* C *be given. The parameter* x *is* OLS-identifiable *in* C *from a measurement* z *of* $\varphi(x)$ *if and only if the NLS problem (4.1) is* Q-wellposed, *that is, if* $\varphi(C)$ *possesses an open neighborhood* ϑ *such that*

(i) Existence and uniqueness: *For every* $z \in \vartheta$*, problem (1.10) has a unique solution* $\hat{x}$

(ii) Unimodality: *For every* $z \in \vartheta$*, the objective function* $x \rightsquigarrow J(x)$ *has no parasitic stationary point*

(iii) Local stability: *The mapping $z \rightsquigarrow \hat{x}$ is locally Lipschitz continuous from $(\vartheta, \|\cdot\|_F)$ to $(C, \|\cdot\|_E)$.*

In the nonlinear least-squares problem (4.1), the distance of the data $z \in F$ to the attainable set $D = \varphi(C)$ represents the modeling and measurement errors, and we would like to be sure that keeping these errors below a certain level r will ensure that z belongs to the set ϑ on which existence, uniqueness, optimizability, and stability hold.

This is why we seek in this chapter sufficient conditions that ensure that ϑ contains an open *enlargement neighborhood* of the attainable set

$$\vartheta \supset \left\{ z \in F \mid d(z, D) < r \right\} \tag{4.4}$$

for some $r > 0$, which then represents the upper limit of the noise and modeling error level for which the NLS problem is well posed.

We recall first in Sect. 4.1 the results for the linear case we would like to generalize to nonlinear problems, and define in Sect. 4.2 the class of FC/LD problems that retain some useful properties of the linear case. We introduce in Sect. 4.3 the notions of linearized identifiability and stability. These ingredients are used in Sect. 4.4 to state a sufficient conditions for Q-wellposedness of NLS problems, based on the results of Chaps. 7 and 8 on strictly quasi-convex (s.q.c.) sets. The case of finite dimensional parameters is studied in Sect. 4.5, where it is shown that Q-wellposedness holds locally as soon as the linearized problem is identifiable. The remaining sections are devoted to examples of Q-wellposed parameter estimation problems in elliptic equations.

4.1 The Linear Case

We recall here the case where φ is linear:

$$\varphi(x) = Bx \quad \text{with } B \in \mathcal{L}(E; F), \tag{4.5}$$

which will serve as a guideline for the study of the nonlinear case.

The output set $\varphi(C) = B.C$ is then a convex set D, and we can take advantage of the good properties of the projection on convex sets:

Proposition 4.1.1 *Let D be a convex set of the Hilbert space F. Then*

(i) Uniqueness: *For any $z \in F$, there exists at most one projection $\hat{X}$ of z on D*

(ii) Unimodality: *If $z \in F$ admits a projection $\hat{X}$ on D, the "distance to z" function has no parasitic stationary point on D (its unique stationary point is $\hat{X}$)*

(iii) Stability: *If $z_0, z_1, \in F$ admit projections $\hat{X}_0, \hat{X}_1$, on D, then*

$$\|\hat{X}_0 - \hat{X}_1\|_F \leq \|z_0 - z_1\|_F,$$

(iv) Existence: *If $z \in F$, any minimizing sequence $X_n \in D$ of the "distance to z" function over D is a Cauchy sequence for the distance $\|X - Y\|_F$, and $X_n \to \hat{X} =$ unique projection of z on the closure $\overline{C}$ of C.*

If moreover D is closed, then $\hat{X} \in C$.

It is then straightforward to combine the above properties of the projection with the hypothesis that the parameter x is *identifiable* (respectively, *stable*) according to Definition 1.3.1:

Proposition 4.1.2 *Let hypothesis (4.2) and (4.5) hold.*

1. *If x is* identifiable, *that is,*

$$By = 0 \Longrightarrow y = 0, \tag{4.6}$$

and if the attainable set $B.C$ is closed, *then x is OLS-identifiable for the* "arc length distance"

$$\delta(x_0, x_1) = \|B(x_0 - x_1)\|_F \tag{4.7}$$

in data space: the linear least squares problem (4.1) is Q-wellposed with $\vartheta = F$, and for any z_0 and z_1 in F, the unique solutions $\hat{x}_0$ and $\hat{x}_1$ of (4.1) satisfy the stability estimate

$$\delta(\hat{x}_0, \hat{x}_1) \leq \|z_0 - z_1\|_F. \tag{4.8}$$

2. *If the estimation of x from Bx is* stable, *that is,*

$$\exists \alpha_m > 0 \ s.t. \quad \alpha_m \|y\|_E \leq \|By\|_F \quad \forall y \in E, \tag{4.9}$$

then $B.C$ is necessarily closed, and x is OLS-identifiable and problem (4.1) Q-wellposed for the norm $\|\cdot\|_E$, with the same $\vartheta = F$, and the stability estimate

$$\alpha_m \|\hat{x}_0 - \hat{x}_1\|_E \leq \|z_0 - z_1\|_F. \tag{4.10}$$

For finite dimensional parameters, the stability inequality (4.9) follows immediately from the identifiability property (4.6) (that is, the injectivity of B).

Proof. The attainable set $B.C$ is closed either by hypothesis (part 1), or by a completeness argument using (4.9) and the fact that C is closed (part 2). It is also convex, as the image of a convex set by a linear operator. The proposition follows then immediately from the properties of the projection on a closed convex sets recalled in Proposition 4.1.1. ∎

The above proposition may seem to be a complicated way to state simple results, but its interest is that it gives precisely the stencil of the results we shall be able to generalize to some nonlinear inverse problems in the rest of this chapter.

Linear inverse problems that do not satisfy the hypotheses of Proposition 4.1.2 may or may not have a solution, and the usual cure is to bring additional information by regularization (Sect. 1.3.4). When the L.M.T. regularization is used, the original problem (4.1) is replaced by (1.25), which satisfies the hypotheses of Proposition 4.1.2 as soon as $\epsilon > 0$, and exhibits nice convergence properties when $\epsilon \to 0$ (this will be studied in detail in Sect. 5.1.1).

We define in the next section a class of nonlinear problems that retains the properties of linear problems recalled in Propositions 4.1.1 and 4.1.2.

4.2 Finite Curvature/Limited Deflection Problems

The good properties of the linear case followed from the convexity of the output set $\varphi(C)$. To generalize these properties to the nonlinear case, we introduce the class of FC/LD problems, which all have in common the fact that their attainable set is *strictly quasiconvex (s.q.c.)* – instead of being *convex* as in the linear case.

The generalization of convex sets to s.q.c. sets is the subject of Chaps. 6–8. These chapters are quite technical, but the only thing we need in the present chapter is the combination of the practical sufficient condition for strict quasiconvexity developed in Chap. 8 with the properties of s.q.c. sets regarding projection developed in Chaps. 6 and 7 and summarized in Theorem 7.2.11. We shall refer to the above chapters for the proofs, and give in this section a self-contained and hopefully intuitive presentation of the definition and properties of FC/LD problems.

We define first the class of FC problems. FC problems were originally called *weakly nonlinear* when they were introduced in [28]; but we prefer here the denomination FC, which describes them better as we shall see below.

To any pair of points x_0, x_1 of C, we associate a *curve* P on $\varphi(C)$ by

$$P \,:\, t \in [0,1] \rightsquigarrow \varphi((1-t)x_0 + tx_1) \in \varphi(C). \tag{4.11}$$

Under the minimum set of hypothesis (4.2), $P(t)$ has a derivative $V(t) = D_t\varphi((1-t)x_0 + tx_1)$, and the arc length of the curve P of the output set $\varphi(C)$ is

$$L(P) = \int_0^1 \|V(t)\|_F \, \mathrm{d}t. \tag{4.12}$$

When $t \rightsquigarrow P(t)$ is constant, then $\|V(t)\|_F = 0 \ \forall t \in [0,1]$ and hence $L(P) = 0$, so that the curve P is reduced to one point of $\varphi(C)$ and does not provide any information on the shape of $\varphi(C)$.

On the contrary, the curves P for which $L(P) > 0$ stay by construction on the attainable set $\varphi(C)$, and so it is understandable that an upper bound on their curvature brings some information on the shape of $\varphi(C)$:

Definition 4.2.1 *Let C and φ satisfy the minimum set of hypothesis (4.2). Problem (4.1) is a* finite curvature *least squares problem (in short:* a FC problem*) if the following conditions are satisfied:*

$$\begin{cases} \text{there exists } R > 0 \text{ such that} \\ \forall x_0, x_1 \in C, \text{ the curve } P : t \rightsquigarrow \varphi((1-x_0)t + tx_1) \text{ satisfies} \\ P \in W^{2,\infty}([0,1]; F) \text{ and } \|A(t)\|_F \leq \dfrac{1}{R}\|V(t)\|_F^2 \text{ for a.e. } t \in [0,1], \\ \text{where } V(t) = P'(t), \ A(t) = P''(t). \end{cases} \tag{4.13}$$

Of course, linear problems are FC problems, as they satisfy obviously the definition with $1/R = 0$. The "finite curvature" name given to property (4.13) of problem (4.1) is justified in Chap. 8 (Propositions 8.2.1 and 8.2.2), where it is proved that, when it holds

- Either $P(t)$ =constant, so that $L(P) = 0$ and the curve P is reduced to one *point* of $\varphi(C)$ – its curvature is not defined

- Or $V(t) \neq 0 \ \forall t \in [0,1]$, so that $L(P) > 0$ and the *radius of curvature* $\rho(t)$ along the *curve* P satisfies

$$\frac{1}{\rho(t)} \leq \frac{\|A(t)\|_F}{\|V(t)\|_F^2} \quad \text{for a.e. } t \text{ in } [0,1], \tag{4.14}$$

so that the *radius of curvature* of the curve P and of the attainable set$\varphi(C)$ (Definition 7.2.8) satisfy

$$\frac{1}{R(P)} \stackrel{\text{def}}{=} \sup_{t\in[0,1]} \frac{1}{\rho(t)} \leq \frac{1}{R(\varphi(C))} \stackrel{\text{def}}{=} \sup_{x_0,x_1\in C\,,\,t\in[0,1]} \frac{1}{\rho(t)} \leq \frac{1}{R} < +\infty, \tag{4.15}$$

which explains the FC name given to this class of problems.

The lower bound R to the radius of curvature $R(\varphi(C))$ of the attainable set is called a *radius of curvature of the inverse problem* (4.1) (Definition 4.2.3 below).

Notice that (4.13) is only a sufficient condition for the radius of curvature along P to be larger than R (compare (4.13) with the formula (8.66) for the curvature, and/or think of $\varphi : [0,1] \rightsquigarrow I\!R^2$ defined by $\varphi(x) = (x^2, x^2)$).

Without further constraints, the attainable set $\varphi(C)$ of a FC problem may

- Fold over itself, which prevents property (i) of Proposition 4.1.1 to hold on any neighborhood of $\varphi(C)$ (think of the greek letter α as attainable set, or consider the simple case where $C = [0, \operatorname{diam} C] \subset I\!R$, and $\varphi(x) = (\cos x, \sin x) \in I\!R^2$, so that $\varphi(C)$ is an arc of circle of radius 1 and arc length $\operatorname{diam} C$, when $\operatorname{diam} C > 2\pi$).

- Or possess no neighborhood on which the unimodality property (ii) of Proposition 4.1.1 holds (think again of the arc of circle with $\pi \leq \operatorname{diam} C < 2\pi$: the point $z = (\cos \operatorname{diam} C, \sin \operatorname{diam} C)$ is on $\varphi(C)$, so the "distance to z" function attains its global minimum on $\varphi(C)$ at $z = \varphi(\operatorname{diam} C)$) (with value 0!), but it has also a stationary point (in fact, a local minimum when $\pi < \operatorname{diam} C$) at $(0,0) = \varphi(0)$ (with a strictly positive value)).

It is hence necessary to constrain further $\varphi(C)$ if the properties of the projection listed in Proposition 4.1.1 for the linear case are to be extended to

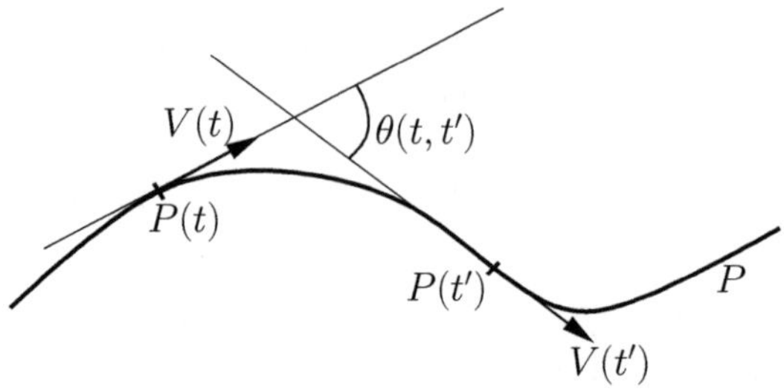

Figure 4.1: The deflection θ between two points of a curve P

the nonlinear case. A natural thing is to impose a constraint on the *deflection* Θ *of the curves* P, defined as *the largest angle* $\theta(t, t') \in [0, \pi]$ *between any two tangent vectors* $V(t)$ and $V(t')$ to P (Fig. 4.1 and Definition 8.0.4). For example, the previous arc-of-circle example with radius $R = 1$ has a deflection bounded by $\Theta = \operatorname{diam} C$. Let us check whether limiting Θ can prevent for this simple case the attainable set to fold over, and can restore the unimodality of the projection:

- If $\Theta < 2\pi$, the arc of circle can obviously not fold over itself!
- If $\Theta \le \pi/2$, all points of $\mathbb{R}^2$ at a distance of the arc of circle strictly smaller than $R = 1$ have the uniqueness and unimodality properties (i) and (ii) of Proposition 4.1.1 (see Fig. 7.3, top).

So the condition $\Theta \le \pi/2$ seems to ensure a nice behavior of the attainable set – at least for the arc-of-circle example. This is in fact general, as we shall see in Proposition 4.2.7 below, so we make a definition:

Definition 4.2.2 *Let* C *and* φ *satisfy the minimum set of hypothesis (4.2). A FC problem (4.1) is a* Limited Deflection *least squares problem (in short, a FC/LD problem) if it satisfies the* Deflection Condition*:*

$$\Theta \le \frac{\pi}{2}. \tag{4.16}$$

The attainable set of an FC/LD problems is s.q.c. The upper bound Θ *to the deflection of the curves* P *is called the* deflection of the inverse problem *(4.1).*

Proof. The strict quasi-convexity property of the attainable set follows from Theorem 8.1.6, which gives a sufficient condition for a set to be s.q.c. ■

We explain now how to estimate the deflection Θ of a FC problem. We remark that when P is an arc of circle, Θ is simply equal to the length of the arc divided by its radius, that is, the product of the size of the arc by its curvature. For a curve P other than a circle, a similar formula holds at the infinitesimal level: the deflection $d\theta$ between the endpoints of the arc corresponding to parameters t and $t + dt$ is bounded (not equal in general) by the length of the arc $\|V(t)\|_F\, dt$ multiplied by its curvature $1/\rho(t)$ (Proposition 8.1.2):

$$d\theta \le \frac{\|V(t)\|_F\, dt}{\rho(t)}, \tag{4.17}$$

(with the equality if P is a plane curve!). Combining with (4.14) gives

$$d\theta \le \frac{\|A(t)\|_F}{\|V(t)\|_F}\, dt \le \frac{\|V(t)\|_F}{R}\, dt. \tag{4.18}$$

If we denote by t_0 and t_1 the values of t corresponding to the points of the curve P where the deflection is maximum, we obtain, for the maximum deflection $\Theta(P)$ along the curve P,

$$\Theta(P) = \int_{t_0}^{t_1} d\theta \le \int_0^1 d\theta \le \int_0^1 \frac{\|A(t)\|_F}{\|V(t)\|_F}\, dt \le L(P)/R \le L/R, \tag{4.19}$$

where $L(P) = \int_0^1 \|V(t)\|_F\, dt$ denotes the arc length of P and L an upper bound to $L(P)$, This shows that any number Θ which satisfies

$$\int_0^1 \frac{\|A(t)\|_F}{\|V(t)\|_F}\, dt \le \Theta \le L/R \quad \text{for all } x_0, x_1 \in C \tag{4.20}$$

is an upper bound to the deflection of all curves P defined in (4.11). Of course, the angle between two vectors belongs always to the $[0, \pi]$ interval, so this upper bound on the deflection constrains actually the deflection of the curves P of $\varphi(C)$ only if Θ happens to be strictly smaller than π!

The above geometric approach to the estimation of the deflection is only meant to provide an intuitive support to formula (4.20). We refer to Theorem 8.2.3 in Chap. 8 for a rigorous proof.

We summarize the geometric quantities of interest introduced so far for the analysis of FC problems:

Definition 4.2.3 *The* geometric attributes *of the FC problem (4.1) are the following:*

1. *Its* radius of curvature $R > 0$*, defined as a lower bound to the radius of curvature along all curves* P *of the form (4.11) with* $L(P) > 0$*. It is estimated by (4.13)*

2. *Its* (arc length) size $L \geq 0$*, defined as an upper bound to the arc length* $L(P)$ *of all curves* P *of the form (4.11). It is estimated by*

$$L(P) = \int_0^1 \|V(t)\|_F \, \mathrm{d}t \leq L \leq \alpha_M \operatorname{diam} C \quad \forall x_0, x_1 \in C, \tag{4.21}$$

where $\operatorname{diam} C$ *is the diameter of* C*:*

$$\operatorname{diam} C = sup_{x,y \in C} \|x - y\|_E, \tag{4.22}$$

3. *Its* deflection $\Theta \geq 0$ *defined as an upper bound to the deflection* $\Theta(P)$ *of all curves* P *of the form (4.11) with* $L(P) > 0$*. It is estimated by (4.23) below*

Then formula (4.20) gives

Proposition 4.2.4 *Let (4.1) be a FC problem with curvature* $1/R$ *and size* $L \leq \alpha_M \operatorname{diam} C$*. Then any angle* Θ *that satisfies*

$$\begin{cases} \int_0^1 \theta(t) \, \mathrm{d}t \leq \Theta \leq L/R, \\ \text{where:} \\ \|A(t)\|_F \leq \theta(t) \|V(t)\|_F \text{ for a.e. } t \in [0,1] \text{ and all } x_0, x_1 \in C \end{cases} \tag{4.23}$$

is an upper bound to the deflection of the FC problem (4.1).

The majoration $\Theta \leq L/R$ in (4.23) is sharp: consider once again the arc-of-circle example. Then $R = 1$ is a lower bound to the radius of curvature of the curves (4.11), and $L = \operatorname{diam} C$ is the upper bound to their arc length. When $\operatorname{diam} C \leq \pi$, the largest deflection of the curves (4.11) is precisely $\Theta = \operatorname{diam} C$, so that $\Theta = L/R$.

Corollary 4.2.5 *The FC problem (4.1) with curvature* $1/R$ *and size* $L \leq \alpha_M \operatorname{diam} C$ *is a FC/LD problem as soon as one of the following* sufficient conditions *is satisfied:*

$$\int_0^1 \frac{\|A(t)\|_F}{\|V(t)\|_F} \, \mathrm{d}t \leq \frac{\pi}{2} \quad \text{for all } x_0, x_1 \in C, \tag{4.24}$$

or

$$\|A(t)\|_F \leq \frac{\pi}{2}\|V(t)\|_F \text{ for a.e. } t \in [0,1] \text{ and all } x_0, x_1 \in C, \tag{4.25}$$

or

$$L \leq \frac{\pi}{2}R. \tag{4.26}$$

Proof. it follows immediately from Proposition 4.2.4. ■

Condition (4.26) shows that the *deflection condition* can be enforced by a *size×curvature condition*, that is, by requiring that the product of the size L of $\varphi(C)$ by its curvature $1/R$ is bounded, here by $\pi/2$.

Though the deflection condition $\Theta \leq \pi/2$ can be satisfied for an unbounded set $\varphi(C)$ (think, e.g., of the graph of a sine function in $I\!R^2$), Corollary 4.2.5 shows that $\Theta \leq \pi/2$ can be ensured by limiting the size of the admissible parameter set C ($\|A(t)\|$ is proportional to $\|x_1 - x_0\|^2$ and $\|V(t)\|$ is proportional to $\|x_1 - x_0\|$!). Hence the deflection condition will act in practice as a *localization constraint*: this is an example of *regularization by size reduction* mentioned in Sect. 1.3.4.

We can now state the nonlinear counterpart of Proposition 4.1.1 concerning the properties of the projection on $\varphi(C)$. The stability of the projection will hold for the "arc length distance" $\delta(X_0, X_1)$ on the attainable set $\varphi(C)$, defined by

$$\forall X_0, X_1 \in \varphi(C), \qquad \delta(X_0, X_1) = sup_{x_0 \in \varphi^{-1}(X_0), x_1 \in \varphi^{-1}(X_1)} L(P), \tag{4.27}$$

where $L(P)$ is the length, defined in (4.12), of the curve P associated to x_0 and x_1 by (4.11).

Remark 4.2.6 *The quantity $\delta(X_0, X_1)$ is positive. But without further hypothesis, it can happen that $\delta(X_0, X_1) = 0$ and $X_0 \neq X_1$. However, for a FC/LD problem, as it is the case in Proposition 4.2.7 below, $\delta(X_0, X_1) = 0$ implies $X_0 = X_1$ (Proposition 6.2.5), so that the first axiom of a distance is satisfied. The second axiom of a distance $\delta(X_0, X_1) = \delta(X_1, X_0)$ is always satisfied, but we do not know whether the third axiom (triangular inequality) is satisfied, this is why we use the word distance between quotes.* ■

Proposition 4.2.7 *Let (4.1) be a FC/LD problem with curvature $1/R < \infty$, and let ϑ be the* enlargement neighborhood *of $\varphi(C)$ defined by*

$$\vartheta = \left\{ z \in F \mid d(z, \varphi(C)) < R \right\}. \tag{4.28}$$

Then the projection on the attainable set $\varphi(C)$ has the following properties:

(i) Uniqueness: *For any $z \in \vartheta$, there exists at most one projection of z on $\varphi(C)$*

(ii) Unimodality: *If $z \in \vartheta$ admits a projection $\widehat{X}$ on $\varphi(C)$, the "distance to z" function has no parasitic stationary point on $\varphi(C)$*

(iii) Local stability: *If $z_0, z_1 \in \vartheta$ admit projections $\widehat{X}_0, \widehat{X}_1$ on $\varphi(C)$ and are close enough so that there exists $d \geq 0$ satisfying*

$$\|z_0 - z_1\|_F + \max_{j=0,1} d(z_j, \varphi(C)) \;\leq\; d < R, \tag{4.29}$$

then one has

$$\|\widehat{X}_0 - \widehat{X}_1\|_F \leq L(\widehat{P}) \leq (1 - d\,/R(\widehat{P}))^{-1}\;\|z_0 - z_1\|_F, \tag{4.30}$$

where $\widehat{P}$ is the curve associated by (4.11) to any a couple $\widehat{x}_0, \widehat{x}_1$ of the preimage of $\widehat{X}_0, \widehat{X}_1$.

This implies, as $L(\widehat{P}) \leq \delta(\widehat{X}_0, \widehat{X}_1)$ and $R(\widehat{P}) \geq R > 0$, a stability property for the arc length "distance" δ in the attainable set

$$\|\widehat{X}_0 - \widehat{X}_1\|_F \leq \delta(\widehat{X}_0, \widehat{X}_1) \leq (1 - d\,/R)^{-1}\;\|z_0 - z_1\|_F. \tag{4.31}$$

(iv) Existence: *If $z \in \vartheta$, any minimizing sequence $X_n \in \varphi(C)$ of the "distance to z" function over $\varphi(C)$ is a Cauchy sequence for both the distance $\|X - Y\|_F$ and the arc length "distance" $\delta(X, Y)$. Hence X_n converges in F to the (unique) projection $\widehat{X}$ of z onto $\overline{\varphi(C)}$.*

If $\varphi(C)$ is closed, then $\widehat{X} \in \varphi(C)$, and $\delta(X_n, \widehat{X}) \to 0$ when $n \to 0$.

Proof. Equation (4.1) is a FC/LD problem, and so its attainable set $\varphi(C)$ is s.q.c. (Definition 4.2.2) and the proposition follows from the properties of the projection on s.q.c. sets summarized in Theorem 7.2.11. ■

We investigate now the shape of the preimage sets, that is, the sets of parameters $x \in C$ that have the same image by φ. For a linear problem $\varphi(x) = B.x$, the preimage of $X \in B.C$ is the intersection of the closed affine subspace $\{x \in E \text{ such that } B.x = X\}$ of E with the admissible parameter set C, it is hence closed and convex. For FC/LD problems, the following result holds:

Proposition 4.2.8 *Let (4.1) be a FC/LD problem. Then the preimage $\varphi^{-1}(X)$ is a closed and convex set for all $X \in \varphi(C)$.*

Proof. let $X \in \varphi(C)$ be given. The finite curvature hypothesis implies that C is closed and φ continuous (second and last properties of (4.2)), and hence that $\varphi^{-1}(X)$ is closed. Then the condition $\Theta \leq \pi/2$ on the deflection implies that $\varphi(C)$ is s.q.c. (Proposition 4.2.7). Let then x_0, x_1 be two pre-image of X, and P the curve (4.11) image by φ of the $[x_0, x_1]$ segment of C. The function $t \rightsquigarrow \|X - P(t)\|^2$ is s.q.c. (use (7.4) with $D = \varphi(C)$ and $z = X$ in the Definition 7.1.2 of s.q.c. sets), positive, and takes the value 0 for $t = 0$ and $t = 1$: this can be possible only if $\|X - P(t)\|^2 = 0 \ \forall t \in [0, 1]$. Hence $x_t = (1 - t)x_0 + tx_1$ belongs to $\varphi^{-1}(X)$ for all $t \in [0, 1]$, which shows that $\varphi^{-1}(X)$ is convex. ■

Propositions 4.2.7 and 4.2.8 (and also 4.3.3 below) show that FC/LD problems are a direct generalization of linear least squares problems.

Remark 4.2.9 *All known examples of FC/LD problems correspond to forward maps φ, which are injective over C, for which $\varphi(X)^{-1} = \{x\}$, which is trivially convex. The existence of non injective function φ that produce FC problems – and hence FC/LD problems by reduction of the size of C – is an open problem.* ■

Remark 4.2.10 *When the deflection Θ is larger than π, we have seen in the arc-of-circle example that the conclusions of Proposition 4.2.7 cannot be true. But what when $\pi/2 < \Theta \leq \pi$? Theorem 8.1.6, which serves to prove Proposition 4.2.7, shows that $\varphi(C)$ is still s.q.c., but with a* smaller regular neighborhood

$$\vartheta = \Big\{ z \in F \mid d(z, \varphi(C)) < R_G \Big\}, \tag{4.32}$$

where $R_G \leq R$ is the global radius of curvature *(Sect. 7.2), as soon as the* geometric attributes $1/R$, L , *and Θ of the FC problem (4.1) (Definition 4.2.3) satisfy the* extended deflection condition

$$R_G \stackrel{\text{def}}{=} R(\sin\Theta + (L/R - \Theta)\cos\Theta) > 0. \tag{4.33}$$

Figure 8.2 shows that the set of deflection Θ and size×curvature product L/R that satisfy (4.33) is of the form

$$\Theta \leq \Theta_{\max}(L/R) \quad \textit{(or < depending on the value of } L/R\textit{)}, \tag{4.34}$$

with $\Theta_{\max}$ given by (8.19). This shows that the range of authorized deflections can be extended beyond $\pi/2$. Examples of simple sets that have a deflection larger than $\pi/2$ but are nevertheless s.q.c. can be seen in Figs. 7.3 and 8.3.

Hence the use of conditions (4.33) or (4.34) allows to enlarge the size of the admissible parameter set C for which $\varphi(C)$ is s.q.c. – so that Propositions 4.2.7 and 4.2.8 above and 4.3.3 below still hold – at the price of reducing the size of the neighborhood ϑ on which the projection is well-behaved.

For example, in the case where only the worst deflection estimate $\Theta = L/R$ (see (4.23)) is available, one sees that (4.33) is satisfied for any $\pi/2 < \Theta < \pi$, but on the smaller neighborhood (4.32) of size $R_G = R \sin\Theta$.

For sake of simplicity, we shall not attempt, in the rest of this chapter and in Chap. 5, which deal with the regularization of inverse problems, to produce the least constraining conditions on the size of C. So we shall not use the extended deflection condition (4.34), but only the simple deflection condition (4.16), which will ensure that the output set $\varphi(C)$ is s.q.c. with a neighborhood (4.28) of size R independent of the deflection Θ. ■

4.3 Identifiability and Stability of the Linearized Problems

We introduce in this section identifiability and stability properties of the linearized problems, and look into their relation with the corresponding properties of the original nonlinear problem (Definition 1.3.1).

Definition 4.3.1 *Let C, φ satisfy the minimum set of hypothesis (4.2). The parameter x is* linearly identifiable *over C if*

$$x_0, x_1 \in C\,,\ t \in [0,1] \text{ and } \|V(t)\|_F = 0 \implies x_0 = x_1, \tag{4.35}$$

where $V(t)$ is defined in (4.13).

Remark 4.3.2 *Identifiability and linear identifiability coincide of course for linear problems, but also for the class of so-called "bilinear" problems where the φ mapping admits a* state-space decomposition *(see (1.33) and (1.34) in Sect. 1.3.5 of Chap. 1) of the form:*

$$\begin{cases} e(x,y) &=& b(x,y)+c(y)+d \text{ with } b \text{ bilinear, } c \text{ linear, } d \text{ constant,} \\ M(y) &=& \text{linear and injective,} \\ b,c,M &=& \text{continuous.} \end{cases}$$

This can be seen easily: given $x_0, x_1 \in C$ *and* $t \in [0,1]$, *subtraction of the state equations written at* x_1 *and* x_0 *gives*

$$b(x_1 - x_0, y_1) + b(x_0, y_1 - y_0) + c(y_1 - y_0) = 0, \tag{4.36}$$

and derivation with respect to t of the equation at $x_t = (1-t)x_0 - tx_1$ *gives*

$$b(x_1 - x_0, y_t) + \quad b(x_t, \frac{\partial y_t}{\partial t}) \quad + \quad c(\frac{\partial y_t}{\partial t}) \quad = 0, \tag{4.37}$$

where y_t *is the solution of* $e(x,y) = 0$ *for* $x = x_t$.

Identifiability *holds as soon as* $\varphi(x_1) - \varphi(x_0) = 0 \Longrightarrow x_1 = x_0$, *that is, using the injectivity of* M, *as soon as* $y_1 = y_0 \Longrightarrow x_1 = x_0$, *or, using (4.36),*

$$b(x_1 - x_0, y_1) = 0 \Longrightarrow x_1 = x_0. \tag{4.38}$$

Similarly, linear identifiability *will hold as soon as* $V = M\,\partial y_t/\partial t = 0 \Longrightarrow x_1 = x_0$, *or, using the injectivity of* M *and (4.37), as soon as*

$$b(x_1 - x_0, y_t) = 0 \Longrightarrow x_1 = x_0,$$

which is equivalent to (4.38).

The class of "bilinear" problems is relatively large, as it contains all problems governed by a linear equation, where the unknown parameter appear as a coefficient, see the examples in Sects. 1.4 and 1.6 of Chap. 1. ■

But for nonlinear problems, identifiability does not imply in general linearized identifiability (think, e.g., of $\varphi(x) = x^3$ on $C = [-1,+1]$, but this is *not* a FC problem), and conversely linearized identifiability does not imply identifiability. However, for FC/LD problems the following result holds:

Proposition 4.3.3 *Let (4.1) be a FC/LD problem. Then*

$$\textit{linear identifiability (4.35)} \quad \Longrightarrow \quad \textit{identifiability (1.14)}$$

Proof. Let $X \in \varphi(C)$ be given, and x_0, x_1 be two preimage of X. We know from Proposition 4.2.8 that the curve P defined in (4.11) satisfies $P(t) = X$ for all $t \in [0,1]$. Derivation with respect to t gives $V(t) = 0\ \forall t \in [0,1]$, and hence $x_0 = x_1$ using (4.35). ■

To define the linearized version of the stability property (4.41), we replace now the length $\|\varphi(x_1) - \varphi(x_0)\|$ of the $[\varphi(x_0), \varphi(x_1]$ segment of F by the arc length L of the curve $P : t \rightsquigarrow \varphi((1-t)x_0 - tx_1)$:

Definition 4.3.4 *Let C, φ satisfy the minimum set of hypothesis (4.2). The parameter x is* linearly stable *on C if*

$$\begin{cases} \exists \alpha_m > 0 \text{ such that } \forall x_0, x_1 \in C : \\ \alpha_m \|x_0 - x_1\|_E \leq L(P) = \int_0^1 \|V(t)\|_F \, \mathrm{d}t, \end{cases} \tag{4.39}$$

where $L(P)$ is the arc length of the curve P image by φ of the $[x_0, x_1]$ segment, as defined by (4.13), (4.11), and (4.21). A sufficient condition for (4.39) to hold is

$$\begin{cases} \exists \alpha_m > 0 \text{ such that } \forall x_0, x_1 \in C, \forall t \in [0,1] : \\ \alpha_m \|x_0 - x_1\|_E \leq \|V(t)\|_F. \end{cases} \tag{4.40}$$

Linear stability is weaker than stability:

Proposition 4.3.5 *Let C, φ satisfy the minimum set of hypothesis (4.2). Then*

$$\textit{stability} \implies \textit{linear stability.}$$

Proof. The stability condition (1.15) implies (4.39) with $\alpha_m = 1/k$, which proves the implication. ■

4.4 A Sufficient Conditions for OLS-Identifiability

We establish in this section the additional condition that ensures that a FC/LD problem is Q-wellposed, or equivalently that x is OLS-identifiable. Finite dimensional problems are studied in Sect. 4.5, and an infinite dimensional example is given in Sect. 4.8 of this chapter.

The use of regularization to produce a FC/LD problem, which meets this condition – and is hence Q-wellposed – is studied in Chap. 5, for FC/LD problems in Sect. 5.1.2, for general nonlinear problems in 5.1.3, and for a special family of infinite curvature problems in Sect. 5.3.

As we have seen in (4.21) and (4.23), the deflection Θ can always be chosen smaller than $L/R \leq \alpha_M \mathrm{diam} C/R$. Hence the projection on the output set of FC problems can be made well-behaved by reducing the size of the admissible parameter set C until the deflection condition $\Theta \leq \pi/2$ is met. But as in the linear case, these nice properties of the projection are not enough to ensure the OLS-identifiability of x in problem (4.1): one has to add the nonlinear counterpart of the stability condition (4.9).

The first and most natural way to do that consists in considering that B in (4.9) is the forward map φ, and to replace (4.9) by the *stability condition* (1.15) of Definition 1.3.1, which we recall here for convenience:

$$\begin{cases} \exists k \geq 0 \text{ such that} \\ \|x_0 - x_1\|_E \leq k \,\|\varphi(x_0) - \varphi(x_1)\|_F \quad \forall x_0, x_1 \in C. \end{cases} \tag{4.41}$$

Of course, combining (4.41) with the stability property (4.30) of the projection onto $\varphi(C)$ would ensure the desired OLS-identifiability property. But it would not take full advantage of (4.30), where the stability of the projection is achieved not only for the usual distance $\|\widehat{X}_0 - \widehat{X}_1\|_F$ of the projections of z_0 and z_1, but also for their larger "arc length distance" $\delta(\widehat{X}_0, \widehat{X}_1)$ measured along the curves $\widehat{P}$. This distance is defined in term of the directional derivative $V(t)$ of φ (see (4.31)). This leads us to consider that B in (4.9) is rather the derivative of the forward map φ, and to replace (4.9) by the *linear stability condition* of Definition 4.3.4:

Theorem 4.4.1 *Let (4.1) be a FC/LD problem with curvature* $1/R < \infty$. *If the* linear stability condition *(4.39) is satisfied, the parameter* x *is OLS-identifiable, that is, the NLS problem (4.1) is Q-wellposed on a neighborhood*

$$\vartheta = \Big\{ z \in F | d(z, \varphi(C)) < R \Big\}, \tag{4.42}$$

and the following local Lipschitz stability result holds for the inverse problem: for all $z_0, z_1 \in \vartheta$ *close enough so that there exists* $d \geq 0$ *satisfying*

$$\|z_0 - z_1\|_F + \max_{j=0,1} d(z_j, \varphi(C)) \leq d < R, \tag{4.43}$$

the corresponding solutions $\widehat{x_0}$, $\widehat{x_1}$ *of (4.1) satisfy*

$$\alpha_m \|\widehat{x_0} - \widehat{x_1}\|_E \leq L(\widehat{P}) \leq (1 - d\,/R(\widehat{P}))^{-1} \,\|z_0 - z_1\|_F, \tag{4.44}$$

where

- $L(\widehat{P})$ *is the arc length of the curve* $\widehat{P} : t \rightsquigarrow \varphi((1-t)\widehat{x_0} + t\widehat{x_1})$ *image by* φ *of the* $[\widehat{x_0}, \widehat{x_1}]$ *segment*
- $R(\widehat{P}) \geq R$ *is a lower bound to the radius of curvature along* $\widehat{P}$

Proof. The linear stability hypothesis implies of course the linear identifiability of x and, because the problem has finite curvature and satisfies the deflection condition, identifiability of x on C (Proposition 4.3.3). Hence φ is injective on C: any $X \in \varphi(C)$ has a unique preimage $x = \varphi^{-1}(X)$ in C.

Let then $X_j \in \varphi(C)$, $j = 0, 1$, be two points on the output set, and $x_j = \varphi^{-1}(X_j)$, $j = 1, 2$, be the corresponding (unique) preimage. Combining the Definition (4.27) of $\delta(X, Y)$ with the linear stability condition (4.39) shows that

$$\delta(X_0, X_1) = L(P) = \int_0^1 \|V(t)\|_F \, dt \geq \alpha_m \|x_0 - x_1\|_E. \tag{4.45}$$

We prove now point 1 of the Definition 4.0.10 of OLS-identifiability. Let $z \in \vartheta$ be given. The uniqueness of the solution of problem (4.1) follows from the injectivity of φ over C (see above) combined with the uniqueness of the projection of z onto $\varphi(C)$ (property i) of Proposition 4.2.7). As for the existence of a solution, let $X_n \in \varphi(C)$ be a minimizing sequence of the "distance to z" function over $\varphi(C)$, and $x_n = \varphi^{-1}(X_n)$ the corresponding sequence of preimage. Because the problem has finite curvature and satisfies the deflection condition (4.16), Proposition 4.2.7 applies, so that X_n is a Cauchy sequence for the "distance" $\delta(X, Y)$ and hence, using (4.45), x_n is a Cauchy sequence in the Banach space E. But C is closed, so there exists $\hat{x} \in C$ such that $x_n \to \hat{x}$.Then the last property of (4.2) implies that φ is Lipschitz continuous, so that $X_n \to \widehat{X} = \varphi(\hat{x})$. This proves that $\widehat{X}$ is the projection of z onto $\varphi(C)$, and that $\hat{x}$ is the (unique) solution of problem (4.1).

Then point 2 of Definition 4.0.10 follows immediately from property (ii) of Proposition 4.2.7, combined with the existence of a (unique) projection on $\varphi(C)$ for all $z \in \vartheta$.

Finally, we notice that the stability property (4.43) (4.44) follows immediately from the property (iii) of the projection on $\varphi(C)$ in Proposition 4.2.7, combined with the stability result (4.45). This implies the local Lipschitz continuity of the $z \rightsquigarrow \hat{x}$ mapping from $(\vartheta, \|\cdot\|_F)$ to $(C, \|\cdot\|_E)$, which shows that point 3 of Definition 4.0.10 is satisfied. ■

4.5 The Case of Finite Dimensional Parameters

Using the results of the previous sections, we derive in this section simple sufficient conditions for Q-wellposedness (Definition 4.0.10) of problem (4.1) when the unknown parameter x is *finite dimensional.* These conditions will allow to check whether the additional information brought to the original ill-posed problem by regularization (Sect. 1.3.4) and, if necessary, discretization is sufficient to restore Q-wellposedness.

We shall suppose that the following *finite dimension (FD) minimum set of hypothesis* holds (compare with (4.2))

$$\left\{\begin{array}{lcl} E & = & \text{finite dimensional vector space, with norm } \|\ \|_E, \\ C & = & \text{closed, convex subset of } E, \\ C_\eta & = & \text{convex open neighborhood of } C \text{ in } E, \\ F & = & \text{Hilbert space, with norm } \|\ \|_F, \\ z & \in & F, \\ \varphi & : & C_\eta \rightsquigarrow F \text{ is twice differentiable along segments of } C_\eta, \\ \text{and} & : & V = \dfrac{\partial}{\partial t}\varphi((1-t)x_0 + tx_1),\ A = \dfrac{\partial^2}{\partial t^2}\varphi((1-t)x_0 + tx_1) \\ & & \text{are continuous functions of } x_0, x_1 \in C_\eta \text{ and } t \in [0,1]. \end{array}\right. \tag{4.46}$$

For example, when C is defined by

$$C = \{x \in E \mid c_\ell(x) \leq 0 \ \ \forall \ell \in L\}, \tag{4.47}$$

where c_ℓ, $\ell \in L$ are continuous and convex constraints, the neighborhood C_η can be defined simply by

$$C_\eta = \{x \in E \mid c_\ell(x) < \eta \ \ \forall \ell \in L\ \} \tag{4.48}$$

for some $\eta > 0$.

The differentiability and continuity results required in (4.46) happen to hold in most of the applications, so that the FD minimum set of hypothesis (4.46) is satisfied for the reduction to finite dimension of a large number of problems. But it is obviously not sufficient to ensure the Q-wellposedness of the inverse problem (4.1) over the finite dimensional set C. The next theorem indicates which additional properties are sufficient:

Theorem 4.5.1 *Let the* FD minimum set of hypothesis *(4.46) hold. Then φ is continuously differentiable over C_η, and the following properties hold:*

1. *If C is* bounded, *then*

   the attainable set $\varphi(C)$ is compact *– and hence closed,*
   the minimum set of hypothesis (4.2) holds on C.

2. *If moreover x is* linearly identifiable *(Definition 4.3.1)* over *C_η, then*

   x is linearly stable over *C (Definition 4.3.4),*
   the NLS problem (4.1) is a FC problem (Definition 4.2.1).

3. *If moreover C* is small enough *for the deflection condition $\Theta \leq \pi/2$ to hold, then*

   x is OLS-identifiable on C, or equivalently:
   the NLS problem (4.1) is Q-wellposed on C.

Proof. Let S denote the unit sphere of E. For any $x_0, x_1 \in C_\eta$, $x_0 \neq x_1$, and for any $t \in [0,1]$, we can define

$$x = (1-t)x_0 + tx_1 \in C_\eta, \qquad h = \frac{x_1 - x_0}{\|x_1 - x_0\|_E} \in S. \tag{4.49}$$

Then V and A defined in (4.46) satisfy

$$V = \|x_1 - x_0\|_E \; D_h\varphi(x), \;\; A = \|x_1 - x_0\|_E^2 \; D^2_{h,h}\varphi(x), \tag{4.50}$$

where $D_h\varphi(x)$ and $D^2_{h,h}\varphi(x)$ are the first and second derivatives of φ at x in the direction h, which exist by hypothesis. This shows that $V/\|x_1 - x_0\|_E$ and $A/\|x_1 - x_0\|_E^2$ depend only on the point $x \in C_\eta$ and on the direction $h \in S$.

Conversely, given $x \in C_\eta$ and $h \in S$, there are many ways to find $x_0, x_1 \in C_\eta$, $x_0 \neq x_1$, and $t \in [0,1]$, which satisfy (4.49), for example,

$$\left\{ \begin{array}{lcll} x_0 & = & x - d(x, E\setminus C_\eta)h/2 & \in \; C_\eta, \\ x_1 & = & x + d(x, E\setminus C_\eta)h/2 & \in \; C_\eta, \\ t & = & 1/2 & \in \; [0,1], \end{array} \right. \tag{4.51}$$

where $d(x, E \setminus C_\eta)$ denotes the distance of x to the complementary set of C_η in E. The function $x \rightsquigarrow d(x, E \setminus C_\eta)$ is continuous, and satisfies

$$d(x, E \setminus C_\eta) > 0 \ \forall x \in C_\eta. \tag{4.52}$$

Hence by composing the continuous mappings $(x, h) \rightsquigarrow (x_0, x_1, t)$ defined by (4.51) and $(x_0, x_1, t) \rightsquigarrow (V/\|x_1 - x_0\|_E, A/\|x_1 - x_0\|_E^2)$ defined by (4.50), we see that the first and second directional derivatives of φ are given by

$$D_h\varphi(x) = \frac{V}{\|x_1 - x_0\|_E}, \ D^2_{h,h}\varphi(x) = \frac{A}{\|x_1 - x_0\|_E^2} \quad \forall (x, h) \in C_\eta \times S, \tag{4.53}$$

and satisfy

$$D_h\varphi(x) \text{ and } D^2_{h,h}\varphi(x) \text{ are continuous functions of } x, h \text{ over } C_\eta \times S. \tag{4.54}$$

In particular, the partial derivatives $\partial\varphi(x)/\partial x_i = D_{e_i}\varphi(x)$ exist and are continuous functions of x over the open set C_η (e_i denotes the ith basis vector of E), which implies that φ is continuously differentiable over C_η.

We begin with point one of the theorem: C is now bounded. As we have just seen, φ is continuous over the open neighborhood C_η of the closed, bounded – and hence compact – set C, which implies that $\varphi(C)$ is compact, and hence closed. As for the minimum set of hypothesis (4.2), the only property that does not follow immediately from (4.46) is the existence of an α_M in the last line. The best (smallest) α_M is by definition

$$\begin{aligned}
\alpha_M &\stackrel{\text{def}}{=} \sup_{x_0, x_1 \in C \ , \ x_0 \neq x_1 \ , \ t \in [0,1]} \frac{\|V\|_F}{\|x_1 - x_0\|_E} \\
&\leq \sup_{x \in C \ , \ h \in S} \|D_h\varphi(x)\|_F \qquad \text{(use (4.53) left)} \\
&< +\infty \qquad \text{(use (4.54) left and } C \times S \text{ compact).}
\end{aligned}$$

We prove now the second point of the theorem. The linear identifiability hypothesis over C_η implies that, for any $x \in C_\eta$ and $h \in S$, the numerator V in $D_h\varphi(x) = V/\|x_1 - x_0\|_E$ is nonzero, as it corresponds to points $x_0, x_1 \in C_\eta$ such that $\|x_0 - x_1\|_E = d(x, E \setminus C_\eta) > 0$, c.f. (4.51) (4.52). The best (largest) α_m that satisfies (4.40) is, by definition,

$$\begin{aligned}
\alpha_m &\stackrel{\text{def}}{=} \inf_{x_0, x_1 \in C \ , \ x_0 \neq x_1 \ , \ t \in [0,1]} \frac{\|V\|_F}{\|x_1 - x_0\|_E} \qquad (4.55) \\
&\geq \inf_{x \in C \ , \ h \in S} \|D_h\varphi(x)\|_F \qquad \text{(use (4.53) left)} \\
&> 0 \qquad \text{(use (4.54) left and } D_h\varphi(x) \neq 0 \text{ over } C \times S \text{ compact),}
\end{aligned}$$

which proves the linear stability of x over C. As for the finite curvature of the problem, the only thing in Definition 4.2.1 that does not follows immediately from the FD minimum set of hypothesis (4.46) is the inequality (4.13). The proof is similar: the best (smallest) $1/R$ that satisfies (4.13) is by definition

$$
\begin{aligned}
1/R &\stackrel{\text{def}}{=} \sup_{x_0,x_1\in C\ ,\ x_0\neq x_1\ ,\ t\in[0,1]} \frac{\|A\|_F}{\|V\|_F^2} && (4.56)\\
&\leq \sup_{x\in C\ ,\ h\in S} \frac{\|D^2_{h,h}\varphi(x)\|}{\|D_h\varphi(x)\|^2} && \text{(use (4.53) left and right)}\\
&< +\infty && \text{(use (4.54) and } D_h\varphi(x)\neq 0 \text{ over } C\times S \text{ compact),}
\end{aligned}
$$

which proves the finite curvature property.

Finally, the last point of the theorem is simply a rewriting of the sufficient condition for OLS-identifiability of Theorem 4.4.1. ■

4.6 Four Questions to Q-Wellposedness

We summarize here the successive steps required to prove Q-wellposedness by application of Proposition 4.2.7, Theorems 4.4.1 or 4.5.1. These steps can (and ideally should) be performed before any attempt to minimize the least square objective function is done. Let the minimum set of hypotheses (4.2) be satisfied, x_0, x_1 be two admissible parameters, and denote by V(t) (for velocity) and A(t) (for acceleration) the first and second derivatives with respect to t of the forward map φ at $x_t = (1-t)x_0 + tx_1$ along the segment $[x_0, x_1]$.

A full analysis requires the answer to four questions:

1. Linear identifiability:

does $V=0$ ***imply*** $x_1 = x_0$***?***

2-a. Closedness:

is the attainable set $\varphi(C)$ ***closed?***

or alternatively,

2-b. Linear stability:

does there exists $\alpha_m > 0$ ***such that, for all*** $x_0, x_1 \in C$***,***

$$\alpha_m \|x_1 - x_0\|_E \leq \int_0^1 \|V(t)\|_F dt \textbf{\textit{?}} \qquad (4.57)$$

3. Deflection condition:

$$\textbf{\textit{does one have, for all }} x_0, x_1 \in C \textbf{\textit{ and }} t \in [0,1], \quad \|A(t)\|_F \le \theta(t)\, \|V(t)\|_F \textbf{\textit{ with }} \Theta = \int_0^1 \theta(t) \le \pi/2\,\textbf{\textit{?}} \tag{4.58}$$

4. Finite curvature:

$$\textbf{\textit{does there exist }} R > 0 \textbf{\textit{ such that, for all }} x_0, x_1 \in C \textbf{\textit{ and }} t \in [0,1], \ \|A(t)\|_F \le 1/R\, \|V(t)\|_F^2\,\textbf{\textit{?}} \tag{4.59}$$

The theoretical and numerical tools available (or not available...) to answer these questions are described in Sect. 4.7 below.

4.6.1 Case of Finite Dimensional Parameters

When the *FD minimum set of hypothesis* (4.46) is satisfied and the *admissible set C is bounded*, then the minimum set of hypothesis (4.2) is verified, and a positive answer to question **1** (linear identifiability) implies a positive answer to questions **2-a** (closedness), **2-b** (linear stability, for any norm on E), and **4** (finite curvature):

$$\left.\begin{array}{l}\text{finite dimensional parameters}\\ \text{bounded parameter set}\\ \text{linear identifiability}\end{array}\right\} \Longrightarrow \text{local OLS-identifiability,}$$

where *local OLS-identifiability* is a short name for *closed attainable set, linearly stable parameter, finite curvature problem.* Local OLS-identifiability implies that x is OLS-identifiable over any convex and closed subset D of C small enough to ensure a positive answer to question **3** (deflection condition):

$$\left.\begin{array}{l}\text{local OLS-identifiability}\\ \text{deflection condition } \Theta \le \pi/2\end{array}\right\} \Longrightarrow \text{OLS-identifiability.}$$

The deflection condition (4.58) provides an *estimation of the size* of D, which ensures OLS-identifiability. It is automatically satisfied if D satisfies

$$\alpha_M \operatorname{diam} D \le \frac{\pi}{2} R, \tag{4.60}$$

but this last estimation is usually less precise than the ones derived directly from (4.58).

4.6.2 Case of Infinite Dimensional Parameters

When the answer to questions **1**, **2-a**, **3**, and **4** above is positive, the NLS problem (4.1) is Q-wellposed for all data z in the neighborhood

$$\vartheta = \{z \in F \mid d(z, \varphi(C)) < R\} \tag{4.61}$$

of the attainable set, for the "distance" δ on C defined by

$$\delta(x_0, x_1) = \int_0^1 \|V(t)\|_F \,\mathrm{d}t \quad \text{(arc length in the attainable set)},$$

with the stability property

$$\delta(x_0, x_1) \leq \left(1 - \frac{d}{R}\right)^{-1} \|z_1 - z_0\|_F \tag{4.62}$$

for any $z_0, z_1 \in \vartheta$ and $d > 0$ such that

$$\|z_1 - z_0\|_F + \max_{j=0,1} d(z_j, \varphi(C)) \leq d < R.$$

In the case where one is able to give a positive answer to question **2-b** for some norm $\|\cdot\|_E$, the *closedness* question **2** is also answered positively, and one has

$$\alpha_m \|\hat{x}_1 - \hat{x}_0\|_E \leq \delta(x_0, x_1), \tag{4.63}$$

so that the the NLS problem (4.1) is still Q-wellposed on the same neighborhood (4.61), but for the stronger norm $\|\cdot\|_E$ on the parameter space for which one has been able to prove linear stability.

Remark 4.6.1 *In practice, the infinite dimensional problem has to be reduced at some point to finite dimension for computational purpose, by searching for x in a finite dimensional subspace* $\mathbf{E}$ *of E (see Chap. 3). When the infinite dimensional parameter x happens to be linearly identifiable over C, the results of previous Sect. 4.6.1 apply automatically to the resulting finite dimensional problem.* ■

4.7 Answering the Four Questions

Most often, the original inverse problem does not satisfy, rigorously or even approximately, any of the sets of sufficient conditions for Q-wellposedness recalled in the previous section. Because the conditions are only sufficient, this does not mean that it is ill-posed, only that one cannot decide between well- and ill-posedness. A reasonable thing to do is then to add information, either by reducing the number of parameters (Chap. 3), or by using L.M.T. regularization or strengthening the observation in the perspective of using state-space regularization (Chap. 5), until the regularized problem satisfies one set of sufficient conditions. Hence these sufficient conditions provide a guideline for the design of a wellposed and optimizable regularized problem, and it is only at that point that numerical attempts to minimize the least squares objective function should be made, now with a reasonable hope of producing meaningful results.

We discuss for the rest of this section the *available tools* for checking these condition under the minimum set of hypotheses (4.2): one can try to answer questions 1–4 either *rigorously*, by *proving* the corresponding property or giving a counterexample, or *approximately*, when a proof is out of reach, by *checking numerically* the property – but this becomes computationally intensive when there are more than a few parameters.

In any cases, it is useful, whenever possible, to estimate even crudely the linear stability constant α_m, and the lower bound R to the radius of curvature of the problem, which provide useful practical information: α_m, R appear in the stability estimate (4.62) and (4.63) of the inverse problem, and R gives the size of the neighborhood of the attainable set on which the inverse problem is stable, and hence provides an upper bound on the size of the admissible measurement and model errors.

For problems of small size, one can combine formula (4.65), (4.70), and (4.73) for the numerical determination of α_m, Θ, and R with Theorem 4.5.1 and the stability property (4.44) to perform a *nonlinear stability analysis*, see, for example, [80].

Checking for Linear Identifiability (Question 1)

This is the first step on the way to OLS-identifiability. For the full nonlinear inverse problem, identifiability has received much attention in the recent years, see, for example, [46, 34, 47]. Unluckily, the availability of a nonlinear identifiability result does not imply automatically linear identifiability ($x \rightsquigarrow x^3$ is injective, but its derivative at $x = 0$ is not injective...), and, when a nonlinear identifiability result is available for the problem at hand, it is necessary to check wether the nonlinear proof goes over to the linearized problem. This suggests that the identifiability studies should be devoted to the linearized problems rather than to the original nonlinear problem.

However, for *bilinear problems*, where the inverse problem is governed by a linear state equation with the unknown parameter being one coefficient of the equation, and with a linear injective observation operator, identifiability coincides with linear identifiabiliy (see Remark 4.3.2), and the existence of a nonlinear identifiability result implies automatically linear identifiability.

If none of this works, one is left with the challenge of proving directly linear identifiability – not necessarily easy, nor necessarily true!

When the theoretical approach fails, one has to resort to numerical determination. This makes sense only after the parameter has been reduced, if necessary, to finite dimension. The approach is similar to that of Sect. 3.2: a nominal parameter $x_{\rm nom} \in C$ is chosen, and the *singular value decomposition* (SVD) (3.9)–(3.11) of the $q \times n$ Jacobian $\varphi'(x_{\rm nom})$ is performed. Linear identifiability is achieved in theory as soon as the number r of strictly positive singular values μ is equal to n for all $x_{\rm nom} \in C$, and it fails if $r < n$ for some $x_{\rm nom} \in C$.

Of course, this is impossible to check rigorously on the computer:

- One has to limit the computation of $\varphi'(x_{\rm nom})$ and its SVD to a finite set C_N of nominal parameters, and then cross fingers that things do not change too much away from the chosen nominal value(s). The number of points in C_N depend on the computational power one can afford, it can be reduced to one for computationally intensive problems...

- Testing that a floating number in the computer is strictly positive is a difficult task. In practice, the test $\mu > 0$ is replaced, for the determination of r, by
$$\mu \geq \mu_{\rm min} > 0,$$
where $\mu_{\rm min}$ is a threshold determined by the level of noise or error on the data, as explained in Sect. 3.2 of Chap. 3.

Remark 4.7.1 *For infinite dimensional parameters, the existence of a theoretical linear identifiability result does not eliminate the need of performing an SVD analysis after the problem has been reduced to finite dimension: some of the singular values may (and usually will...) be zero or below the threshold* $\mu_{\rm min}$. ■

Checking for Closedness (Question 2-a)

The attainable set is closed as soon as one can equip the parameter space E with a norm that makes the admissible parameter set C compact and the forward map φ continuous. This holds true in particular as soon as E is finite dimensional and C bounded.

Also, the attainable set of a linearly stable FC/LD problem is closed as soon as the parameter space E is complete.

Checking for Linear Stability (Question 2-b)

Analytical proofs of (4.57) are rare, and when they exist, give a pessimistic (too small) estimate of the constant $\alpha_m > 0$, or even show only its existence, without any information on its value.

So one has to evaluate α_m numerically, which will necessarily produce an optimistic (too large) value. There are two ways to do that:

- One can go along the line used to check numerically for linear identifiability: discretize the admissible set C into a finite set C_N, and perform the SVD decomposition (3.9)–(3.11) of $\varphi'(x_{\mathrm{nom}})$ for all $x_{\mathrm{nom}} \in C_{\mathrm{N}}$. When linear identifiability holds, one has $r = n \ \forall x_{\mathrm{nom}} \in C_N$, and α_m can be estimated by
$$\alpha_m = \min_{x_{\mathrm{nom}} \in C_{\mathrm{N}}} \mu_{n,\mathrm{nom}}. \tag{4.64}$$
The estimate (4.64) is optimistic because one performs the SVD only at a finite number of points of C, but it can also be pessimistic because performing the SVD amounts to investigate the stability in all directions of $\mathbb{R}^n$, some of them pointing outside of C when the dimension of C is strictly smaller than n.

- One can also go along the line of the next sections on the deflection and finite curvature conditions: one discretizes both points and directions of the admissible set C. Figure 4.2 shows two naturals way to do that:

  – On the left, a very coarse coverage where only one (or a few) point(s) and the directions of the coordinate axis are investigated

  – On the right a more comprehensive coverage, based on a discretization of the boundary ∂C into a finite set ∂C_N (circles), where each $[x_0, x_1]$ interval is in turn discretized into N intervals by $N+1$ points $x_{k/N}, k = 0, 1 \cdots N$ (black dots)

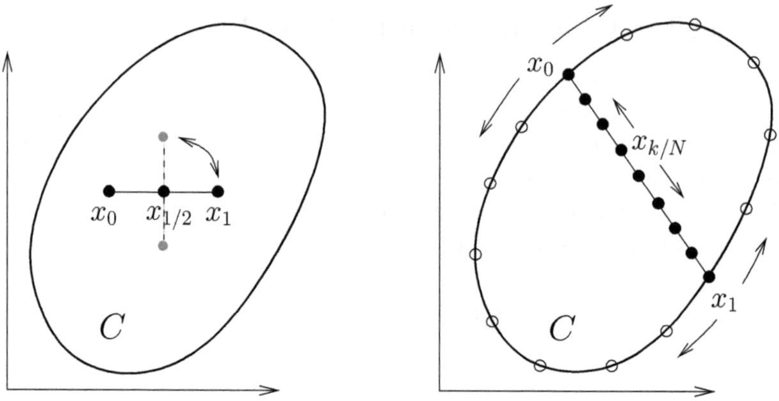

Figure 4.2: Very scarce (*left*) and comprehensive (*right*) coverage of the points and directions in C

Then α_m is estimated by

$$\alpha_m = \min_{x_0, x_1 \in \partial C_N,\ k=1\cdots N} \frac{||V_{k-1/2}||}{||x_1 - x_0||}, \tag{4.65}$$

where $V_{k-1/2}$ (see (4.13)) is the derivative of the forward map φ at $x_{(k-1/2)/N} \stackrel{\text{def}}{=} (x_{(k-1)/N} + x_{k/N})/2$ along the vector $x_1 - x_0$. The derivative $V_{k-1/2}$ is evaluated either analytically, if the corresponding code is available, or numerically if not, for example,

$$\frac{V_{k-1/2}}{||x_1 - x_0||} = \frac{\varphi(x_{k/N}) - \varphi(x_{(k-1)/N})}{||x_{k/N} - x_{(k-1)/N}||}. \tag{4.66}$$

Checking the Deflection Condition (Question 3)

The deflection Θ of the inverse problem (4.1) can sometimes be estimated analytically, in particular for bilinear problems (see Remark 4.3.2) with full observation. The state equation for bilinear problems is of the form:

$$e(x, y) = b(x, y) + c(y) + d = 0, \tag{4.67}$$

where b is bilinear and c linear, and by full observation property we mean that the observation operator M is linear and satisfies, for some $\kappa_M \geq \kappa_m > 0$,

$$\kappa_m \|y\|_Y \leq \|My\|_F \leq \kappa_M \|y\|_Y \text{ for all } y \in Y. \tag{4.68}$$

The velocity and acceleration are then given by

$$V(t) = M\,\frac{\partial y_t}{\partial t} \quad \text{and} \quad A(t) = M\,\frac{\partial^2 y_t}{\partial t^2},$$

where y_t is the solution of the state equation (4.67) for $x = x_t = (1-t)x_0 + tx_1$. Deriving (4.67) twice gives the equations for $\partial y_t/\partial t$ and $\partial^2 y_t/\partial t^2$:

$$b\Big(x_t, \frac{\partial y_t}{\partial t}\Big) + c\Big(\frac{\partial y_t}{\partial t}\Big) + b(x_1 - x_0, y_t) = 0,$$

$$b\Big(x_t, \frac{\partial^2 y_t}{\partial t^2}\Big) + c\Big(\frac{\partial^2 y_t}{\partial t^2}\Big) + 2b\Big(x_1 - x_0, \frac{\partial y_t}{\partial t}\Big) = 0. \tag{4.69}$$

The same arguments that ensure that the state equation (4.67) has a unique solution depending continuously on c will, in general, imply that (4.69) has a unique solution, with

$$\left\| \frac{\partial^2 y_t}{\partial t^2} \right\|_Y \leq \kappa(C) \left\| \frac{\partial y_t}{\partial t} \right\|_Y,$$

where the constant $\kappa(C)$ can be expressed in terms of the bounds imposed on x in C. The deflection condition is then satisfied as soon as

$$\frac{\kappa_M}{\kappa_m}\kappa(C) \leq \pi/2.$$

An example of this situation can be found in the deflection estimates of Proposition 4.9.3 (respectively, 5.4.5) for the estimation of the diffusion coefficient in a two-dimensional elliptic equation with H^1-observation (respectively, H^1-observation with adapted regularization).

As for the numerical evaluation of the deflection Θ, it can be performed on a coverage of points and directions of C that takes into account the size of C, as the one at the right of Fig. 4.2:

$$\Theta = \max_{x_0, x_1 \in \partial C_N,\ k,k'=0,1\cdots N,\ k \neq k'} \theta(k, k'), \tag{4.70}$$

where $\theta(k,k')$ is the deflection between the points $x_{k/N}$ and $x_{k'/N}$:

$$\theta(k,k') = \cos^{-1}\frac{\langle V_k, V_{k'}\rangle}{\|V_k\|\,\|V_{k'}\|}.$$

Here V_k is the velocity at x_k along $x_1 - x_0$. It can be evaluated either analytically, or numerically by

$$V_k = (V_{k+1/2} + V_{k-1/2})/2, \tag{4.71}$$

for example, where $V_{k-1/2}$ is defined in (4.66).

Formula (4.70), which is based on the definition of Θ, is the more precise one, but is computationally expensive, as it requires the comparison of $\theta(k,k')$ for all *couples* k, k'. A formula with a simple summation in k only, which is based on the majoration of Proposition 4.2.4 but requires an estimation of the acceleration A, is given in the next section in (4.74).

Remark 4.7.2 *The deflection condition $\Theta \le \pi/2$ is only a sufficient condition for the more precise* extended deflection condition $R_G > 0$ *on the* global radius of curvature, *which characterizes s.q.c. sets (see Remark 4.2.10 and Chap. 7).*

So in a situation where one can afford to compute the deflection by (4.70), one should rather compute, at a similar computational burden, the global radius of curvature R_G using Proposition 7.3.1

$$R_G = \max_{x_0,x_1\in\partial C_N,\ k,k'=0,1\cdots N,\ k\neq k'} \rho_G^{ep}(k,k'),$$

where $\rho_G^{ep}(k,k')$ is the global radius of curvature at $x_{k/N}$ seen from $x_{k'/N}$ ($\rho_G^{ep}(k,k') \neq \rho_G^{ep}(k',k)$!), given by

$$\rho_G^{ep}(k,k') = \begin{cases} \max\{0,N\}/D & \text{if } \langle V,V'\rangle \ge 0, \\ \max\{0,N\} & \text{if } \langle V,V'\rangle \le 0, \end{cases} \tag{4.72}$$

with (Proposition 7.2.6)

$$\begin{cases} X = \varphi(x_{k/N}), & X' = \varphi(x_{k'/N}), \\ v = V/\|V\|, & v' = V'/\|V'\|, \\ N = \operatorname{sgn}(k'-k)\langle X'-X, v'\rangle, & \\ D = (1-\langle v,v'\rangle^2)^{1/2}. & \end{cases}$$

When $R_G > 0$, all stability results of this chapter apply on a neighborhood of size $R_G \le R$ (Remark 4.2.10). ■

Checking for Finite Curvature (Question 4)

Constructive proofs of finite curvature for an inverse problem (4.1), which provide an estimation of the *radius of curvature R of the inverse problem*, are very rare, the author knows of two cases only: the 1D elliptic parameter estimation problem of Sect. 1.4 in the case of an H^1-observation, analyzed in Sect. 4.8, and the 2D elliptic nonlinear source estimation problem of Sect. 1.5 in the case of an H^1-observation, analyzed in Sect. 5.2.

Hence one has to resort in practice to numerical estimation if one wants to assign a value to R. Given one coverage of points and directions of C as in Fig. 4.2, one can estimate R by

$$R = \min_{x_0, x_1 \in \partial C_N,\ k=1\cdots N-1} \frac{\|V_k\|^2}{\|A_k\|}, \tag{4.73}$$

and, using Proposition 4.2.4, the deflection Θ by

$$\Theta = \sum_{k=1}^{N-1} \frac{\|A_k\|}{\|V_k\|} \times \frac{\|x_{(k+1/2)/N} - x_{(k-1/2)/N}\|}{\|x_1 - x_0\|}, \tag{4.74}$$

where $x_{(k-1/2)/N} \stackrel{\text{def}}{=} (x_{(k-1)/N} + x_{k/N})/2$, and V_k and A_k are the velocity and acceleration at $x_{k/N}$ along the vector $x_1 - x_0$. They can be determined analytically if the codes computing the first and second directional derivative are available. If not, they can be approximated numerically, for example, using formula (4.71) for V_k, and for A_k,

$$\frac{A_k}{\|x_1 - x_0\|} = \frac{V_{k+1/2} - V_{k-1/2}}{\|x_{(k+1/2)/N} - x_{(k-1/2)/N}\|}, \qquad k = 1 \cdots N-1.$$

With this approximation, the formula for Θ becomes

$$\Theta = 2 \sum_{k=1}^{N-1} \frac{\|V_{k+1/2} - V_{k-1/2}\|}{\|V_{k+1/2} + V_{k-1/2}\|}.$$

4.8 Application to Example 2: 1D Parameter Estimation with H^1 Observation

The estimation of the diffusion coefficient a from temperature measurements is a long-time classic of inverse problems (see, e.g.,[4, 49, 47, 46]). We discuss in this section the OLS-identifiability of the diffusion coefficient a in a one-dimensional elliptic problem (1.38) and (1.39) with an H^1 observation (1.50) and (1.51), that is, when a measurement z of the space derivative u' of its solution u is available, as described in Sect. 1.4. This material is adapted from the original paper [25], with a simpler proof and sharper estimations. The case of an L^2 observation of u will be considered in Chap. 5 using state-space regularization. The results of this section can be generalized to problems with distributed sources, at the price of more technical complexities. We refer the interested reader to Sect. 6 of reference [25].

As explained in Sect. 1.4, we shall search for $b = a^{-1}$ rather than for a. The state equation over the domain $\Omega = [0, 1]$ is then, with the notations (1.38),

$$-(b^{-1}u_\xi)_\xi = \sum_{j\in J} g_j\,\delta(\xi - \xi_j), \qquad \xi \in \Omega, \tag{4.75}$$

with the Dirichlet boundary conditions

$$u(0) = 0, \qquad u(1) = 0. \tag{4.76}$$

We choose as parameter space $E = L^2(\Omega)$ for b, with the admissible parameter set C defined in (1.47). In fact we shall not be able to achieve L^2 stability on this parameter set, but only on a smaller set $D \subset C$ to be defined later.

For any $b \in C$, (4.75) and (4.76) have a unique solution u, whose space derivative is given by

$$u_\xi = -b\,q_b,$$

where the heat flux profile q_b defined by (1.41)–(1.43) depend on b through the b-weighted mean $\widetilde{H}$.

Following (1.50), we choose $F = L^2(\Omega)$ as data space, and the estimation of $b \in C$ from a measurement $z \in L^2(\Omega)$ of u_ξ corresponds to the inversion of the forward map φ defined in (1.51), which is given here by the particularly simple closed form formula (1.44):

$$\varphi(b) = -b\,q_b. \tag{4.77}$$

For any $b_0, b_1 \in C$ and $t \in [0, 1]$, define $b = (1-t)b_0 + tb_1 \in C$.

Derivation of the state equations (4.75) and (4.76) or the closed form formula (4.77) with respect to t shows that the velocity $V = \mathrm{d}u_\xi/dt = \mathrm{d}\varphi(b)/dt$ is given either by

$$V = \eta_\xi, \quad \text{where } \eta \text{ is the solution of}$$

$$-(b^{-1}\eta_\xi)_\xi = -\Big(\frac{b_1 - b_0}{b} q_b\Big)_\xi, \qquad \xi \in \Omega, \tag{4.78}$$

$$\eta(0) = 0, \qquad \eta(1) = 0. \tag{4.79}$$

or by

$$V = \frac{b}{\int_0^1 b} \int_0^1 (b_1 - b_0)q_b - (b_1 - b_0)q_b. \tag{4.80}$$

Derivation of (4.80) gives then the acceleration $A = \mathrm{d}^2\varphi(b)/\mathrm{d}t^2$:

$$A = 2\frac{b}{\int_0^1 b} \int_0^1 (b_1 - b_0)q_b \left\{ \frac{\int_0^1 (b_1 - b_0)}{\int_0^1 b} - \frac{b_1 - b_0}{b} \right\}, \tag{4.81}$$

or equivalently, using (4.80),

$$A = 2\Big(\frac{V}{b} + \frac{b_1 - b_0}{b} q_b\Big) \left\{ \frac{b}{\int_0^1 b} \int_0^1 (b_1 - b_0) - (b_1 - b_0) \right\}. \tag{4.82}$$

Formula (4.82) will be used to estimate the deflection Θ, and (4.81) will serve to estimate the curvature $1/R$.

We check first that the *minimum set of hypothesis* (4.2) is satisfied: the only part that needs a proof is the last property. Multiplication of (4.78) by η_ξ and integration over $\Omega = [0, 1]$ shows that

$$|V|_{L^2} = |\eta_\xi|_{L^2} \leq b_M \Big|\frac{b_1 - b_0}{b} q_b\Big|_{L^2},$$

and (4.2) holds with

$$\alpha_M = \frac{b_M}{b_m} q_M.$$

4.8.1 Linear Stability

Linear stability on C means that we can find $\alpha_m > 0$ such that

$$\alpha_m\, |b_1 - b_0|_{L^2} \leq |\eta_\xi|_{L^2} = |V|_{L^2} \quad \forall b_0, b_1 \in C\ ,\ \ \forall t \in [0, 1]. \tag{4.83}$$

With the notation

$$d = (b_1 - b_0)/b,$$

(4.78) and (4.79), which define η, rewrite

$$- b^{-1}\eta_\xi = -d\, q_b + \text{unknown constant}, \qquad \xi \in \Omega, \tag{4.84}$$

$$\eta(0) = 0, \qquad \eta(1) = 0.$$

The flux function q_b is constant between consecutive source points. If one of these constants is zero, parameters b_0 and b_1 that differ only on the corresponding interval produce $\eta = 0$, that is, $V = 0$, and linear stability cannot hold. It is hence necessary, if we want stability, to suppose that the flux satisfies

$$0 < q_m \leq q_b \leq q_M \quad \forall b \tag{4.85}$$

for some $0 < q_m \leq q_M$ (notice that (4.85) will be automatically satisfied if, e.g., a finite number of sources and sinks of the same amplitude are located in an alternate way on the $[0, 1]$ interval).

But now q_b^{-1} is finite, and so it can happen that $d = (b_1 - b_0)/b$ becomes proportional to q_b^{-1}. In this case, $d\, q_b$ is a constant, and (4.78) and (4.79) imply $\eta = 0$. Hence $V = 0$, and linear stability fails once again! So we have to prevent $d\, q_b$ to become close to a constant function. To quantify this, we decompose $L^2(\Omega)$ into the direct sum of two orthogonal subspaces

$$L^2(\Omega) = L^2(\Omega)/I\!R \oplus I\!R, \tag{4.86}$$

where $L^2(\Omega)/I\!R$ is the quotient space of L^2 by the equivalence relation "v is equivalent to w if and only if $v - w$ is a constant," and $I\!R$ is the space of constant functions. Let then $\gamma \in [0, \pi/2]$ be the *indetermination angle* between the direction of $d\, q_b$ and that of constant functions (see Fig. 4.3).

The angle γ measure the "distance" of $d\, q_b$ to constants. We proceed in two steps:

– First, we find a linear stability constant that depends on the angle γ

– Second, we reduce the admissible parameter set to $D \subset C$ by adding constraints that ensure that $\gamma \geq \gamma_m > 0$ all over D

Step 1: The decomposition (4.86) is orthogonal, hence for any function v of $L^2(\Omega)$, one has

$$|v|^2_{L^2} = |v|^2_{L^2/I\!R} + |v|^2_{I\!R},$$

where

$$|v|_{L^2/I\!R} = \inf_{c \in I\!R} |v + c|_{L^2} = |v - \int_\Omega v\, |_{L^2} \leq |v|_{L^2}, \qquad |v|_{I\!R} = |\int_\Omega v\, |. \tag{4.87}$$

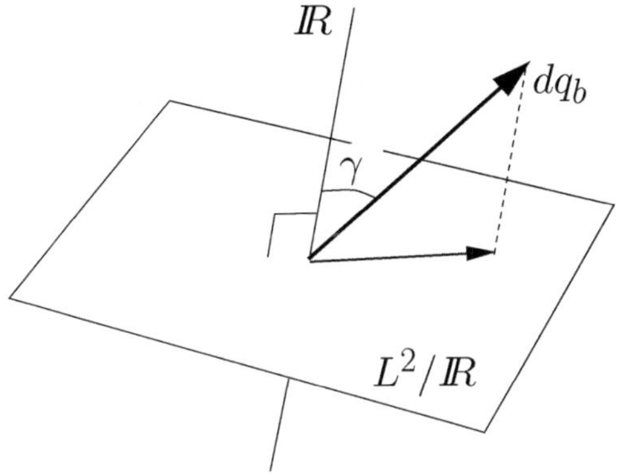

Figure 4.3: The decomposition $L^2(\Omega) = L^2(\Omega)/\mathbb{R} \oplus \mathbb{R}$ and the indetermination angle γ

Then (4.84) shows that $b^{-1}\eta_\xi = b^{-1}V$ and $d\,q_b$ are equivalent, so that

$$|d\,q_b|^2_{L^2/\mathbb{R}} = |b^{-1}V|^2_{L^2/\mathbb{R}} \le |b^{-1}V|_{L^2} \le b_m^{-1}|V|_{L^2}. \tag{4.88}$$

But, by definition of γ, one has

$$|d\,q_b|_{L^2/\mathbb{R}} = \sin\gamma |d\,q_b|_{L^2},$$

which gives the linear stability estimate depending on γ:

$$b_m \sin\gamma |d\,q_b|_{L^2} \le |V|_{L^2}. \tag{4.89}$$

Step 2: To keep the angle γ away from zero, we remark that q_b^{-1} is constant between two source points, and is discontinuous at each active source point, where $g_i \neq 0$. So if we require that the admissible coefficients b (and hence the vector $d = (b_1 - b_0/b!)$) are regular, say, for example, constant, on some neighborhood of at least one active source, then $d\,q_b$, which is also discontinuous at active source points, cannot be constant, and we can expect that γ remains larger than some $\gamma_m > 0$.

So we define a smaller admissible parameter set

$$D = \{b \in C \mid \forall j \in J,\ b(x) = b_j = \text{unknown constant on } I_j\}, \tag{4.90}$$

where

$$I_j =]\xi_j - \eta_j^-, \xi_j + \eta_j^+[\quad \forall j \in J \quad (\text{with } \eta_j^- > 0 \ , \ \eta_j^+ > 0) \tag{4.91}$$

are intervals surrounding the sources ξ_j. Of course, the intervals are chosen such that

$$I_j \subset \Omega =]0,1[\quad \forall j \in J, \qquad I_j \cap I_k = \emptyset \quad \forall j,k \in J, \ j \neq k. \tag{4.92}$$

Let v and e be unit vectors in the directions of $d\,q_b$ and of constant functions

$$v = \pm d\,q_b / |d\,q_b|_{L^2}, \qquad e(\xi) = 1 \quad \forall \xi \in \Omega,$$

where the sign is chosen such that $\langle e, v \rangle \geq 0$. With this convention, the angle γ between the directions of v and e is given by the median theorem:

$$0 \leq \cos\gamma = \langle e, v \rangle = 1 - \frac{1}{2}|e - v|_{L^2}^2, \qquad \gamma \in [0, \pi/2]. \tag{4.93}$$

To minor γ when $b_0, b_1 \in D$ and $t \in [0,1]$, we have to search for a lower bound to $|e - v|_{L^2}^2$:

$$\begin{aligned} |e - v|_{L^2}^2 &= \int_0^1 |e - v|^2 \\ &\geq \sum_{j \in J} \int_{I_j} |1 - v|^2 \\ &= \sum_{j \in J} \{\eta_j^-(1 - \widetilde{d}_j q_b(\xi_j^-))^2 + \eta_j^+(1 - \widetilde{d}_j q_b(\xi_j^+))^2\}, \end{aligned} \tag{4.94}$$

where $\widetilde{d}_j = d_j / |d\,q_b|_{L^2} \in I\!R$. We have used the fact that, on each interval I_j, e is equal to 1, and v takes constant values $\widetilde{d}_j q_b(\xi_j^-)$ left from ξ_j and $\widetilde{d}_j q_b(\xi_j^+)$ right from ξ_j. Taking the infimum over $\widetilde{d}_j \in I\!R$ for each interval gives

$$|e - v|_{L^2}^2 \geq \sum_{j \in J} \frac{\eta_j^- \eta_j^+}{\eta_j^- q_b(\xi_j^-)^2 + \eta_j^+ q_b(\xi_j^+)^2} \, g_j^2 \tag{4.95}$$

$$\geq \frac{1}{q_M^2} \sum_{j \in J} \frac{\eta_j^- \eta_j^+}{\eta_j^- + \eta_j^+} \, g_j^2 \tag{4.96}$$

Combining (4.95) with (4.93) shows that

$$\forall b \in D, \qquad 0 \leq \cos\gamma \leq \cos\gamma_m \leq 1, \tag{4.97}$$

where the *minimum indetermination angle* γ_m is defined by

$$\cos\gamma_m = 1 - \frac{1}{2q_M^2}\sum_{j\in J}\frac{\eta_j^-\eta_j^+}{\eta_j^- + \eta_j^+}\, g_j^2 \geq 0, \qquad 0 \leq \gamma_m \leq \pi/2. \tag{4.98}$$

The sum term in the right-hand side of (4.98) depends only on the strength of the sources (last factor), and on the disposition of the intervals I_j surrounding the sources (first factor). It is strictly positive if at least one active source is interior to the corresponding interval, that is,

$$\exists j \in J \ \text{ such that } \ \eta_j^-\eta_j^+ g_j \neq 0. \tag{4.99}$$

Combining (4.89) (step 1) and (4.97) (step 2) give then the "flux-weighted relative stability estimate"

$$b_m \sin\gamma_m \left|\frac{b_1 - b_0}{b}\, q_b\right|_{L^2} \leq |V|_{L^2}. \tag{4.100}$$

Thus we have proved the

Proposition 4.8.1 *Let hypothesis (4.85) and (4.99) hold. The estimation of $b \in L^2$ from $u_\xi \in L^2$ is then* linearly stable *(inequality (4.83)) on the admissible parameter set D defined by (4.90)–(4.92), with a stability constant given by*

$$\alpha_m = q_m \frac{b_m}{b_M} \sin\gamma_m > 0, \qquad \textit{with } \gamma_m \textit{ defined in (4.98).} \tag{4.101}$$

Remark 4.8.2 *It is possible to obtain numerically a better (i.e., larger) estimation of γ_m – and hence of α_m. Equation (4.95) can be rewritten, using the definition (1.41) of q_b,*

$$|e - v|_{L^2}^2 \geq g(H_b).$$

where $g : I\!R \rightsquigarrow I\!R$ and $H_b \in I\!R$ are defined by (see (1.43))

$$g(h) = \sum_{j\in J}\frac{\eta_j^-\eta_j^+}{\eta_j^-(h - H(\xi_j^-))^2 + \eta_j^+(h - H(\xi_j^+))^2}\, g_j^2,$$

$$H_b = \frac{\int_0^1 b\, H(\xi)\, \mathrm{d}\xi}{\int_0^1 b\, \mathrm{d}\xi} \in I\!R.$$

Let Ω^+ (resp. Ω^-) be the subsets of $\Omega = [0,1]$, where H (defined in (1.43)) is positive (respectively, negative). Then the b-weighted mean H_b satisfies

$$H_{b,m} \le H_b \le H_{b,M},$$

where $H_{b,m}$ and $H_{b,M}$ are defined by

$$\begin{aligned} H_{b,m} &= \Big\{ b_m \int_{\Omega^+} H(\xi)\,\mathrm{d}\xi - b_M \int_{\Omega^-} H(\xi)\,\mathrm{d}\xi \Big\} \Big/ \big\{ b_m|\Omega^+| + b_M|\Omega^-| \big\}, \\ H_{b,M} &= \Big\{ b_M \int_{\Omega^+} H(\xi)\,\mathrm{d}\xi - b_m \int_{\Omega^-} H(\xi)\,\mathrm{d}\xi \Big\} \Big/ \big\{ b_M|\Omega^+| + b_m|\Omega^-| \big\}. \end{aligned}$$

A better estimation $\widetilde{\gamma}_m$ of γ_m is then

$$\cos \widetilde{\gamma}_m = 1 - \frac{1}{2} \inf_{H_{b,m} \le h \le H_{b,M}} g(h) \le \cos \gamma_m,$$

where the infimum can be determined numerically, for example, by plotting the function g. ■

4.8.2 Deflection Estimate

We start from (4.82). It can be rewritten, with the notations $c = b_1 - b_0$ and $d = (b_1 - b_0)/b$,

$$A = 2\Big(\frac{V}{b} + d\, q_b\Big) \Big\{ \frac{b - \int_0^1 b}{\int_0^1 b} \int_0^1 c - \Big(c - \int_0^1 c\Big) \Big\}.$$

This gives, using the stability estimate (4.100) and the property (4.87) of the norm in $L^2/I\!R$,

$$|A|_{L^2} \le \frac{2}{b_m}\Big(1 + \frac{1}{\sin \gamma_m}\Big)|V|_{L^2} \left\{ \frac{|b|_{L^2/I\!R}}{b_m}(b_M - b_m) + |c|_{L^2/I\!R} \right\}. \quad (4.102)$$

Let us denote by $\max \in J$ the index of the source for which $|I_j| = \eta_j^- + \eta_j^+$ is maximum, and by $c_{\max}$ and $d_{\max}$ the constant values of c and d on $I_{\max}$. Using (4.87) again gives, as $c - c_{\max}$ and $d - d_{\max}$ vanish over $I_{\max}$,

$$\begin{aligned} |c|_{L^2/I\!R} &\le |c - c_{\max}|_{L^2} \le (1 - |I_{\max}|)^{1/2}\, 2(b_M - b_m), \\ |d|_{L^2/I\!R} &\le |d - d_{\max}|_{L^2} \le (1 - |I_{\max}|)^{1/2}\, (b_M - b_m). \end{aligned}$$

Hence (4.102) becomes

$$|A|_{L^2} \leq \frac{2}{b_m}\Big(1+\frac{1}{\sin\gamma_m}\Big)|V|_{L^2}(1-|I_{\max}|)^{1/2}\,(b_M-b_m)\Big\{2+\frac{b_M-b_m}{b_m}\Big\},$$

$$|A|_{L^2} \leq 2\Big(\Big(\frac{b_M}{b_m}\Big)^2-1\Big)\Big(1+\frac{1}{\sin\gamma_m}\Big)(1-|I_{\max}|)^{1/2}\,|V|_{L^2}.$$

Thus we have proved the

Proposition 4.8.3 *Let hypothesis (4.85) and (4.99) hold. The deflection* Θ *for the identification of* $b \in L^2([0,1])$ *in the admissible parameter set D, defined in (4.90), from a measurement of* $u_\xi \in L^2$ *is*

$$\Theta = 2\Big(\Big(\frac{b_M}{b_m}\Big)^2-1\Big)\Big(1+\frac{1}{\sin\gamma_m}\Big)(1-|I_{max}|)^{1/2}. \tag{4.103}$$

When $|I_{\max}| \to 1$, that is, when the coefficients tend to become constant over Ω, we see that

– Last factor goes to zero

– $\sin\gamma$ increases – likely to one –

so that $\Theta \to 0$, which corresponds to the fact that the problem becomes "more linear."

4.8.3 Curvature Estimate

Equation (4.82) we have used for the deflection estimate is no more suited for the estimation of the curvature: we want to major $|A|_{L^2}$ by $|V|^2_{L^2}$, but the right-hand side of (4.82) contains the term $V\,(b_1-b_0)$, and there is no hope of majorating $|V\,(b_1-b_0)|_{L^2}$ by $|V|_{L^2}|b_1-b_0|_{L^2} \leq \alpha_m^{-1}|V|^2_{L^2}$!

So we start instead from (4.81). The stability estimate (4.100) gives

$$\begin{aligned}\Big|\int_0^1 (b_1-b_0)q_b\Big| &\leq |(b_1-b_0)q_b|_{L^1} \leq |(b_1-b_0)q_b|_{L^2} \leq \frac{b_M}{b_m\,\sin\gamma_m}|V|_{L^2},\\ \Big|\frac{b_1-b_0}{b}\Big|_{L^2} &\leq \frac{1}{b_m q_m\,\sin\gamma_m}|V|_{L^2},\end{aligned}$$

which, combined with (4.81), gives

$$|A|_{L^2} \leq \frac{2}{b_m q_m\,\sin^2\gamma_m}\Big(\frac{b_M}{b_m}\Big)^2\Big\{\frac{b_M}{b_m}+1\Big\}|V|^2_{L^2}.$$

Proposition 4.8.4 *(Same hypotheses as Proposition 4.8.3). The estimation of $b \in L^2([0,1])$ in the admissible parameter set D defined by (4.90)–(4.92) from a measurement of $u_\xi \in L^2$ is a FC problem. Its radius of curvature is given by*

$$R = \frac{1}{2} b_m q_m \sin^2 \gamma_m \frac{(b_m/b_M)^3}{1 + b_m/b_M} > 0. \tag{4.104}$$

The question of the behavior of the curvature of the problem when $|I_{\max}| \to 1$ is open. One could expect that it tends towards zero, but all attempts to prove it have failed....

4.8.4 Conclusion: OLS-Identifiability

We can now combine the above results with Theorem 4.4.1: the parameter b in (4.75) and (4.76) is OLS-Identifiable in $L^2([0,1]$, on the admissible set D defined in (4.90), from a measurement of u_ξ in $L^2([0,1]$, as soon as the deflection (4.103) satisfies $\Theta \le \pi/2$. The size of the wellposedness neighborhood of the attainable set is then R given by (4.104), and the linear stability constant α_m is given by (4.101).

4.9 Application to Example 4: 2D Parameter Estimation, with H^1 Observation

We discuss in this section the OLS-identifiability of the diffusion coefficient a in the two-dimensional elliptic problem described in Sect. 1.6 in the case of an H^1 observation, that is, when a measurement z of the gradient ∇u of its solution u is available. This material is adapted from the original paper [30]. The case of an L^2 observation of u will be considered in Chap. 5 using state-space regularization.

There are no distributed source terms inside Ω or on the Neumann boundary $\partial\Omega_N$ in this example. All sources or sinks are modeled by holes with boundaries $\partial\Omega_i, i = 1, \ldots, N$, equipped with a given injection or production rate condition (see (1.64)).

When the diffusion coefficient is only constrained to stay within a lower and an upper bound, as it was the case for the set C defined in (1.66), the coefficient is allowed to oscillate wildly, and the homogeneization theory [71] shows that the least squares function does not in general attain its minimum over C.

So we regularize the problem a first time by adding the information that we are only seeking

- *...smooth coefficients:* This will be achieved by requiring that the diffusion coefficient belongs to the space $C^{0,1}(\overline{\Omega})$ of functions which are bounded and uniformly Lipschitz continuous over $\overline{\Omega}$
- *...with limited oscillations:* this can be implemented by requiring that the Lipschitz constants of all coefficients a of C are uniformly bounded by $b_M \geq 0$
- *...and which take (unknown but) constant values* on the source or sink boundaries $\partial\Omega_i$: this is the 2D generalization of the hypothesis that a is constant on some neighborhood of each Dirac source term, which was required in the 1D case to ensure OLS-identifiability (see Sect. 4.8 above and [25]). It is also physically not too restrictive, as one can assume that the sizes of the $\partial\Omega_i$'s, which model the well boundaries, are small compared to the size of Ω and to the scale at which the coefficient a is expected to vary.

This leads us to define a vector space

$$\mathcal{E} = \{a \in C^{0,1}(\overline{\Omega}) \colon a|_{\partial\Omega_i} = \text{ unknown constant } a_i = i = 1, \cdots, N\}, \quad (4.105)$$

and to replace the admissible set C defined in (1.66) by the smaller one

$$\begin{aligned} C = \{ \, a \in \mathcal{E} \quad | \quad & a_m \leq a(\xi) \leq a_M \qquad \forall \xi \in \overline{\Omega}, \\ & |a(\xi_1) - a(\xi_0)| \leq b_M \|\xi_1 - \xi_0\| \quad \forall \xi_0, \xi_1 \in \overline{\Omega} \, \}, \end{aligned} \quad (4.106)$$

where a_m, a_M, and b_M are given numbers that satisfy

$$a_M \geq a_m > 0 \quad \text{and} \quad b_M \geq 0, \quad (4.107)$$

(this corresponds to the technique of **Regularization by size reduction of** C of Sect. 1.3.4).

To be able to infer the smoothness properties (4.109) below the solution u of (1.64) from the smoothness (4.105) and (4.106) of the coefficient a, we will also to suppose that the domain Ω itself is smooth, in the sense that its boundary $\partial\Omega$ is smooth, and that the subparts $\partial\Omega_D$, $\partial\Omega_N$, and $\partial\Omega_i$, $i = 1, \cdots, N$ do not intersect,

$$\begin{cases} \Omega \subset I\!R^2 \ \text{ has a } \ C^{1,1} \text{ boundary } \partial\Omega, \\ \overline{\partial\Omega_{\rm N}}, \overline{\partial\Omega_{\rm D}} \text{ and } \overline{\partial\Omega_i}, i = 1, \cdots, N \text{ are pairwise disjoint} \end{cases} \quad (4.108)$$

With these hypotheses, the regularity theorem in [76], page 180, implies that $\{|u_a|_{W^{2,p}} : a \in C\}$ is bounded for any $p > 2$. Since $W^{2,p}(\Omega)$ is continuously embedded into $C^1(\overline{\Omega})$ for every $p > 2$, there exists u_M and γ_M such that

$$|u_a|_{L^\infty(\Omega)} \leq u_M, |\nabla u_a|_{\mathbb{L}^\infty(\Omega)} \leq \gamma_M \text{ for all } a \in C. \tag{4.109}$$

The results in this section are derived under the assumption (1.65) that the Dirichlet condition $u = 0$ holds on a nonvoid part $\partial\Omega_{\rm D}$ of the boundary $\partial\Omega$, but all results can be extended to the case where meas $(\partial\Omega_D) = 0$, see [30].

We have now to choose the Banach parameter space E for which one hopes to prove the stability of the inverse problem. As we will end up in this section by further reducing the problem to finite dimension, all norms will be equivalent, and we shall choose simply for E the space that makes the $a \rightsquigarrow u_a$ mapping naturally regular:

$$E = \mathcal{C}^0(\overline{\Omega}) \quad \text{with the norm } \|v\|_{\mathcal{C}^0} = \sup_{x\in\overline{\Omega}} |u(x)|. \tag{4.110}$$

The admissible set C defined in (4.106) is clearly a closed and convex subset of E – but with a void interior!

To any $a \in C$, we associate the solution u_a of the state equation (1.64), which we write here for convenience in its variational form. We incorporate for this the boundary conditions of lower order in the state-space Y:

$$\begin{cases} Y = \{v \in H^1(\Omega) : v|_{\partial\Omega_{\rm D}} = 0,\ v|_{\partial\Omega_i} = v_i = \text{ const },\ i = 1, \cdots, N\} \\ \|v\|_Y = |\nabla v|_{\mathbb{L}^2}, \end{cases}$$

(compare to (1.67)), and define u_a as the solution of

$$\text{find } u \in Y \text{ such that } \int_\Omega a\nabla u \nabla v = \sum_{i=1}^N Q_i v_i \text{ for all } v \in Y, \tag{4.111}$$

where the production or injection rates $Q_i \in \mathbb{R}$, $i = 1, \cdots, N$, through the source or sink boundaries $\partial\Omega_i$ are given.

The observation operator corresponding to the H^1 observation considered in this section is (cf. (1.70))

$$\begin{cases} M\ :\ w \in Y \rightsquigarrow \nabla w\ \in F, \text{ where:} \\ F = \mathbb{L}^2(\Omega) \text{ is equipped with the norm } \|v\|_F = |v|_{\mathbb{L}^2(\Omega)}. \end{cases}$$

Hence we are concerned with the inversion of the mapping:

$$\varphi : a \in C \subset E = \mathcal{C}^0(\overline{\Omega}) \rightsquigarrow \nabla u_a \in F = \mathbb{L}^2(\Omega) \tag{4.112}$$

in the least-squares sense

$$\hat{a} \quad \text{minimizes } \tfrac{1}{2}|\nabla u_a - z|^2_{\mathbb{L}^2} \text{ over } C, \tag{4.113}$$

where $z \in \mathbb{L}^2(\Omega)$ is a given observation.

We analyze now the Q-wellposedness of this problem using the approach of Sect. 4.4 summarized in Sect. 4.6. We have to evaluate the velocity $V(t)$ and acceleration $A(t)$ along the curve $t \in [0,1] \rightsquigarrow \nabla u_{a(t)} \in \mathbb{L}^2(\Omega)$ associated to couples a_0, a_1 of parameters of C by (4.11) and (4.13). These curves stay by construction in the range of the mapping $a \rightsquigarrow \nabla u_a$. One finds

$$\|V(t)\|_F = |\nabla \eta(t)|_{\mathbb{L}^2}, \qquad \|A(t)\|_F = |\nabla \zeta(t)|_{\mathbb{L}^2},$$

where $\eta(t)$ and $\zeta(t)$ denote the first and second derivatives of $u_{a(t)}$ with respect to t. The equations for $\eta(t)$ and $\zeta(t)$ are simply obtained by derivating one and two times with respect to t the state equation (4.111) written at point $a = (1-t)a_0 + ta_1$:

$$\int_\Omega a\nabla\eta \cdot \nabla v = -\int_\Omega (a_1 - a_0)\nabla u_a \cdot \nabla v, \quad \text{for all } v \in Y, \tag{4.114}$$

$$\int_\Omega a\nabla\zeta \cdot \nabla v = -2\int_\Omega (a_1 - a_0)\nabla\eta \cdot \nabla v, \quad \text{for all } v \in Y. \tag{4.115}$$

We investigate first the *linear identifiability* of a (Definition 4.3.1).

Lemma 4.9.1 *For $a_0, a_1 \in C$ and $t \in [0,1]$, define $a = (1-t)a_0 + ta_1 \in C$, $h = a_1 - a_0 \in \mathcal{E}$, $u = u_a$, and $v = \frac{hu}{a}$. Then $v \in Y$ and*

$$\int_\Omega h\nabla u \cdot \nabla v = \frac{1}{2}\int_\Omega \frac{h^2}{a}|\nabla u|^2 + \sum_{i=1}^{N} \frac{h_i^2}{a_i^2}\, u_i Q_i.$$

Proof. Since $C \subset \mathcal{E} \subset C^{0,1}(\overline{\Omega})$, one has $v = \frac{hu}{a} \in H^1(\Omega)$. Moreover, v satisfies the boundary conditions defining Y and hence $v \in Y$. It follows that

$$\int_\Omega h\nabla u \cdot \nabla v = \int_\Omega \tfrac{h^2}{a}|\nabla u|^2 + \tfrac{1}{2}\int_\Omega \tfrac{u}{a}\nabla u \cdot \nabla h^2$$

$$-\int_\Omega \tfrac{h^2 u}{a^2}\nabla a \cdot \nabla u.$$

Integrating by parts the second term on the right hand side implies

$$\int_\Omega h\nabla u\cdot\nabla v = \frac{1}{2}\int_\Omega \frac{h^2}{a}|\nabla u|^2 - \frac{1}{2}\int_\Omega h^2\left(\frac{u}{a}\,\Delta u + \frac{u}{a^2}\nabla a\cdot\nabla u\right) + \frac{1}{2}\sum_{i=1}^{N} u_i Q_i\,\frac{h_i^2}{a_i^2},$$

which, utilizing $-a\Delta u - \nabla a\cdot\nabla u = 0$, gives the desired result. ■

Proposition 4.9.2 *Let notations and hypothesis (1.65) and (4.105) through (4.107) hold. Then a is* linearly identifiable *over C as soon as*

$$Q_i,\ i = 1,\cdots,N \text{ are not all zero,} \tag{4.116}$$

and

$$|\partial\Omega_i|,\ i = 1,\cdots,N, \text{ are sufficiently small.} \tag{4.117}$$

Proof. Let $a_0, a_1 \in C$ and $t \in [0,1]$ be such that the velocity η defined by (4.114) is zero. Linear identifiability will be proved if we show that $h = a_1 - a_0 = 0$. By construction, h satisfies

$$\int_\Omega h\,\nabla u_a\cdot\nabla v = 0 \quad\text{for all}\quad v\in Y,$$

where u_a is the solution of (4.111) for $a = (1-t)a_0 + ta_1$. Lemma 4.9.1 implies then

$$\frac{1}{2}\int_\Omega \frac{h^2}{a}|\nabla u_a|^2 + \sum_{i=1}^{N} \frac{h_i^2}{a_i^2}\,u_i Q_i = 0. \tag{4.118}$$

We argue first that the second term in 4.118 can be made positive using (4.118). Suppose that $\partial\Omega_i$ surrounds for each $i = 1,\cdots,N$, a fixed source/sink location x_i. If $|\partial\Omega_i| \to 0$, for all $i = 1,\cdots,N$, the solution u_a converges towards the weak solution associated to a right-hand side with Dirac source term $\sum_{i=1}^{N} Q_i\delta(x-x_i)$, which is singular at x_i. Hence $u_a|_{\partial\Omega_i} = u_{a,i} \to \infty$ if $Q_i > 0$ and $u_a|_{\partial\Omega_i} \to -\infty$ if $Q_i < 0$. Since C is compact in E by the Ascoli theorem, and $a \to u_{a,i}$ is continuous on E, we conclude that, when $|\partial\Omega_i|$ satisfies (4.117), the solution u_a satisfies

$$u_{a,i}\,Q_i \geq 0 \quad\text{for}\quad i = 1,\cdots,N, \quad\text{and all}\quad a\in C,$$

so that (4.118) implies $h\nabla u = 0$ a.e. on Ω.

We argue now that $\nabla u(a)$ cannot vanish on a set I of positive measure. Let γ denote a curve in Ω connecting the inner boundaries $\partial\Omega_i$ to $\partial\Omega_{\mathrm{D}} \cup \partial\Omega_{\mathrm{N}}$, such that $\Omega \setminus \gamma$ is simply connected and meas $\gamma = 0$. Then $I_\gamma = (\Omega \setminus \gamma) \cap I$ satisfies meas $I_\gamma > 0$. From [5], Theorem 2.1, and Remark, it follows that either u_a is constant on $\Omega \setminus \gamma$ and hence on Ω or u_a has only isolated critical points, that is, points z satisfying $\nabla u(z) = 0$. But u_a cannot equal a constant over Ω as this violates the boundary conditions at the wells $\partial\Omega_i$ at which $Q_i \neq 0$. On the other hand, the number of isolated critical points in I_γ can be at most countable, and hence meas $I_\gamma = 0$. Consequently, meas$\{x : \nabla u_a(x) = 0\} = 0$.

Hence $h = 0$ a.e. in Ω, and a is linearly identifiable over C, which ends the proof. ■

We turn now to the *deflection condition* (4.16):

Proposition 4.9.3 *Let notations and hypotheses (1.65) and (4.105) through (4.107) hold. Then the* deflection condition $\Theta \leq \pi/2$ *is satisfied for problem (4.113) as soon as*

$$a_M - a_m \leq \frac{\pi}{4}\, a_m. \tag{4.119}$$

Proof: Taking $v = \zeta$ in (4.115) gives, using (4.107), the Cauchy–Schwarz inequality and (4.112):

$$\begin{aligned} a_m \|\nabla\zeta\|_F &\leq 2\|a_1 - a_0\|_{C^0} \|\nabla\eta\|_F \\ &\leq 2(a_M - a_m)\|\nabla\eta\|_F, \end{aligned} \tag{4.120}$$

that is, using (4.119),

$$\|\nabla\zeta\|_F \leq \frac{\pi}{2}\|\nabla\eta\|_F.$$

This shows that (4.25) is satisfied, and Corollary 4.2.5 ensures that the deflection condition is satisfied. ■

The next step toward OLS-identifiability would be to prove a linear stability property (4.39) or (4.57) and a finite curvature condition (4.13) or (4.59), as this would imply OLS-identifiability using Theorem 4.4.1. However, there are some hints that already linear stability does not hold for the infinite dimensional set C defined in (4.106) (see Remark 5.4.1 in Sect. 5.4 below). Hence the problem needs once again to be regularized. We describe in Sect. 5.4 how to regularize this problem in a way that is specifically adapted to its nature. But we conclude the present section by the simplest regularization, which is to add the information that we search for the parameter in a *finite dimensional subspace* of E. So we define

$$\begin{cases} \boldsymbol{E} &= \text{finite dimensional subspace of } \mathcal{E}, \\ \boldsymbol{C} &= C \cap \boldsymbol{E}, \end{cases} \tag{4.121}$$

(one can, e.g., construct $\boldsymbol{E}$ using finite elements or splines). The result follows then immediately from Theorem 4.5.1:

Theorem 4.9.4 *Let notations and hypothesis (1.65) and (4.105) through (4.110) hold, as well as (4.116), (4.117), and (4.121). Then*

1. $\mathbf{a}$ *is* linearly stable *over* $\boldsymbol{C}$*:*

$$\alpha_m \|\mathbf{a}_0 - \mathbf{a}_1\|_{C^0(\overline{\Omega})} \leq |\nabla u_0 - \nabla u_1|_{\mathbb{L}^2(\Omega)} \quad \forall \mathbf{a}_0, \mathbf{a}_1 \in \boldsymbol{C},$$

where the constant α_m *is given by*

$$\alpha_m = \inf_{\mathbf{a}_0, \mathbf{a}_1 \in \boldsymbol{C}\ ,\ \mathbf{a}_0 \neq \mathbf{a}_1\ ,\ t \in [0,1]} \frac{|\nabla u|_{\mathbb{L}^2(\Omega)}}{\|\mathbf{a}_1 - \mathbf{a}_0\|_{C^0(\overline{\Omega})}} > 0$$

2. *The estimation of* $\mathbf{a} \in \boldsymbol{C}$ *from* $\nabla u \in \mathbb{L}^2(\Omega)$ *is a* finite curvature *problem, with a curvature*

$$\frac{1}{R} = \frac{2}{\alpha_m a_m} < +\infty \tag{4.122}$$

3. *If moreover the admissible set* C *satisfies* $a_M/a_m \leq 1 + \pi/4$ *(condition (4.119)), the diffusion coefficient* $\mathbf{a}$ *is* OLS-identifiable *in the finite dimensional subset* $\boldsymbol{C}$ *of* C *from a measurement* $z \in \vartheta$ *of* $\nabla u \in \mathbb{L}^2(\Omega)$*, where* ϑ *is the neighborhood of the attainable set defined by*

$$\vartheta = \{z \in \mathbb{L}^2(\Omega) \mid \inf_{\mathbf{a} \in \boldsymbol{C}} |z - \nabla u_a|_{\mathbb{L}^2(\Omega)} < R\},$$

and the Lipschitz constant of the $z \rightsquigarrow \hat{a}$ *mapping is given by (4.43) and (4.44) with* x *replaced by* a*.*

Proof. To apply Theorem 4.5.1, we define $\epsilon = a_m/2 > 0$, and, according to (4.47) and (4.48), C_η by

$$C_\eta = \{\, a \in \mathcal{E} \mid a_m - \epsilon < a(x) < a_M + \epsilon \quad \forall x \in \overline{\Omega}, \\ |a(x_1) - a(x_0)| < (b_M + \epsilon)\|x_1 - x_0\| \quad \forall x_0, x_1 \in \overline{\Omega}\,\}.$$

The FD-minimum set of hypotheses (4.46) is then verified, the set C – and hence $\boldsymbol{C}$ – is obviously bounded, and Proposition 4.9.2 applied to C_η instead of C shows that a is linearly identifiable over C_η – and hence over $\boldsymbol{C}_\eta$. Then points 1 and 2 of Theorem 4.9.4 follow immediately from points 1 and 2 of Theorem 4.5.1, and point 3 follows from Proposition 4.9.3 and point 3 of Theorem 4.5.1. ■

Remark 4.9.5 *When* $\mathbf{E}$ *is one member of a family of embedded subspace filling out* E*, the stability constant* α_m *and , following (4.122), the radius of curvature* R *decrease – most likely to zero – when the dimension of* $\mathbf{E}$ *goes to infinity, that is, when the discretization of the diffusion coefficient is refined. However, the size condition (4.119) which ensures that the estimation problem is Q-wellposed remains unchanged.*

To figure out the shape of the attainable set for the diffusion problem, one can take a look at the attainable sets $\varphi(\boldsymbol{C}^0)$ *and* $\varphi(\boldsymbol{C}^1)$ *at scale* 0 *and* 1 *for the prototype "nicely nonlinear" example of Fig. 3.3, Sect. 3.6: they both have the same size, are obviously s.q.c., with the second one having a stronger curvature and hence a much smaller neighborhood* ϑ *on which the projection is Q-wellposed.*

This leads us to conjecture that the $a \rightsquigarrow \nabla u$ *map under study in the present section is also a "nicely nonlinear" problem in the sense of Definition 3.6.1: directions of coarse perturbations should correspond to large sensibilities (i.e., large velocities* $\|V\|$*) and small nonlinearities (i.e., small accelerations* $\|A\|$*), whereas directions of fine perturbations should correspond to small sensibilities and large nonlinearities. This "nice nonlinearity" property has been shown to hold in [57] for the estimation of a one-dimensional diffusion parameter, but further research is needed for the present two dimensional case.*

Another hint in favor of this conjecture is the recognized ability of multiscale parameterizations to overcome stationary points while searching for diffusion coefficients, which is shared with "nicely nonlinear" problem (see paragraph 4 of Sect. 3.6.3 in Chap. 3).

Hence Fig. 3.2 (bottom) is a good mental representation of the attainable set for the diffusion inverse problem . ■

Chapter 5

Regularization of Nonlinear Least Squares Problems

We consider in this chapter various approaches for the regularization of the general NLS problem (1.10), recalled here for convenience:

$$\hat{x} \quad \text{minimizes} \quad J(x) = \frac{1}{2}\|\varphi(x) - z\|_F^2 \quad \text{over} \quad C. \tag{5.1}$$

and we suppose throughout the chapter that it satisfies the minimum set of hypothesis (1.12) or (4.2).

We develop three of the five approaches described in Sect. 1.3.4 of the introduction: Levenberg–Marquardt–Tychonov (LMT), state-space, and adapted regularization. The two remaining approaches, regularization by parameterization and regularization by size reduction of the admissible parameter set, have been already addressed in Chaps. 3 and 4, respectively.

5.1 Levenberg–Marquardt–Tychonov (LMT) Regularization

The LMT regularization of problem (5.1) is, as announced in (1.25),

$$\hat{x}_\epsilon \text{ minimizes } J_\epsilon(x) = J(x) + \frac{\epsilon^2}{2}\|x - x_0\|_E^2 \text{ over } C, \tag{5.2}$$

where $\epsilon > 0$ denotes the regularization parameter and x_0 represents an a-priori estimate to a solution of (5.1). The paper by Levenberg [54] goes

G. Chavent, *Nonlinear Least Squares for Inverse Problems: Theoretical Foundations and Step-by-Step Guide for Applications*, Scientific Computation,
DOI 10.1007/978-90-481-2785-6_5,

back to the forties, and that of Marquardt [62] to the sixties. The approach was popularized in the seventies by Tikhonov and the Russian school [75, 63].

In practice, the available data are always corrupted by noise. So we retain the letter z to denote the noise free data, and suppose that a sequence z_n of noisy measurement of increasing quality is available:

$$|z_n - z| \leq \delta_n, \qquad \text{with } \delta_n \to 0. \tag{5.3}$$

One associates to these data a sequence of regularization parameters:

$$\epsilon_n > 0, \qquad \text{with } \epsilon_n \to 0, \tag{5.4}$$

and the sequence of regularized problems:

$$\hat{x}_n \text{ minimizes } J_n(x) = \frac{1}{2}\|\varphi(x) - z_n\|_F^2 + \frac{\epsilon_n^2}{2}\|x - x_0\|_E^2 \text{ over } C. \tag{5.5}$$

- For the class of *linear problems*, where $\varphi(x) = Ax$ with $A \in \mathcal{L}(E, F)$, the theory is rather well developed, see, for example, the monographs by Baumeister, Groetsch, Louis, and Morozov [8, 43, 59, 63], and the papers [65, 66, 27]. The main results are the following:

    1. The regularized problem (5.2) is Q-wellposed as soon as $\epsilon > 0$, hence problems (5.5) are Q-wellposed for all n
    2. When the original problem (5.1) admits a solution, $\hat{x}_n$ converges to the x_0-minimum-norm solution $\hat{x}$ when $\epsilon \to 0$, provided the regularization parameter goes to zero slower than the noise on the data
    3. When the x_0-minimum-norm solution $\hat{x}$ is regular, convergence rates for $\hat{x}_n - \hat{x}$ and of $A\hat{x}_n - A\hat{x}$ are available, provided the regularization parameter goes to zero as the square root of the noise on the data

    We give in Sect. 5.1.1 a direct hard analysis proof of these results, which will serve as guideline for the nonlinear case.

- The results of the linear case are generalized completely in Sect. 5.1.2 to the class of *finite curvature/limited deflection (FC/LD)* problems, provided that the true data z is close enough to the attainable set $\varphi(C)$.

This class contains NLS problems that satisfy both the finite curvature property of Definition 4.2.1 and the deflection condition (4.25) of Corollary 4.2.5. We follow [28] for the proof, but use sharper estimations that allow to obtain the convergence results even for unattainable data and without the need for the a-priori guess x_0 to be close to the minimal norm solution $\hat{x}$.

- For *general nonlinear problems,* however, where the parameter $\rightsquigarrow$ output mapping φ exhibits no interesting mathematical properties except regularity, $\epsilon > 0$ does not necessarily ensure wellposedness of the regularized problem (5.2) any more, in particular for the small values of ϵ. We give first in Sect. 5.1.3 an estimation of a minimum value $\epsilon_{\min} > 0$ of ϵ, which restores Q-wellposedness of (5.2) [18]. Then we study the convergence of a sequence of (non-necessarily unique) minimizers $\hat{x}_n$ of (5.5) when the data z is attainable and the a-priori guess x_0 is close enough to a minimum-norm solution $\hat{x}$ of (5.1) [27]. Convergence results for unconstrained nonlinear problems are also available in [37, 67, 45].

5.1.1 Linear Problems

We follow in this section the presentation of [27]. We consider here the case where

$$\varphi(x) = Ax, \ A \in \mathcal{L}(E, F), \qquad C \subset F \text{ closed, convex} \quad \text{and} \quad z \in F, \tag{5.6}$$

so that the minimum set of hypothesis (1.12) is satisfied. We define

$$\hat{z} = \text{ projection of } z \text{ on } \overline{A(C)}. \tag{5.7}$$

The x_0-Minimum-Norm Solution

When the unregularized problem (5.1) admits solution(s), one has

$$\hat{z} \in A(C) \qquad \text{(always true when } A(C) \text{ is closed !)}, \tag{5.8}$$

and the *solution set* of problem (5.1) is then

$$X = \{x \in E \ : \ Ax = \hat{z}\} \cap C \qquad \text{closed, convex .} \tag{5.9}$$

The choice of an a-priori guess

$$x_0 \in E \tag{5.10}$$

implies the selection of a specific solution $\hat{x}$ in X. It is the element closest to x_0, and is found by solving

$$\hat{x} \in X \quad \text{minimizes} \quad \frac{1}{2}\|x - x_0\|_E^2 \quad \text{over} \quad X. \tag{5.11}$$

Clearly (5.11) has a unique solution, which will be referred to as the x_0-*minimum-norm solution* of (5.1).

To exhibit some of its properties, we shall utilize the following notions. For any convex set $K \subset E$ and $x \in K$, the *tangent cone* and *negative polar cone* to K at x are defined by

$$T(K,x) = \{y \in E : \exists x_n \in K, \lambda_n > 0 \text{ with } \lambda_n(x_n - x) \to y\}, \tag{5.12}$$
$$T(K,x)^- = \{y \in E \ : \ \langle y, \bar{y}\rangle \le 0 \text{ for all } \bar{y} \in T(K,x)\}. \tag{5.13}$$

$T(K,x)$ and $T(K,x)^-$ are closed convex cones. The tangent cone to X satisfies

Lemma 5.1.1

$$\forall x \in X, \ T(X,x) \subset \operatorname{Ker} A \cap T(C,x).$$

Proof. Let $y \in T(X,x)$, and $\{x_n\}$ in X, $\{\lambda_n\}$ with $\lambda_n > 0$ be sequences such that $y = \lim \lambda_n(x_n - x)$. It follows that $y \in \operatorname{Ker} A$ and, since $x_n \in X \subset C$, we also have that $y \in T(C,x)$. ■

Unluckily, the converse inclusion is not true in general: for example, if C is a closed ball and X is reduced to one single point $\{x\}$, then $T(X,x) = \{0\}$ and $\operatorname{Ker} A \cap T(C,x) = \operatorname{Ker} A$, which do not coincide as soon as $\operatorname{Ker} A$ is not trivial. So we make the following definition:

Definition 5.1.2 *An element $x \in X$ is said to be* qualified *if*

$$T(X,x) = \operatorname{Ker} A \cap T(C,x). \tag{5.14}$$

Lemma 5.1.3 *Let (5.1) admit a solution and x be* identifiable *over C (Definition 1.3.1). Then the solution set X contains one single element $\hat{x}$, and $\hat{x}$* is qualified.

Proof. The identifiability property implies (1) that $X = \{\hat{x}\}$, so that $T(X,\hat{x}) = \{0\}$, and (2) that $\operatorname{Ker} A = \{0\}$ and (5.14) is proved. ■

For the case of linear constraints, one has the following result:

Lemma 5.1.4 *Let C be defined by a finite number N_C of linear constraints:*

$$C = \{x \in E \ : \ M_i x \leq b_i \ , \ i \in I \stackrel{\text{def}}{=} \{1 \dots N_C\}\},$$

where M_i are bounded linear functionals on E and $b_i \in \mathbb{R}$. Then all points x of X are qualified.

Proof. Step 1: for any $\tilde{x} \in C$, we prove that

$$\begin{aligned} K &= \{y \in E \ : \ \exists x \in C, \lambda > 0 \text{ such that } y = \lambda(x - \tilde{x}\} \\ &= \cup_{\lambda > 0} \lambda(C - \tilde{x}) \end{aligned}$$

coincides with $T(C, \tilde{x})$. By definition of $T(C, \tilde{x})$, one has $T(C, \tilde{x}) = \overline{K}$, and hence it suffices to show that K is closed. Let $y_n \in K$ and $y \in E$ be such that $y_n \to y$. As $y_n \in K$, there exist $\lambda_n > 0$ and $x_n \in C$ such that $y_n = \lambda_n(x_n - \tilde{x})$. Let us denote by $I(\tilde{x})$ the set of active indices at $\tilde{x}$: $I(\tilde{x}) = \{i \in I \ : \ M_i \tilde{x} = b_i\}$. Then we find $M_i y_n = \lambda_n(M_i x_n - M_i \tilde{x}) = \lambda_n(M_i x_n - b_i) \leq 0$ for all $i \in I(\tilde{x})$ and $n = 1, 2, \dots$. Hence $M_i y \leq 0$ for all $i \in I(\tilde{x})$. Next we choose $\lambda > 0$ small enough so that $M_i \tilde{x} + \lambda M_i y \leq b_i$ for all $i \notin I(\tilde{x})$. It is simple to check that $\tilde{x} + \lambda y \in C$ and hence $y \in K$ and K is closed.

Step 2: we prove that $\operatorname{Ker} A \cap K \subset T(X, \tilde{x})$, for any $\tilde{x} \in X$. Let $y \in \operatorname{Ker} A \cap K$ be given. Then $y = \lambda(x - \tilde{x})$, where $\lambda > 0$ and $x \in C$, and $A y = 0$. Hence $A x = A \tilde{x}$, so that $x \in X$ and $y = \lambda(x - \tilde{x}) \in T(X, \tilde{x})$. ■

The following property will be useful in the convergence study of the regularized solutions x_n to the x_0-minimum-norm solution $\hat{x}$:

Lemma 5.1.5 *Hypothesis and notations (5.6) through (5.11).*

1. *If the x_0-minimum-norm $\hat{x}$ solution of (5.1) is qualified, then*

$$x_0 - \hat{x} \in \overline{Rg\, A^* + T(C, \hat{x})^-}. \tag{5.15}$$

2. *Conversely, if (5.15) holds, then $\hat{x}$ is the x_0-minimum-norm solution.*

Remark that in the case of an unconstrained problem, where $C = E = T(C, \hat{x})$, one has $T(C, \hat{x})^- = \{0\}$, and property (5.15) reduces to

$$x_0 - \hat{x} \in \overline{Rg\, A^*} = \operatorname{Ker} A^\perp,$$

which expresses the fact that $x_0 - \hat{x}$ is orthogonal to $\operatorname{Ker} A$.

Proof. The Euler condition applied to problem (5.11) shows that the x_0-minimum-norm solution $\hat{x}$ satisfies

$$\langle x_0 - \hat{x}, x - \hat{x} \rangle \leq 0 \quad \forall x \in X,$$

and, using the definition (5.12) of tangent cones

$$\langle x_0 - \hat{x}, y \rangle \leq 0 \quad \forall y \in T(X, \hat{x}),$$

or in terms of the negative polar cone

$$x_0 - \hat{x} \in T(X, \hat{x})^-. \tag{5.16}$$

The qualification hypothesis for $\hat{x}$ implies that $T(X, \hat{x})$ is given by (5.14), and using a property of polar cones

$$T(X, \hat{x})^- = \overline{\operatorname{Ker} A^- + T(C, \hat{x})^-}.$$

But $\operatorname{Ker} A^- = \overline{Rg\, A^*}$, and (5.16) becomes

$$x_0 - \hat{x} \in \overline{\overline{Rg\, A^*} + T(C, \hat{x})^-}.$$

Let $\eta > 0$ be given. We can first find $x_1 \in T(C, \hat{x})^-$ and $x_2 \in \overline{Rg\, A^*}$ such that $|x_0 - \hat{x} - x_1 - x_2| \leq \eta/2$. Then we can find $x_3 \in Rg\, A^*$ such that $|x_3 - x_2| \leq \eta/2$. Hence we obtain

$$|x_0 - \hat{x} - x_1 - x_3| \leq \eta,$$

which proves (5.15).

The second part of the lemma is proved by reversing the order of the previous calculations, using lemma 5.1.1 instead of the qualification hypothesis. ■

The Regularity Condition

To obtain convergence rates for the regularized problems (5.5), we shall require the

Definition 5.1.6 *The x_0-minimum-norm solution $\hat{x}$ satisfies the* regularity condition *if*

$$x_0 - \hat{x} \in Rg\, A^* + T(C, \hat{x})^-. \tag{5.17}$$

The name given to condition (5.17) will be explained after Definition 5.1.14.

The regularity condition is equivalent to the existence of a Lagrange multiplier for the optimization problem (5.11), which can be rewritten as a constrained optimization problem:

$$\hat{x} \in C \ \text{ minimizes } \frac{1}{2}\|x - x_0\|_E^2 \ \text{ over } C \ \text{ under the constraint } Ax = \hat{z}.$$

The Lagrangian for this problem is

$$\mathcal{L}(x, w) = \frac{1}{2}\|x - x_0\|_E^2 + \langle w, A\,x - \hat{z}\rangle_F,$$

and $\hat{w}$ is a Lagrange multiplier at $\hat{x}$ if

$$\mathcal{L}(\hat{x}, \hat{w}) \leq \mathcal{L}(x, \hat{w}) \quad \text{for all } x \in C$$

or equivalently, as $\mathcal{L}$ is convex in x

$$\frac{\partial \mathcal{L}}{\partial x}(\hat{x}, \hat{w})(x - \hat{x}) \geq 0 \quad \text{for all } x \in C,$$

$$\langle \hat{x} - x_0, x - \hat{x}\rangle + \langle \hat{w}, A\,(x - \hat{x})\rangle \geq 0 \quad \text{for all } x \in C. \tag{5.18}$$

If we define $\mu \in E$ by

$$x_0 - \hat{x} = A^*\,\hat{w} + \mu,$$

we can rewrite (5.18) as

$$\langle \mu, x - \hat{x}\rangle \leq 0 \quad \text{for all } x \in C.$$

Hence $\mu \in T(C, \hat{x})^-$, and $\hat{x}$ satisfies the regularity condition (5.17), and the equivalence is proved.

Remark 5.1.7 *An alternative formulation of the regularity condition (5.17) is*

$$\hat{x} \in P_C\big(Rg\,A^* + \{x_0\}\big),$$

where P_C denote the metric projection in E onto C. This formulation is used in [65, 67] to obtain the convergence results of part 3 of Theorem 5.1.8 below. ■

Convergence of Regularized Problems

Theorem 5.1.8 *Hypothesis and notations (5.3), (5.4), (5.6), (5.7), (5.9), and (5.11).*

1. *Without further hypothesis on z, the unregularized problem (5.1) has generally no solution, but all regularized problems (5.5) have a unique solution $\hat{x}_n$, which satisfies, when $\epsilon_n \to 0$ and $\delta_n \to 0$,*

$$\epsilon_n \hat{x}_n \to 0. \tag{5.19}$$

$$A\hat{x}_n \to \hat{z}. \tag{5.20}$$

2. *Let the data z satisfy (5.8), which means that (5.1) has solution(s). Then if the x_0-minimum-norm solution $\hat{x}$ satisfies*

$$x_0 - \hat{x} \in \overline{Rg\, A^* + T(C,\hat{x})^-}, \tag{5.21}$$

(e.g., if $\hat{x}$ is qualified), one has, when $\epsilon_n \to 0$ and $\delta_n/\epsilon_n \to 0$,

$$x_n - \hat{x} \to 0, \tag{5.22}$$

$$\|A\, x_n - \hat{z}\|_F = O(\epsilon_n). \tag{5.23}$$

3. *If moreover $\hat{x}$ satisfies the regularity condition*

$$x_0 - \hat{x} \in Rg\, A^* + T(C,\hat{x})^-, \tag{5.24}$$

one has, when $\epsilon_n \to 0$ and $\delta_n \sim \epsilon_n^2$,

$$\|x_n - \hat{x}\|_E = O(\epsilon_n) = O(\delta_n^{\frac{1}{2}}), \tag{5.25}$$

$$\|A\, x_n - \hat{z}\|_F = O(\epsilon_n^2) = O(\delta_n). \tag{5.26}$$

Proof.

Part 1 Let $\eta > 0$ be given. Using (5.7), we can find $\tilde{x} \in C$ such that

$$\|A\tilde{x} - \hat{z}\|_F \le \eta/4. \tag{5.27}$$

Then by definition of $\hat{x}_n$ we have

$$\|A\,\hat{x}_n - z_n\|_F^2 + \epsilon_n^2 \|\hat{x}_n - x_0\|_E^2 \le \|A\,\tilde{x} - z_n\|_F^2 + \epsilon_n^2 \|\tilde{x} - x_0\|_E^2.$$

To estimate $\|A\,\hat{x}_n - A\,\tilde{x}\|$ and $\|\hat{x}_n - \tilde{x}\|$, we rewrite the above inequality as

$$\begin{aligned}\|A(\hat{x}_n - \tilde{x})\|_F^2 + \epsilon_n^2\|\hat{x}_n - \tilde{x}\|_E^2 \leq\ & \|A(\hat{x}_n - \tilde{x})\|_F^2 + \epsilon_n^2\|\hat{x}_n - \tilde{x}\|_E^2 \\ & +\|A\,\tilde{x} - z_n\|_F^2 + \epsilon_n^2\|\tilde{x} - x_0\|_E^2 \\ & -\|A\,\hat{x}_n - z_n\|_F^2 - \epsilon_n^2\|\hat{x}_n - x_0\|_E^2.\end{aligned} \tag{5.28}$$

Using identity $a^2 + b^2 - (a+b)^2 = -2ab$, we obtain

$$\begin{aligned}\|A(\hat{x}_n - \tilde{x})\|_F^2 + \epsilon_n^2\|\hat{x}_n - \tilde{x}\|_E^2 \leq\ & 2\langle A(\hat{x}_n - \tilde{x}), z_n - A\,\tilde{x}\rangle \\ & +2\epsilon_n^2\langle \hat{x}_n - \tilde{x}, x_0 - \tilde{x}\rangle.\end{aligned} \tag{5.29}$$

But $z - \hat{z}$ is orthogonal to $A(\hat{x}_n - \tilde{x}) \in Rg\,A$, and so (5.29) can be rewritten as

$$\begin{aligned}\|A(\hat{x}_n - \tilde{x})\|_F^2 + \epsilon_n^2\|\hat{x}_n - \tilde{x}\|_E^2 \leq\ & 2\langle A(\hat{x}_n - \tilde{x}), z_n - z + \hat{z} - A\,\tilde{x}\rangle \\ & +2\epsilon_n^2\langle \hat{x}_n - \tilde{x}, x_0 - \tilde{x}\rangle,\end{aligned} \tag{5.30}$$

and, using (5.3) and (5.27) and the Cauchy–Schwarz inequality

$$\begin{aligned}\|A(\hat{x}_n - \tilde{x})\|_F^2 + \epsilon_n^2\|\hat{x}_n - \tilde{x}\|_E^2 \leq\ & 2\|A(\hat{x}_n - \tilde{x})\|_F\big(\delta_n + \frac{\eta}{4}\big) \\ & +2\epsilon_n^2\|\hat{x}_n - \tilde{x}\|_E\,\|x_0 - \tilde{x}\|_E,\end{aligned} \tag{5.31}$$

which is of the form, with obvious notations,

$$a^2 + b^2 \leq 2a\alpha + 2b\beta.$$

Hence,

$$(a-\alpha)^2 + (b-\beta)^2 \leq \alpha^2 + \beta^2 \leq (\alpha+\beta)^2$$

and finally

$$\left\{\begin{array}{lcrcr} a & \leq & 2\alpha & + & \beta, \\ b & \leq & \alpha & + & 2\beta. \end{array}\right. \tag{5.32}$$

Hence we deduce from (5.31) that

$$\|A(\hat{x}_n - \tilde{x})\|_F \ \leq\ 2\Big(\delta_n + \frac{\eta}{4}\Big) + \epsilon_n\|x_0 - \tilde{x}\|_E, \tag{5.33}$$

$$\epsilon_n\|\hat{x}_n - \tilde{x}\|_E \ \leq\ \delta_n + \frac{\eta}{4} + 2\epsilon_n\|x_0 - \tilde{x}\|_E. \tag{5.34}$$

From the first inequality we obtain, using (5.27),

$$\|A\hat{x}_n - \hat{z}\|_F \leq 2\delta_n + \frac{3\eta}{4} + \epsilon_n\|x_0 - \tilde{x}\|_E,$$

so that $\|A\hat{x}_n - \hat{z}\|_F \leq \eta$ for n large enough, and (5.20) is proved. Then we deduce from (5.34) that

$$\epsilon_n\|\hat{x}_n\|_E \leq \epsilon_n\|\tilde{x}\|_E + \delta_n + \frac{\eta}{4} + 2\epsilon_n\|x_0 - \tilde{x}\|_E,$$

so that $\epsilon_n\|\hat{x}_n\|_E \leq \eta$ for n large enough, and (5.19) is proved.

Part 2 From now on we suppose that (5.1) has a solution, and hence an x_0-minimum-norm solution $\hat{x}$. Hence we can choose $\eta = 0$ and $\tilde{x} = \hat{x}$ in (5.27), so that (5.30), (5.33), and (5.34) become

$$\|A\,\hat{x}_n - \hat{z}\|_F^2 + \epsilon_n^2\|\hat{x}_n - \hat{x}\|_E^2 \leq 2\langle A\,\hat{x}_n - \hat{z}, z_n - z\rangle + 2\epsilon_n^2\langle \hat{x}_n - \hat{x}, x_0 - \hat{x}\rangle \tag{5.35}$$

and

$$\|A\,\hat{x}_n - \hat{z}\|_F \leq 2\delta_n + \epsilon_n\|x_0 - \hat{x}\|_E, \tag{5.36}$$
$$\epsilon_n\|\hat{x}_n - \hat{x}\|_E \leq \delta_n + 2\epsilon_n\|x_0 - \hat{x}\|_E, \tag{5.37}$$

where (5.36) proves (5.23) as now $\delta_n/\epsilon_n \to 0$. But (5.37) gives no information on the convergence of $\hat{x}_n$ to $\hat{x}$, it is necessary for that to use hypothesis (5.21) on $\hat{x}$: let again $\eta > 0$ be given, and $w \in F, \mu \in T(C,\hat{x})^-$ be such that

$$\|x_0 - \hat{x} - (A^*\,w + \mu)\|_F \leq \eta/3.$$

Then we can rewrite (5.35) as

$$\begin{aligned}\|A\,\hat{x}_n - \hat{z}\|_F^2 + \epsilon_n^2\|\hat{x}_n - \hat{x}\|_E^2 \leq\ & 2\langle A\,\hat{x}_n - \hat{z}, z_n - z\rangle \\ & +2\epsilon_n^2\langle \hat{x}_n - \hat{x}, x_0 - \hat{x} - (A^*\,w + \mu)\rangle \\ & +2\epsilon_n^2\langle \hat{x}_n - \hat{x}, A^*\,w\rangle \\ & +2\epsilon_n^2\langle \hat{x}_n - \hat{x}, \mu\rangle,\end{aligned}$$

and, transposing A^* in the right-hand side and using the fact that $\langle \hat{x}_n - \hat{x}, \mu\rangle \leq 0$

$$\begin{aligned}\|A\,\hat{x}_n - \hat{z}\|_F^2 + \epsilon_n^2\|\hat{x}_n - \hat{x}\|_E^2 \leq\ & 2\langle A\,\hat{x}_n - \hat{z}, z_n - z + \epsilon_n^2 w\rangle \\ & +2\epsilon_n^2\langle \hat{x}_n - \hat{x}, x_0 - \hat{x} - (A^*\,w + \mu)\rangle.\end{aligned}$$

The Cauchy–Schwarz inequality and the formula (5.32) give now

$$\begin{aligned} \|A\hat{x}_n - \hat{z}\|_F &\leq 2(\delta_n + \epsilon_n^2\|w\|_F), \qquad (5.38) \\ \epsilon_n\|\hat{x}_n - \hat{x}\|_E &\leq \delta_n + \epsilon_n^2\|w\|_F + \frac{2\eta}{3}\epsilon_n. \qquad (5.39) \end{aligned}$$

When $\delta_n/\epsilon_n \to 0$, the last inequality shows that $\|\hat{x}_n - \hat{x}\|_E \leq \eta$ for n large enough, which proves (5.22).

Part 3 Now $\hat{x}$ satisfies the regularity condition (5.24), and so we can choose $\eta = 0$ in **Part 2**, and estimations (5.38), and (5.39) simplify to

$$\begin{aligned} \|A\hat{x}_n - \hat{z}\|_F &\leq 2(\delta_n + \epsilon_n^2\|w\|_F), \qquad (5.40) \\ \|\hat{x}_n - \hat{x}\|_E &\leq \delta_n/\epsilon_n + \epsilon_n\|w\|_F, \qquad (5.41) \end{aligned}$$

which prove (5.25) and (5.26). ■

5.1.2 Finite Curvature/Limited Deflection (FC/LD) Problems

We consider here the application of LMT-regularization to the class of FC/LD least squares problems introduced in Definition 4.2.2, which we summarize here:

Definition 5.1.9 *The NLS problem (5.1) is a FC/LD problem if it satisfies, beside the minimum set of hypothesis (4.2), the* finite curvature condition *of Definition 4.2.1:*

$$\begin{cases} \text{there exists } R > 0 \text{ such that} \\ \forall x_0, x_1 \in C, \text{ the curve } P : t \rightsquigarrow \varphi((1-x_0)t + tx_1) \text{ satisfies} \\ P \in W^{2,\infty}([0,1]; F) \text{ and } \|A(t)\|_F \leq \dfrac{1}{R}\|V(t)\|_F^2 \text{ for a.e. } t \in [0,1], \\ \text{where } V(t) = P'(t),\ A(t) = P''(t), \end{cases} \qquad (5.42)$$

and the deflection condition *of Corollary 4.2.5*

$$\|A(t)\|_F \leq \frac{\pi}{2}\|V(t)\|_F \text{ for a.e. } t \in [0,1] \text{ and all } x_0, x_1 \in C. \qquad (5.43)$$

A set of geometric attributes (Definition 4.2.3) of problem (5.1) is then made of the *radius of curvature* R given by (5.42), the *deflection* $\Theta = \pi/2$ given by (5.43), and the (arc length) size $L = \alpha_M \text{diam}\, C$.

Combining (5.42) and (5.43) shows that condition (5.43) is met as soon as $L \leq \frac{\pi}{2} R$. Hence, for a FC problem, the deflection condition (5.43) can always be satisfied by diminishing the size of the admissible set C.

FC/LD problems have two nice properties:

- The forward map φ is not necessarily injective, but the preimages $\varphi^{-1}(X)$ of any $X \in \varphi(C)$ are always closed and convex (Proposition 4.2.8), despite the fact that φ itself is not linear. This will simplify the handling of the x_0-minimum-norm solution.

- The attainable set $\varphi(C)$ is not necessarily convex nor closed, but Proposition 4.2.7 shows that the projection of z onto $\varphi(C)$ is unique and stable (when it exists) as soon as z belongs to the neighborhood:

$$\vartheta = \Big\{ z \in F \mid d(z, \varphi(C)) < R \Big\}. \tag{5.44}$$

  This property will allow to handle the case of nonattainable data, provided the error level is smaller than R.

Let

$$\hat{z} = \text{ projection of } z \text{ on } \overline{\varphi(C)}. \tag{5.45}$$

The unregularized problem (5.1) admits solution(s) as soon as

$$\hat{z} \in \varphi(C) \qquad \text{(always true when } \varphi(C) \text{ is closed !)}, \tag{5.46}$$

and the *solution set* of problem (5.1) is then

$$X = \varphi(\hat{z})^{-1} \cap C,$$

which is *closed and convex* as shown in Proposition 4.2.8.

The x_0-Minimum-Norm Solution

Given an a-priori guess $x_0 \in E$, the x_0-*minimum-norm solution* $\hat{x}$ of (5.1) is the element of the solution set X closest to x_0, that is, the unique solution of the minimization problem (5.11). We shall suppose throughout this section that

$$\varphi \text{ admits a Gâteaux derivative } \varphi'(\hat{x}) \text{ at } \hat{x}. \tag{5.47}$$

and we follow the presentation of Sect. 5.1.1 on linear problems, with the necessary adaptations.

The situation concerning the tangent cone to X at the x_0-minimum-norm solution $\hat{x}$ is similar to that of the linear case:

Lemma 5.1.10

$$T(X,\hat{x}) \subset \operatorname{Ker} \varphi'(\hat{x}) \cap T(C,\hat{x}). \tag{5.48}$$

Proof. Let $y \in T(X,\hat{x})$, and $\{x_n\}$ in X, $\{\lambda_n\}$ with $\lambda_n > 0$ be such that $y = \lim \lambda_n(x_n - \hat{x})$. For n given, the point $x_n(t) = (1-t)\hat{x} + tx_n$ belongs to $X \subset \varphi(\hat{z})^{-1}$ for $0 \leq t \leq 1$ because of the convexity of X. The existence of a Gâteaux derivative of φ at $\hat{x}$ gives, for each n,

$$\lim_{t\to 0} \underbrace{\left(\varphi(x_n(t)) - \varphi(\hat{x})\right)}_{=0}/t = \varphi'(\hat{x})(x_n - \hat{x}) = 0,$$

and hence

$$0 = \lim_{n\to\infty} \lambda_n \varphi'(\hat{x})(x_n - \hat{x}) = \varphi'(\hat{x})y.$$

It follows that $y \in \operatorname{Ker} \varphi'(\hat{x})$ and, since $x_n \in X \subset C$, we have also $y \in T(C,\hat{x})$. ∎

We make the following definition:

Definition 5.1.11 *An element $x \in X$ is said to be* qualified *if*

$$T(X,x) = \operatorname{Ker} \varphi'(x) \cap T(C,x). \tag{5.49}$$

Lemma 5.1.3 generalizes to the nonlinear case as follows:

Lemma 5.1.12 *Let (5.1) admit a solution, φ be defined over a convex open neighborhood C_η of C (see Sect. 4.5 for an example), and x be* linearly identifiable *over C_η (Definition 4.3.1). Then the solution set X contains one single element $\hat{x}$ and $\hat{x}$* is qualified.

Proof. Let $\hat{x} \in X$ denote the minimum-norm solution, and x denote an arbitrary element of X. Then the curve $P : t \rightsquigarrow (1-t)\hat{x} + tx$ satisfies $P(t) = \hat{z}\ \forall t \in [0,1]$, so that $V(t) = P'(t) = 0\ \forall t \in [0,1]$, and $x = \hat{x}$ using the linear identifiability property. Hence $X = \{\hat{x}\}$, and $T(X,\hat{x}) = \{0\}$.

Let now $y \in \operatorname{Ker} \varphi'(\hat{x})$ be given. One can always suppose y is small enough so that $y = x - \hat{x}$ for some $x \in C_\eta$. Then the velocity along the curve $P : t \rightsquigarrow (1-t)\hat{x} + tx$ satisfies $V(0) = P'(0) = \varphi'(0)y = 0$, which again implies $x = \hat{x}$ using the linear identifiability property. Hence $y = 0$, which proves that $\operatorname{Ker} \varphi'(\hat{x}) = \{0\}$, and the lemma is proved. ■

Lemma 5.1.13 *Hypothesis and notations (5.42), (5.43), (5.46), (5.47), (5.10), and (5.11).*

1. *If the x_0-minimum-norm $\hat{x}$ solution of (5.1) is qualified then*

$$x_0 - \hat{x} \in \overline{Rg\, \varphi'(\hat{x})^* + T(C, \hat{x})^-}, \tag{5.50}$$

2. *Conversely, if (5.50) holds, then $\hat{x}$ is the x_0-minimum-norm solution.*

The proof is identical to that of Lemma 5.1.5. In the case of an unconstrained problem, where $C = E = T(C, \hat{x})$, property (5.50) reduces to

$$x_0 - \hat{x} \in \overline{Rg\, \varphi'(\hat{x})^*} = (\operatorname{Ker} \varphi'(\hat{x}))^\perp,$$

which expresses the fact that $\hat{x}$ is a local minimum of the distance to x_0 over $X = \varphi^{-1}(\hat{z})$.

The Regularity Condition

Definition 5.1.14 *The x_0-minimum-norm solution $\hat{x}$ is said to satisfy a* regularity condition *if*

$$x_0 - \hat{x} \in Rg\, \varphi'(\hat{x})^* + T(C, \hat{x})^-. \tag{5.51}$$

The name given to condition (5.51) comes from the distributer parameter case: when E is an infinite dimension function space, $Rg\, \varphi'(\hat{x})^*$ can be a dense subspace of E made of regular functions, in which case (5.51) can be satisfied only if $x_0 - \hat{x}$, and hence necessarily the data z, are smooth. This is illustrated by (5.129) in Sect. 5.2 on the LMT regularization of the nonlinear 2D source problem.

As in the linear case, we shall obtain convergence rates for the solutions $\hat{x}_n$ of regularized problems (5.5) when this condition is satisfied.

Convergence of Regularized Problems

Theorem 5.1.15 *Let (5.1) be a FC/LD problem (hypothesis and notations (4.2)) plus (5.42) through (5.45)), and let $x_0 \in E$ be a given a-priori guess.*

1. *When the data z satisfies only*

$$z \in \vartheta = \Big\{ z \in F \mid d(z, \varphi(C)) < R \Big\}, \tag{5.52}$$

the unregularized problem (5.1) has generally no solution, but the regularized problems (5.5) are Q-wellposed for n large enough when $\epsilon_n \to 0$ and $\delta_n \to 0$, and

$$\epsilon_n \hat{x}_n \;\to\; 0 \tag{5.53}$$
$$\varphi(\hat{x}_n) \;\to\; \hat{z}. \tag{5.54}$$

2. *Let in addition φ satisfy condition (5.47), and suppose that the data z satisfy (5.46), which means that (5.1) has a convex nonempty solution set, and a unique x_0-minimum-norm solution $\hat{x}$. If this latter satisfies*

$$x_0 - \hat{x} \in \overline{Rg\, \varphi'(\hat{x})^* + T(C, \hat{x})^-}, \tag{5.55}$$

(e.g., if $\hat{x}$ is qualified), one has, when $\epsilon_n \to 0$ and $\delta_n/\epsilon_n \to 0$,

$$\hat{x}_n \;\to\; \hat{x}\;, \tag{5.56}$$
$$\|\varphi(\hat{x}_n) - \hat{z}\|_F \;=\; O(\epsilon_n). \tag{5.57}$$

3. *If moreover $\hat{x}$ satisfies the regularity condition*

$$x_0 - \hat{x} \in Rg\, \varphi'(\hat{x})^* + T(C, \hat{x})^-, \tag{5.58}$$

one has, when $\epsilon_n \to 0$ and $\delta_n \sim \epsilon_n^2$,

$$\|\hat{x}_n - \hat{x}\|_E \;=\; O(\epsilon_n) = O(\delta_n^{\frac{1}{2}}), \tag{5.59}$$
$$\|\varphi(\hat{x}_n) - \hat{z}\|_F \;=\; O(\epsilon_n^2) = O(\delta_n). \tag{5.60}$$

Proof.

Part 1 The regularized problem(5.5) is equivalent to

$$\hat{x}_n \text{ minimizes } \frac{1}{2}\| \underbrace{(\varphi(x), \epsilon_n x)}_{\stackrel{\text{def}}{=} \varphi_n(x)} - (z_n, \epsilon_n x_0)\|^2_{F\times E} \text{ over } C. \tag{5.61}$$

Along the curves $P_n : t \rightsquigarrow \varphi_n((1-t)x_0 + tx_1)$, the velocity and acceleration

$$V_n(t) = (V(t), \epsilon_n(x_1 - x_0)), \qquad A_n(t) = (A(t), 0),$$

satisfy, using (4.2),

$$\|V_n(t)\|_{F\times E} \geq (1 + \epsilon_n^2/\alpha_M^2)^{1/2}\|V(t)\|_F, \qquad \|A_n(t)\|_{F\times E} = \|A(t)\|_F,$$

so that φ_n and C satisfies (5.42) and (5.43) with

$$R_n = R(1 + \epsilon_n^2/\alpha_M^2) > R \quad \text{and} \quad \Theta_n = \Theta/(1 + \epsilon_n^2/\alpha_M^2)^{1/2} < \Theta \leq \frac{\pi}{2}. \tag{5.62}$$

Hence (5.61) is a finite curvature problem that satisfies the deflection condition $\Theta \leq \pi/2$, as well as the linear stability property (Definition 4.3.4):

$$\|V_n(t)\|_{F\times E} \geq \epsilon_n \|x_1 - x_0\| \quad \text{with} \quad \epsilon_n > 0.$$

Then Theorem 4.4.1 proves that (5.61) is a Q-wellposed problem as soon as

$$d_{F\times E}((z_n, \epsilon_n x_0), \varphi_n(C)) < R_n = R(1 + \epsilon_n^2/\alpha_M^2). \tag{5.63}$$

But

$$\begin{aligned} d_{F\times E}((z_n, \epsilon_n x_0), \varphi_n(C)) &\leq d_{F\times E}((z, \epsilon_n x_0), \varphi_n(C)) + \delta_n \\ &\leq \inf_{x\in C} \{(\|z - \varphi(x)\|^2 + \epsilon_n^2\|x_0 - x\|^2)^{1/2}\} + \delta_n \\ &\leq \inf_{x\in C} \{\|z - \varphi(x)\| + \epsilon_n\|x_0 - x\|\} + \delta_n \\ &\leq \inf_{x\in C} \{\|z - \varphi(x)\| + \epsilon_n Rad_{x_0} C\} + \delta_n \\ &\leq d(z, \varphi(C)) + \epsilon_n Rad_{x_0} C + \delta_n. \end{aligned}$$

Hence, we se that (5.63) will hold as soon as

$$d(z, \varphi(C)) + \epsilon_n \text{Rad}_{x_0} C + \delta_n < R_n = R(1 + \epsilon_n^2/\alpha_M^2),$$

which is true for n large enough as $\epsilon_n \to 0$, $\delta_n \to 0$, and $d(z, \varphi(C) < R$ because of (5.52). So the Q-wellposedness of the regularized problems is proved for n large enough.

We turn now to the proof of (5.53) and (5.54). Let ϵ, d, and η be chosen such that

$$\epsilon > 0, \tag{5.64}$$
$$0 \leq d(z, \varphi(C)) < d < R, \tag{5.65}$$
$$0 < \eta \leq d - d(z, \varphi(C)), \tag{5.66}$$
$$\eta + 2\eta^{\frac{1}{2}} d^{\frac{1}{2}} \leq \epsilon, \tag{5.67}$$

which is always possible because of (5.52). The points of ϑ which have a projection on $\varphi(C)$ are dense in ϑ, and so one can find $\tilde{z} \in \vartheta$ (Fig. 5.1) such that

$$\begin{cases} \|\tilde{z} - z\|_F < \dfrac{\eta}{4}, \\ \tilde{z} \text{ admits a projection } \varphi(\tilde{x}) \text{ on } \varphi(C). \end{cases} \tag{5.68}$$

Let $\tilde{L}_n \geq 0$ be the length of the curve $\tilde{P}_n : t \to \varphi((1-t)\tilde{x} + t\hat{x}_n)$, which satisfies

$$\|\varphi(\hat{x}_n) - \varphi(\tilde{x})\|_F \leq \tilde{L}_n.$$

Then by definition of $\hat{x}_n$ we have

$$\|\varphi(\hat{x}_n) - z_n\|_F^2 + \epsilon_n^2 \|\hat{x}_n - x_0\|_E^2 \leq \|\varphi(\tilde{x}) - z_n\|_F^2 + \epsilon_n^2 \|\tilde{x} - x_0\|_E^2, \tag{5.69}$$

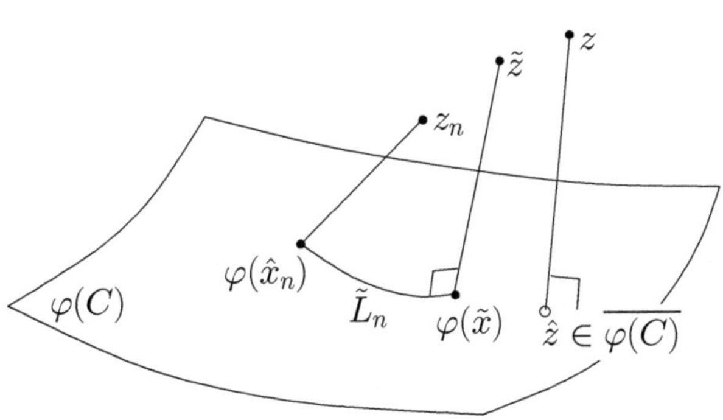

Figure 5.1: Notations for the nonlinear case

and we proceed from here as in the linear case, replacing the estimation of $\|A(\hat{x}_n - \tilde{x})\|_F$ by that of the arc length distance $\tilde{L}_n$ of $\varphi(\tilde{x})$ to $\varphi(\hat{x}_n)$ in $\varphi(C)$. So we rewrite (5.69) as (compare to (5.28))

$$\begin{aligned}\Big(1-\frac{d}{R}\Big)\tilde{L}_n^2 + \epsilon_n^2\|\hat{x}_n - \tilde{x}\|_E^2 \le \Big(1-\frac{d}{R}\Big)\tilde{L}_n^2 &+ \epsilon_n^2\|\hat{x}_n - \tilde{x}\|_E^2 \\ &+\|\varphi(\tilde{x}) - z_n\|_F^2 + \epsilon_n^2\|\tilde{x} - x_0\|_E^2 \\ &-\|\varphi(\hat{x}_n) - z_n\|_F^2 - \epsilon_n^2\|\hat{x}_n - x_0\|_E^2.\end{aligned} \tag{5.70}$$

But the distance of $\tilde{z}$ to $\tilde{P}_n$ has a local minimum at $t = 0$ by definition of $\varphi(\tilde{x})$ = projection of $\tilde{z}$ on $\varphi(C)$, and we can apply the obtuse angle lemma 6.2.9

$$(1 - k(\tilde{z}, \tilde{P}_n))\tilde{L}_n^2 \le \|\tilde{z} - \varphi(\hat{x}_n)\|^2 - \|\tilde{z} - \varphi(\tilde{x})\|^2. \tag{5.71}$$

But (5.68) and the triangular inequality give

$$\begin{aligned}\|\tilde{z} - \varphi(\hat{x}_n)\| &\le \|\tilde{z} - z\| + \|z - z_n\| + \|z_n - \varphi(\hat{x}_n)\| \\ &= \|\tilde{z} - z\| + \|z - z_n\| + \|z_n - \varphi(\tilde{x})\| + \epsilon_n\|\tilde{x} - x_0\| \\ &\le 2\|\tilde{z} - z\| + 2\|z - z_n\| + \|\tilde{z} - \varphi(\tilde{x})\| + \epsilon_n\|\tilde{x} - x_0\| \\ &= 2\|\tilde{z} - z\| + 2\|z - z_n\| + d(\tilde{z}, \varphi(C)) + \epsilon_n\|\tilde{x} - x_0\| \\ &\le 3\|\tilde{z} - z\| + 2\|z - z_n\| + d(z, \varphi(C)) + \epsilon_n\|\tilde{x} - x_0\| \\ &\le d(z, \varphi(C)) + 3\eta/4 + 2\delta_n + \epsilon_n\|\tilde{x} - x_0\| \\ &\le d \quad \text{for } n \text{ large enough}, \\ \|\tilde{z} - \varphi(\tilde{x})\| &= d(\tilde{z}, \varphi(C)) \\ &\le \|\tilde{z} - z\| + d(z, \varphi(C)) \\ &\le d(z, \varphi(C)) + \eta/4 \le d.\end{aligned}$$

This implies, as in the proof of Theorem 7.2.10, that $k(\tilde{z}, \tilde{P}_n) \le d/R$, so that (5.71) gives

$$\Big(1-\frac{d}{R}\Big)\tilde{L}_n^2 \le \|\tilde{z} - \varphi(\hat{x}_n)\|^2 - \|\tilde{z} - \varphi(\tilde{x})\|^2 \text{ for } n \text{ large enough.} \tag{5.72}$$

Substitution of (5.72) in the right-hand side of (5.70) and addition and substraction of $\|\tilde{z} - z_n\|_F^2$ gives

$$\begin{aligned}\Big(1-\frac{d}{R}\Big)\tilde{L}_n^2 + \epsilon_n^2\|\hat{x}_n - \tilde{x}\|_E^2 \le\ &\|\varphi(\hat{x}_n) - \tilde{z}\|_F^2 - \|\varphi(\tilde{x}) - \tilde{z}\|_F^2 \\ &+\|\tilde{z} - z_n\|_F^2 - \|\tilde{z} - z_n\|_F^2\end{aligned}$$

$$
\begin{aligned}
&-\|\varphi(\hat{x}_n) - z_n\|_F^2 + \| \varphi(\tilde{x}) - z_n\|_F^2 \\
&\quad + \epsilon_n^2\|\hat{x}_n - \tilde{x}\|_E^2 \\
&\quad + \epsilon_n^2\|\tilde{x} - x_0\|_E^2 \\
&\quad - \epsilon_n^2\|\hat{x}_n - x_0\|_E^2.
\end{aligned}
$$

Using three times the identity $a^2 + b^2 - (a+b)^2 = -2ab$ gives

$$
\begin{aligned}
\Big(1 - \frac{d}{R}\Big)\tilde{L}_n^2 + \epsilon_n^2\|\hat{x}_n - \tilde{x}\|_E^2 \le\ & 2\langle\varphi(\hat{x}_n) - \tilde{z}, z_n - \tilde{z}\rangle_F \\
& -2\langle\varphi(\tilde{x}) - \tilde{z}, z_n - \tilde{z}\rangle_F \\
& +2\epsilon_n^2\langle\hat{x}_n - \tilde{x}, x_0 - \tilde{x}\rangle_E,
\end{aligned}
$$

and hence

$$
\begin{aligned}
\Big(1 - \frac{d}{R}\Big)\tilde{L}_n^2 + \epsilon_n^2\|\hat{x}_n - \tilde{x}\|_E^2 \le\ & 2\langle\varphi(\hat{x}_n) - \varphi(\tilde{x}), z_n - \tilde{z}\rangle_F \\
& +2\epsilon_n^2\langle\hat{x}_n - \tilde{x}, x_0 - \tilde{x}\rangle_E.
\end{aligned}
\tag{5.73}
$$

Using the Cauchy–Schwarz inequality and the formula (5.32) gives

$$
\Big(1 - \frac{d}{R}\Big)^{1/2}\tilde{L}_n \ \le\ 2\Big(1 - \frac{d}{R}\Big)^{-1/2}\Big(\delta_n + \frac{\eta}{4}\Big) + \epsilon_n\|x_0 - \tilde{x}\|_E, \tag{5.74}
$$

$$
\epsilon_n\|\hat{x}_n - \tilde{x}\|_E \ \le\ \Big(1 - \frac{d}{R}\Big)^{-1/2}\Big(\delta_n + \frac{\eta}{4}\Big) + 2\epsilon_n\|x_0 - \tilde{x}\|_E. \tag{5.75}
$$

Formula (5.74) gives an estimation of $\|\varphi(\hat{x}_n) - \varphi(\tilde{x})\|_F \le \tilde{L}_n$, and we need to estimate $\|\varphi(\tilde{x}) - \hat{z}\|_F$ to prove (5.54). We shall distinguish two cases:

• If $d(z, \varphi(C)) = 0$, one has $z = \hat{z} \in \overline{\varphi(C)}$, so one can choose $\tilde{z} = \varphi(\tilde{x}) \in \varphi(C)$. Then

$$
\Big(1 - \frac{d}{R}\Big)\|\varphi(\tilde{x}) - \hat{z}\|_F = \Big(1 - \frac{d}{R}\Big)\|\tilde{z} - z\|_F \le \Big(1 - \frac{d}{R}\Big)\frac{\eta}{4} \le \frac{\eta}{4},
$$

which, combined with (5.74), gives

$$
\begin{aligned}
\Big(1 - \frac{d}{R}\Big)\|\varphi(\hat{x}_n) - \hat{z}\|_F \ &\le\ 2\delta_n + \frac{3\eta}{4} + \Big(1 - \frac{d}{R}\Big)^{1/2}\epsilon_n\|x_0 - \tilde{x}\|_E, \\
&\le\ 2\delta_n + \frac{3\epsilon}{4} + \Big(1 - \frac{d}{R}\Big)^{1/2}\epsilon_n\|x_0 - \tilde{x}\|_E, \\
&\le\ \epsilon \quad \text{for } n \text{ large enough,}
\end{aligned}
$$

and (5.54) is proved.

- If $d(z,\varphi(C)) > 0$, let us choose η such that

$$0 < \frac{\eta}{4} \leq d(z,\varphi(C)).$$

By construction, $\hat{z} \in \overline{\varphi(C)}$, so there exists $y_p \in C, p = 1, 2 \ldots$ such that

$$\|\varphi(y_p) - \hat{z}\|_F \leq \frac{1}{p}.$$

The function $d_p : t \in [0,1] \rightsquigarrow \|\tilde{z} - \varphi((1-t)\tilde{x} + ty_p)\|_F$ has a local minimum at $t = 0$, as $\varphi(\tilde{x})$ is the projection of $\tilde{z}$ on $\varphi(C)$, and satisfies moreover

$$d_p(0) = d(\tilde{z},\varphi(C)) \leq d(z,\varphi(C)) + \frac{\eta}{4} \leq d,$$

$$d_p(1) = \|\tilde{z} - \varphi(y_p)\|_F \leq \frac{1}{p} + d(z,\varphi(C)) + \frac{\eta}{4} \leq d \text{ for } n \text{ large enough.}$$

So we can apply once again the obtuse angle lemma 6.2.9

$$\Big(1 - \frac{d}{R}\Big)\|\varphi(\tilde{x}) - \varphi(y_p)\|_F^2 + \|\tilde{z} - \varphi(\tilde{x})\|_F^2 \leq \|\tilde{z} - \varphi(y_p)\|_F^2. \tag{5.76}$$

But

$$\begin{aligned}\|\tilde{z} - \varphi(\tilde{x})\|_F = d(\tilde{z},\varphi(C)) &\geq d(z,\varphi(C)) - \frac{\eta}{4} \geq 0,\\ \|\tilde{z} - \varphi(y_p)\|_F &\leq \frac{1}{p} + d(z,\varphi(C)) + \frac{\eta}{4},\end{aligned}$$

which, combined to (5.76), give

$$\begin{aligned}\Big(1 - \frac{d}{R}\Big)\|\varphi(\tilde{x}) - \varphi(y_p)\|_F^2 &\leq \Big(\frac{1}{p} + d(z,\varphi(C)) + \frac{\eta}{4}\Big)^2 - \Big(d(z,\varphi(C)) - \frac{\eta}{4}\Big)^2\\ &= \Big(\frac{1}{p} + \frac{\eta}{2}\Big)\Big(\frac{1}{p} + 2d(z,\varphi(C))\Big),\end{aligned}$$

and, passing to the limit when $p \to \infty$,

$$\Big(1 - \frac{d}{R}\Big)\|\varphi(\tilde{x}) - \hat{z}\|_F^2 \leq \eta\, d(z,\varphi(C)) \leq \eta\, d. \tag{5.77}$$

Combining (5.77) with (5.74) gives

$$\begin{aligned}\Big(1-\frac{d}{R}\Big)\|\varphi(\hat{x}_n)-\hat{z}\|_F &\le 2\delta_n+\frac{\eta}{2}+\epsilon_n\Big(1-\frac{d}{R}\Big)^{\frac12}\|x_0-\tilde{x}\|_E\\ &\quad +\eta^{\frac12}\Big(1-\frac{d}{R}\Big)^{\frac12}d^{\frac12}\\ &\le \frac{\eta}{2}+\eta^{\frac12}d^{\frac12}+2\delta_n+\epsilon_n\|x_0-\tilde{x}\|_E\\ &\le \frac{\epsilon}{2}+2\delta_n+\epsilon_n\|x_0-\tilde{x}\|_E,\end{aligned}$$

so that $(1-d/R)\|\varphi(\hat{x}_n)-\hat{z}\|_F \le \epsilon$ for n large enough, and (5.20) is proved.
Last, (5.53) follows from (5.75) as in the linear case

$$\epsilon_n\|\hat{x}_n\|_E \le \epsilon_n\|\tilde{x}\|_E+\delta_n+\frac{\eta}{4}+2\epsilon_n\|x_0-\tilde{x}\|_E \le \eta \text{ for } n \text{ large enough.}$$

Part 2 We suppose now that (5.1) has a solution, and hence an x_0-minimum-norm solution $\hat{x}$. By definition of $\hat{z}$ one has $\varphi(\hat{x})=\hat{z}$, and so we can choose $\tilde{z}=z$, $\eta=0$, and $\tilde{x}=\hat{x}$ in (5.68). Then (5.73), (5.74), and (5.75) become (with $\tilde{L}_n$ replaced by $\hat{L}_n$)

$$\Big(1-\frac{d}{R}\Big)\hat{L}_n^2+\epsilon_n^2\|\hat{x}_n-\hat{x}\|_E^2 \le \begin{array}{l}2\langle\varphi(\hat{x}_n)-\varphi(\hat{x}),z_n-z\rangle_F\\ +2\epsilon_n^2\langle\hat{x}_n-\hat{x},x_0-\hat{x}\rangle_E\end{array} \quad (5.78)$$

and

$$\Big(1-\frac{d}{R}\Big)^{1/2}\hat{L}_n \le 2\Big(1-\frac{d}{R}\Big)^{-1/2}\delta_n+\epsilon_n\|x_0-\hat{x}\|_E, \quad (5.79)$$

$$\epsilon_n\|\hat{x}_n-\hat{x}\|_E \le \Big(1-\frac{d}{R}\Big)^{-1/2}\delta_n+2\epsilon_n\|x_0-\hat{x}\|_E, \quad (5.80)$$

where (5.79) proves (5.57) as now $\delta_n/\epsilon_n \to 0$. But (5.80) gives no information on the convergence of $\hat{x}_n$ to $\hat{x}$, it is necessary for that to use hypothesis (5.50) on $\hat{x}$: let again $\eta>0$ be given, and $w\in F, \mu\in T(C,\hat{x})^-$ be such that

$$\|x_0-\hat{x}-(\varphi'(\hat{x})^*w+\mu)\|_F \le \eta/3.$$

Equation (5.78) becomes

$$\begin{aligned}\Big(1-\frac{d}{R}\Big)\hat{L}_n^2 + \epsilon_n^2\|\hat{x}_n-\hat{x}\|_E^2 \le\ & 2\langle\varphi(\hat{x}_n)-\varphi(\hat{x}), z_n - z\rangle\\ &+ 2\epsilon_n^2\langle\hat{x}_n-\hat{x}, x_0-\hat{x}-(\varphi'(\hat{x})^*\, w+\mu)\rangle\\ &+ 2\epsilon_n^2\langle\hat{x}_n-\hat{x}, \varphi'(\hat{x})^*\, w\rangle\\ &+ 2\epsilon_n^2\langle\hat{x}_n-\hat{x}, \mu\rangle,\end{aligned}$$

and, transposing $\varphi'(\hat{x})^*$ in the right-hand side and using the fact that $\langle\hat{x}_n - \hat{x}, \mu\rangle \le 0$

$$\begin{aligned}\Big(1-\frac{d}{R}\Big)\hat{L}_n^2 + \epsilon_n^2\|\hat{x}_n-\hat{x}\|_E^2 \le\ & 2\langle\varphi(\hat{x}_n)-\varphi(\hat{x}), z_n - z\rangle\\ &+ 2\epsilon_n^2\langle\hat{x}_n-\hat{x}, x_0-\hat{x}-(\varphi'(\hat{x})^*\, w+\mu)\rangle\\ &+ 2\epsilon_n^2\langle\varphi'(\hat{x})(\hat{x}_n-\hat{x}), w\rangle.\end{aligned}$$

A second-order Taylor expansion for $\hat{P}_n : t \rightsquigarrow \varphi((1-t)\hat{x}+t\hat{x}_n)$ gives

$$\varphi(\hat{x}_n)-\varphi(\hat{x}) = \varphi'(\hat{x})(\hat{x}_n-\hat{x}) + \int_0^1 \hat{P}_n''(t)(1-t)\,\mathrm{d}t,$$

so that

$$\begin{aligned}\Big(1-\frac{d}{R}\Big)\hat{L}_n^2 + \epsilon_n^2\|\hat{x}_n-\hat{x}\|_E^2 \le\ & 2\langle\varphi(\hat{x}_n)-\varphi(\hat{x}), z_n - z + \epsilon_n^2 w\rangle \qquad (5.81)\\ &+ 2\epsilon_n^2\langle\hat{x}_n-\hat{x}, x_0-\hat{x}-(\varphi'(\hat{x})^*\, w+\mu)\rangle\\ &- 2\epsilon_n^2\langle\int_0^1 \hat{P}_n''(t)(1-t)\,\mathrm{d}t, w\rangle.\end{aligned}$$

But the deflection condition (5.43) gives

$$\|\int_0^1 \hat{P}_n''(t)(1-t)\,\mathrm{d}t\|_F \le \int_0^1 \|\hat{P}_n''(t)\|_F\,\mathrm{d}t \le \frac{\pi}{2}\int_0^1 \|\hat{P}_n'(t)\|_F\,\mathrm{d}t = \frac{\pi}{2}\hat{L}_n,$$

and, using the Cauchy–Schwarz inequality, (5.81) becomes

$$\begin{aligned}\Big(1-\frac{d}{R}\Big)\hat{L}_n^2 + \epsilon_n^2\|\hat{x}_n-\hat{x}\|_E^2 \le\ & 2\|\varphi(\hat{x}_n)-\varphi(\hat{x})\|\,(\delta_n+\epsilon_n^2\|w\|)\\ &+ \frac{2}{3}\epsilon_n^2\eta\|\hat{x}_n-\hat{x}\|_E\\ &+ \pi\epsilon_n^2\hat{L}_n\,\|w\|_F,\end{aligned}$$

which in turn gives, using formula (5.32),

$$\begin{aligned}\left(1-\frac{d}{R}\right)^{1/2}\hat{L}_n &\leq 2\left(1-\frac{d}{R}\right)^{-1/2}\left(\delta_n+\epsilon_n^2\left(1+\frac{\pi}{2}\right)\|w\|_F\right)+\frac{\eta}{3}\epsilon_n, &(5.82)\\ \epsilon_n\|\hat{x}_n-\hat{x}\|_E &\leq \left(1-\frac{d}{R}\right)^{-1/2}\left(\delta_n+\epsilon_n^2\left(1+\frac{\pi}{2}\right)\|w\|_F\right)+\frac{2\eta}{3}\epsilon_n. &(5.83)\end{aligned}$$

When $\delta_n/\epsilon_n \to 0$, the last inequality shows that $\|\hat{x}_n-\hat{x}\|_E \leq \eta$ for n large enough, which proves (5.56).

Part 3 Now $\hat{x}$ satisfies the regularity condition (5.58), and so we can choose $\eta=0$ in **Part 2**, and estimations (5.82) and (5.83) simplify to

$$\begin{aligned}\left(1-\frac{d}{R}\right)^{1/2}\hat{L}_n &\leq 2\left(1-\frac{d}{R}\right)^{-1/2}\left(\delta_n+\epsilon_n^2\left(1+\frac{\pi}{2}\right)\|w\|_F\right),\\ \epsilon_n\|\hat{x}_n-\hat{x}\|_E &\leq \left(1-\frac{d}{R}\right)^{-1/2}\left(\delta_n+\epsilon_n^2\left(1+\frac{\pi}{2}\right)\|w\|_F\right),\end{aligned}$$

which prove (5.59) and (5.60). ■

The previous theorem reduces exactly to the Theorem 5.1.8 when the forward map φ is linear.

5.1.3 General Nonlinear Problems

We investigate in this section the Q-wellposedness of the LMT-regularized problem (5.2) when the unregularized problem (5.1) satisfies only the minimum set of hypothesis (4.2) and φ has a *bounded second derivative*

$$\left\{\begin{array}{l}\text{there exists } \beta \geq 0 \text{ such that}\\ \forall x_0, x_1 \in C, \text{ the curve } P\ :\ t \rightsquigarrow \varphi((1-x_0)t+tx_1) \text{ satisfies}\\ P \in W^{2,\infty}([0,1];F) \text{ and } \|A(t)\|_F \leq \beta\|x_1-x_0\|_E^2,\\ \text{for a.e. } t\in[0,1], \text{where } A(t)=P''(t).\end{array}\right. \tag{5.84}$$

Hypothesis (4.2) and (5.84) are verified as soon as the map φ to be inverted is smooth, which is the case in most of the situations. Hence the results of this section apply to all examples of Chap. 1 for which no stronger FC/LD property can be proved:

- The Knott–Zoeppritz equations of Sect. 1.1, all types of observation,
- The 1D elliptic parameter estimation problem of section 1.4 with L^2 or partial or boundary observation (for $H1$ observation the problem has been shown to be Q-wellposed in section 4.8),
- The 2D elliptic nonlinear source estimation problem of Sect. 1.5 for boundary or partial observation (the case of H^1 or L^2 observation leads to a FC/LD problem, see Sect. 5.2 below),
- The 2D elliptic parameter estimation problem of Sect. 1.6 for L^2 or partial or boundary observation (for H^1 observation, Q-wellposedness results after reduction to finite dimension are available in Sect. 4.9).

However, under these sole hypothesis, nothing is known concerning the existence and uniqueness of solution(s) to the NLS problem (5.1) and its LMT-regularized version (5.2). This is a main difference with the linear case of Sect. 5.1.1 (which corresponds to $\beta = 0$) and the case of finite curvature limited deflection (FC/LD) problems (Sect. 5.1.2), where (5.2) has a solution as soon as $\epsilon > 0$, possibly small enough!

We use first the Q-wellposedness conditions of Chap. 4 to quantify the natural intuition that "a minimum amount of LMT-regularization" should be added to compensate for the nonlinearity of φ and restore Q-wellposedness of (5.2) as in the linear case [18].

The size of the parameter set and the position of the a-priori guess in the admissible parameter set will play a role through the following quantities:

$$\operatorname{diam} C = \sup_{x,y\in C} \|x-y\|_E, \qquad \operatorname{rad}_{x_0} C = \sup_{x\in C} \|x - x_0\|_E.$$

The regularized problem(5.2) is equivalent to

$$\hat{x}_\epsilon \text{ minimizes } \frac{1}{2}\| \underbrace{(\varphi(x), \epsilon x)}_{\stackrel{\text{def}}{=}\ \varphi_\epsilon(x)} - (z, \epsilon x_0)\|^2_{F\times E} \text{ over } C.$$

Along the curves $P_\epsilon \ :\ t \rightsquigarrow \varphi_\epsilon((1-t)x_0 + t x_1)$, the velocity and acceleration

$$V_\epsilon(t) = (V(t), \epsilon(x_1 - x_0)), \qquad A_\epsilon(t) = (A(t), 0)$$

satisfy, using (4.2) (5.84),

$$\begin{aligned} \epsilon\|x_1 - x_0\|_E \le \|V_\epsilon(t)\|_{F\times E} &\le (\alpha_M^2 + \epsilon^2)^{1/2}\|x_1 - x_0\|_E, \\ \|A_\epsilon(t)\|_{F\times E} &= \|A(t)\|_F \le \beta\|x_1 - x_0\|^2_E, \end{aligned}$$

and hence

$$\|A_\epsilon(t)\|_{F\times E} \leq \frac{\beta}{\epsilon^2}\|V_\epsilon(t)\|^2_{F\times E}, \quad \|A_\epsilon(t)\|_{F\times E} \leq \frac{\beta}{\epsilon}\operatorname{diam} C\|V_\epsilon(t)\|_{F\times E}.$$

This shows that the the regularized problem (5.2) is linearly stable (Definition 4.3.4), and has a curvature and a deflection bounded by (compare with (5.42) and (5.43) for the case of FC/LD problems)

$$\frac{1}{R_\epsilon} = \frac{\beta}{\epsilon^2}, \qquad \Theta_\epsilon = \frac{\beta}{\epsilon}\operatorname{diam} C.$$

As expected, this upper bound to the curvature blows up to infinity when the regularization parameter goes to zero. Application of Theorem 4.4.1 shows that (5.2) is Q-wellposed in $E \times F$ on the neighborhood

$$\Big\{(z, x_0) \in F \times E \mid d_{F\times E}((z, x_0), \varphi_\epsilon(C)) < R_\epsilon = \frac{\epsilon^2}{\beta}\Big\}$$

of $\varphi_\epsilon(C)$ as soon as the deflection condition

$$\Theta_\epsilon = (\beta/\epsilon)\operatorname{diam} C \leq \pi/2$$

is satisfied. But one checks easily that

$$d_{F\times E}((z, x_0), \varphi_\epsilon(C))^2 \leq d(z, \varphi(C))^2 + \epsilon^2 \operatorname{rad}^2_{x_0} C,$$

which proves the

Proposition 5.1.16 *Hypothesis (4.2) and (5.84). If the regularization parameter ϵ satisfies*

$$\epsilon > \epsilon_{\min} \stackrel{\text{def}}{=} \beta \max\big\{\operatorname{rad}_{x_0} C,\ (2/\pi)\operatorname{diam} C\big\},$$

then $\epsilon^2/\beta = R_\epsilon > \epsilon \operatorname{rad}_{x_0} C$ and $(\beta/\epsilon)\operatorname{diam} C = \Theta_\epsilon \leq \pi/2$, and the regularized problem (5.2) is Q-wellposed in F on the neighborhood

$$\vartheta_\epsilon = \Big\{z \in F \mid d_F(z, \varphi(C)) < d_\epsilon\Big\}, \tag{5.85}$$

where d_ϵ is defined by

$$d_\epsilon^2 + \epsilon^2 \operatorname{rad}^2_{x_0} C = R_\epsilon^2. \tag{5.86}$$

When the data $z_0, z_1 \in \vartheta_\epsilon$ are close enough, one can choose $d < d_\epsilon$ such that

$$\begin{aligned} \|z_0 - z_1\|_F \;&+\; \big(\max_{j=0,1} d(z_j, \varphi(C))^2 + \epsilon^2 \mathrm{rad}^2_{x_0} C\big)^{\frac{1}{2}} \\ &\leq \big(d^2 + \epsilon^2 \mathrm{rad}^2_{x_0} C\big)^{1/2} < R_\epsilon, \end{aligned} \tag{5.87}$$

and the corresponding regularized solutions $\hat{x}_{\epsilon,j}, j = 0, 1$ satisfy

$$\epsilon \, \|\hat{x}_{\epsilon,0} - \hat{x}_{\epsilon,1}\|_E \leq \Big(1 - \frac{(d^2 + \epsilon^2 \mathrm{rad}^2_{x_0} C)^{\frac{1}{2}}}{R_\epsilon}\Big)^{-1} \|z_0 - z_1\|_F. \tag{5.88}$$

This proposition can be used, for example, to determine, given an estimation of an upper bound $d_{\max}$ of the measurement and model errors, a minimum amount of regularization $\epsilon_{\min}$, which ensures Q-wellposedness of the regularized NLS problem on a neighborhood of $\varphi(C)$ large enough to contain the expected data.

Remark 5.1.17 *Formula (5.85) through (5.88) describe also the stability property of the regularized problem (5.5) for a fixed n, both in the linear case of Theorem 5.1.8 (take $1/R_\epsilon = \Theta_\epsilon = 0$, so that $d_\epsilon = +\infty$) and in the case of FC/LD problems of Theorem 5.1.15 (replace $R_\epsilon, \Theta_\epsilon$ by R_n, Θ_n defined in (5.62)).* ■

Next we study the *convergence of solutions* $\hat{x}_n$ to (5.5) when $n \to \infty$ [27]. Because of the possibly infinite curvature of the attainable set, we consider only the case where

$$z \in \varphi(C) \qquad \text{(attainable data),} \tag{5.89}$$

which eliminates the need of projecting z onto $\varphi(C)$. The following additional properties of φ will be needed:

$$\varphi \text{ is sequentially closed,} \tag{5.90}$$

$$\varphi \text{ has a Fréchet derivative } \varphi'(\hat{x}) \text{ at } \hat{x}. \tag{5.91}$$

Condition (5.90) ensures that the solution set $X = \varphi(z)^{-1}$ of (5.1) is nonvoid, and so we can define an x_0-minimum-norm solution $\hat{x}$ of (5.1) as one solution of the minimization problem (5.11) (there may be more than one x_0-minimum-norm solution as X is not necessarily convex).

The characterization of $\hat{x}$ follows the same line of arguments as in Sect. 5.1.2, with some adaptations: the tangent cone to X at $\hat{x}$ is now defined by (compare with (5.12))

$$T(X, \hat{x}) = \{y \in E : \exists x_n \in X, \lambda_n > 0, x_n \to \hat{x} \text{ s.t. } \lambda_n(x_n - \hat{x}) \to y\},$$

and a short calculation using hypothesis (5.91) shows that Lemma 5.1.10 still holds

$$T(X, \hat{x}) \subset \mathrm{Ker}\varphi'(\hat{x}) \cap T(C, \hat{x}) \ ,$$

and, when $\hat{x}$ is *qualified* in the sense of definition 5.1.11, that lemma 5.1.13 also holds:

$$x_0 - \hat{x} \in \overline{Rg\, \varphi'(\hat{x})^* + T(C, \hat{x})^-} \ .$$

The LMT-regularized problems (5.5) are not in general Q-wellposed, but they all have at least one solution because of hypothesis (5.90).

Theorem 5.1.18 *Let hypothesis (4.2), (5.84), and (5.89) through (5.91) hold, and let $\hat{x}$ be an x_0-minimum-norm solution to (5.1) that satisfies*

$$x_0 - \hat{x} \in Rg\, \varphi'(\hat{x})^* + T(C, \hat{x})^- \quad \textit{(Regularity Condition)}, \tag{5.92}$$

$$\beta \|w\|_F \leq 1, \tag{5.93}$$

where β is defined in (5.84), and $w \in F$ is an element satisfying

$$x_0 - \hat{x} = \varphi'(\hat{x})^* \, w + \mu \quad \textit{where } \mu \in T(C, \hat{x})^-.$$

Then any sequence $\{x_n\}$ of solutions to (5.5) satisfies, when $\epsilon_n \to 0$ and $\delta_n \sim \epsilon_n^2$,

$$\begin{aligned} \|\hat{x}_n - \hat{x}\|_E &= O(\epsilon_n) = O(\delta_n^{\frac{1}{2}}), \\ \|\varphi(\hat{x}_n) - z\|_F &= O(\epsilon_n^2) = O(\delta_n). \end{aligned}$$

Proof. By definition, $\hat{x}_n$ satisfies

$$\|\varphi(\hat{x}_n) - z_n\|_F^2 + \epsilon_n^2 \|\hat{x}_n - x_0\|_E^2 \leq \|\varphi(\hat{x}) - z_n\|_F^2 + \epsilon_n^2 \|\hat{x} - x_0\|_E^2.$$

Adding and subtracting $\|\varphi(\hat{x}_n) - z\|_F^2 + \epsilon_n^2 \|\hat{x}_n - \hat{x}\|_E^2$ gives, as in the proof of Theorems 5.1.8 or 5.1.15,

$$\begin{aligned} \|\varphi(\hat{x}_n) - z\|_F^2 + \epsilon_n^2 \|\hat{x}_n - \hat{x}\|_E^2 \leq{} & 2\langle \varphi(\hat{x}_n) - z, z_n - z\rangle_F \\ & + 2\epsilon_n^2 \langle \hat{x}_n - \hat{x}, x_0 - \hat{x}\rangle_E. \end{aligned}$$

By (5.92) this implies

$$\begin{aligned}\|\varphi(\hat{x}_n) - z\|_F^2 + \epsilon_n^2\|\hat{x}_n - \hat{x}\|_E^2 \leq 2\langle\varphi(\hat{x}_n) - z, z_n - z\rangle \\ +2\epsilon_n^2\langle\varphi'(\hat{x})(\hat{x}_n - \hat{x}), w\rangle.\end{aligned} \quad (5.94)$$

A second order Taylor expansion for $P : t \rightsquigarrow \varphi((1-t)\hat{x} + t\hat{x}_n)$ gives

$$\varphi(\hat{x}_n) - \varphi(\hat{x}) = \varphi'(\hat{x})(\hat{x}_n - \hat{x}) + \int_0^1 P''(t)(1-t)\,\mathrm{d}t,$$

and, using (5.84)

$$\| \int_0^1 P''(t)(1-t)\,\mathrm{d}t\,\|_F \leq \frac{1}{2}\beta\,\|\hat{x}_n - \hat{x}\|_E^2. \quad (5.95)$$

Hence (5.94) becomes

$$\begin{aligned}\|\varphi(\hat{x}_n) - z\|_F^2 + \epsilon_n^2\|\hat{x}_n - \hat{x}\|_E^2 \leq 2\langle\varphi(\hat{x}_n) - z, z_n - z + \epsilon_n^2 w\rangle \\ -2\epsilon_n^2\langle\int_0^1 P''(t)(1-t)dt, w\rangle.\end{aligned} \quad (5.96)$$

Formula (5.95) and the Cauchy–Schwarz inequality give, rearranging terms,

$$\|\varphi(\hat{x}_n) - z\|_F^2 + \epsilon_n^2(1 - \beta\|w\|_F)\|\hat{x}_n - \hat{x}\|_E^2 \leq 2\|\varphi(\hat{x}_n) - z\|_F(\delta_n + \epsilon_n^2 w), \quad (5.97)$$

and, using (5.32),

$$\begin{aligned}\|\varphi(\hat{x}_n) - z\|_F^2 &\leq 2(\delta_n + \epsilon_n^2\|w\|_F), \\ (1 - \beta\|w\|_F)^{\frac{1}{2}}\epsilon_n\|\hat{x}_n - \hat{x}\|_E &\leq (\delta_n + \epsilon_n^2\|w\|_F),\end{aligned}$$

and the desired result follows. ■

Remark 5.1.19 *Using weak subsequential arguments, one can show that any sequence $\{\hat{x}_n\}$ of solutions to (5.5) contains a subsequence, which converges strongly to an X_0-minimum-norm solution of (5.1), provided that $\epsilon_n \to 0$ and $\delta_n/\epsilon_n \to 0$.* ■

5.2 Application to the Nonlinear 2D Source Problem

We give in this section an example of nonlinear infinite dimensional FC/LD problem to which Theorem 5.1.15 on regularization applies. It is the source estimation problem in a nonlinear elliptic equation described in Sect. 1.5:

$$\begin{cases} -\Delta u + k(u) \;=\; f \quad \text{in } \Omega, \\ u \;=\; 0 \quad \text{on a part } \partial\Omega_{\rm D} \; \text{ of } \partial\Omega, \\ \dfrac{\partial u}{\partial \nu} \;=\; g \quad \text{on } \partial\Omega_{\rm N} \;=\; \partial\Omega \setminus \partial\Omega_{\rm D}, \end{cases} \tag{5.98}$$

where the objective is to estimate the right-hand sides f, g from a measurement z of the solution u in $L^2(\Omega)$. With the notation of Sect. 1.5 and

$$x = (f, g) \in E = L^2(\Omega) \times L^2(\partial\Omega_N), \tag{5.99}$$

the NLS problem (1.61) coincides with (5.1) and its regularized versions (1.63) coincides with (5.5).

We show in this section that (1.61) is a finite curvature problem, and estimate the size of C, which ensures a deflection smaller than $\pi/2$, making thus (1.61) a FC/LD problem to which Theorem 5.1.15 can be applied. We follow for the proof reference [28], where the result was proved for a slightly more complex model involving a convection term.

Remark 5.2.1 *When the* nonlinearity is in the higher order term, *as, for example, in*

$$-\nabla.(a(u)\nabla u) = f \quad + \textit{boundary conditions},$$

the finite curvature property is lost. However, the least square objective function still possesses minimizer(s) over a suitable admissible set, is derivable, and the parameter a can be retrieved numerically over the range of u (see [31]). ■

We begin with the *FC/LD properties* of (1.61). The variational formulation of (5.98) is, with the definition (1.58) for Y,

$$\begin{cases} \text{find } u \in Y \text{ such that} \\ \langle \nabla u, \nabla w\rangle_{L^2(\Omega)} + \langle k(u), w\rangle_{L^2(\Omega)} = \langle f, w\rangle_{L^2(\Omega)} + \langle g, w\rangle_{L^2(\partial\Omega_{\rm N})} \\ \text{for all } w \in Y. \end{cases} \tag{5.100}$$

We endow Y with the norm

$$\|w\|_Y = |\nabla w|_{L^2(\Omega)}.$$

Since $\partial\Omega_{\rm D}$ is assumed to be nonempty, this norm is equivalent to the usual H^1 norm using the Poincaré inequality:

$$|w|_{L^2(\Omega)} \leq C_P |\nabla w|_{L^2(\Omega)}, \quad \text{where } C_P \text{ is the Poincaré constant.} \tag{5.101}$$

The right-hand side of (5.100) is a continuous linear form L on Y:

$$\begin{aligned} L(w) &= \int_\Omega fw + \int_{\partial\Omega_N} gw \\ &\leq \big(C_P|f|_{L^2(\Omega)} + C_N|g|_{L^2(\partial\Omega_N)}\big)\|w\|_Y \\ &\leq \underbrace{(C_P^2 + C_N^2)^{\frac{1}{2}}}_{M} \|(f,g)\|_E \|w\|_Y, \end{aligned} \tag{5.102}$$

where C_N denotes the continuity constant of the trace operator τ_N from Y to $L^2(\partial\Omega_N)$:

$$|w|_{L^2(\partial\Omega_N)} \leq C_N \|w\|_Y.$$

Lemma 5.2.2 *(Hypothesis and notations of Sect. 1.5). Then (5.100) has a unique solution $u \in Y$, which satisfies the a-priori estimate*

$$\|u_1 - u_0\|_Y \leq M\|(f_1, g_1) - (f_0, g_0)\|_E, \quad M \text{ defined in (5.102)}. \tag{5.103}$$

Proof. The left-hand side of (5.100) defines an operator $\mathcal{A} : Y \rightsquigarrow Y'$:

$$\begin{cases} \forall v \in Y\ ,\ \mathcal{A}(v) \in Y' \ \text{ is defined by} \\ \langle \mathcal{A}(v), w\rangle_{Y'Y} = \langle \nabla v, \nabla w\rangle_{L^2(\Omega)} + \langle k(v), w\rangle_{L^2(\Omega)} \ \forall w \in Y. \end{cases}$$

Using the properties imposed on k, one can check that $\mathcal{A}$ maps bounded sets to bounded sets, and that $\mathcal{A}$ is hemicontinuous (i.e., for all the function $\lambda \rightsquigarrow \langle \mathcal{A}(u + \lambda v), w\rangle_{Y'Y}$ is continuous from $I\!R$ to $I\!R$). For every $v, w \in Y$, moreover,

$$\begin{cases} \langle \mathcal{A}(v) - \mathcal{A}(w), v - w\rangle_{Y'Y} = |\nabla(v - w)|^2_{L^2(\Omega)} \\ \qquad\qquad + \int_\Omega \int_0^1 k'((1-t)w + tv)(v-w)^2 \\ \qquad\qquad \geq \|v - w\|_Y^2\ , \end{cases}$$

hence $\mathcal{A}$ is strictly monotone, and the lemma is proved [56, page 171]. ■

Lemma 5.2.3 *(Hypothesis and notations of Sect. 1.5). Let* $x_j = (f_j, g_j) \in C$, $j = 0, 1$ *be given. Then* $P : t \in [0,1] \rightsquigarrow \varphi((1-t)x_0 + tx_1)$ *is in* $W^{2,\infty}([0,1], Y)$*, and its first and second derivatives* $V(t)$ *and* $A(t)$ *are the variational solutions in* Y *of*

$$\begin{cases} -\Delta V(t) + k'(P(t))V(t) \;=\; f_1 - f_0 \quad \text{in } \Omega, \\ V(t) \;=\; 0 \quad \text{on } \partial\Omega_{\mathrm{D}}, \\ \dfrac{\partial V(t)}{\partial \nu} \;=\; g_1 - g_0 \quad \text{on } \partial\Omega_{\mathrm{N}}, \end{cases} \tag{5.104}$$

$$\begin{cases} -\Delta A(t) + k'(P(t))A(t) \;= -k''(P(t))V(t)^2 \quad \text{in } \Omega, \\ A(t) \;=\; 0 \quad \text{on } \partial\Omega_{\mathrm{D}}, \\ \dfrac{\partial A(t)}{\partial \nu} = 0 \quad \text{on } \partial\Omega_{\mathrm{N}}. \end{cases} \tag{5.105}$$

Proof. Because of (5.103), P is Lipschitz continuous and hence a.e. differentiable with

$$\|V(t)\|_Y \leq M\|(f_1, g_1) - (f_0, g_0)\|_E \quad \text{a.e. on } [0,1], \tag{5.106}$$

so that $P \in W^{1,\infty}([0,1], Y)$. By definition, $P(t)$ is the solution of

$$\begin{cases} \langle \nabla P(t), \nabla w \rangle_{L^2(\Omega)} + \langle k(P(t)), w \rangle_{L^2(\Omega)} = \\ \qquad \langle (1-t)f_0 + tf_1, w \rangle_{L^2(\Omega)} + \langle (1-y)g_0 + tg_1, w \rangle_{L^2(\partial\Omega_{\mathrm{N}})} \\ \text{for all } w \in Y. \end{cases} \tag{5.107}$$

Differentiation of (5.107) with respect to t shows that $V(t)$ satisfies

$$\begin{cases} \langle \nabla V(t), \nabla w \rangle_{L^2(\Omega)} + \langle k'(P(t))V(t), w \rangle_{L^2(\Omega)} = \\ \qquad \langle f_1 - f_0, w \rangle_{L^2(\Omega)} + \langle g_1 - g_0, w \rangle_{L^2(\partial\Omega_{\mathrm{N}})} \\ \text{for all } w \in Y, \end{cases} \tag{5.108}$$

which is the variational formulation of (5.104). We next argue that $t \rightsquigarrow V(t)$ is Lipschitzian from $[0,1]$ to Y. For $t, \tau \in [0,1]$ we have

$$\begin{cases} |\nabla(V(t) - V(\tau))|^2_{L^2(\Omega)} + \langle k'(P(t)), (V(t) - V(\tau))^2 \rangle_{L^2(\Omega)} \\ \qquad = -\langle (k'(P(t) - k'(P(\tau))V(\tau), V(t) - V(\tau) \rangle_{L^2(\Omega)} \\ \text{for all } w \in Y, \end{cases}$$

and hence

$$\|V(t) - V(\tau)\|_Y^2 \leq \|k'(P(t) - k'(P(\tau)\|_{L^3(\Omega)} \|V(\tau)\|_{L^3(\Omega)} \|V(t) - V(\tau)\|_{L^3(\Omega)}.$$

But $L^3(\Omega)$ imbeds continuously into $H^1(\Omega)$ if the dimension m of Ω is smaller than 6, which is the case as we have supposed $m \leq 3$ in (1.57). Hence if we denote by const the corresponding embedding constant we obtain,

$$\begin{aligned} \|V(t) - V(\tau)\|_Y &\leq \text{const}\|k''\|_{L^\infty}\|P(t) - P(\tau)\|_{L^3(\Omega)}\|V(\tau)\|_{L^3(\Omega)} \\ &\leq \text{const}^3\|k''\|_{L^\infty}\|P(t) - P(\tau)\|_Y\|V(\tau)\|_Y \\ &\leq \text{const}^3 M\|k''\|_{L^\infty}\|P(t) - P(\tau)\|_Y\|(f_1, g_1) - (f_0, g_0)\|_E, \end{aligned}$$

where we have used (5.106) to obtain the last inequality. Since $t \rightsquigarrow P(t)$ is Lipschitz continuous, it follows that $t \rightsquigarrow A(t)$ is Lipschitz continuous as well. Hence V is a.e. differentiable, and using (5.103),

$$\|A(t)\|_Y \leq \text{const}^3 M^2\|k''\|_{L^\infty}\|(f_1, g_1) - (f_0, g_0)\|_E^2 \quad \text{a.e. in } [0, 1],$$

which proves that $P \in W^{2,\infty}([0, 1], Y)$. Derivation of (5.108) with respect to t gives then

$$\left\{ \begin{array}{l} \langle \nabla A(t), \nabla w\rangle_{L^2(\Omega)} + \langle k'(P(t))A(t), w\rangle_{L^2(\Omega)} = \\ \qquad\qquad -\langle k''(P(t))V(t)^2, w\rangle_{L^2(\Omega)} \\ \text{for all } w \in Y, \end{array} \right. \tag{5.109}$$

which is the variational formulation of (5.105). ■

Proposition 5.2.4 *(Hypothesis and notations of Sect. 1.5). There exists constants $\alpha_M > 0$, $C_R > 0$, and $C_\Theta > 0$ such that*

$$\|V(t)\|_F \leq \alpha_M\|(f_1, g_1) - (f_0, g_0)\|_E \tag{5.110}$$
$$\|A(t)\|_F \leq C_R\|k''\|_{L^\infty(I\!R)}\,\|V(t)\|_F^2 \tag{5.111}$$
$$\|A(t)\|_F \leq C_\Theta\|k''\|_{L^\infty(I\!R)}\,\text{diam}\, C\,\|V\|_F, \tag{5.112}$$

where F is the data space $L^2(\Omega)$ defined in (1.60). Hence (1.61) is a FC problem, and its curvature and deflection of (1.61) are bounded by

$$1/R = C_R\,\|k''\|_{L^\infty(I\!R)}, \qquad \Theta = C_\Theta\,\|k''\|_{L^\infty(I\!R)}\,\text{diam}\, C. \tag{5.113}$$

It follows that (1.61) is also a FC/LD problem as soon as

$$\Theta = C_\Theta\,\|k''\|_{L^\infty(I\!R)}\,\text{diam}\, C \leq \frac{\pi}{2}. \tag{5.114}$$

Proof. From (5.106) and the Poincaré inequality (5.101), (5.110) follows with $\alpha_M = C_P M = C_P(C_P^2 + C_N^2)^{1/2}$.

To prove (5.111), we introduce, for fixed but arbitrary $t \in [0,1]$ and $(f_0, g_0), (f_1, g_1) \in C$, an operator $B : D(B) \rightsquigarrow L^2(\Omega)$ defined by

$$D(B) = \{\psi \in H^2(\Omega) : \psi = 0 \text{ on } \partial\Omega_{\mathrm{D}} \text{ and } \frac{\partial\psi}{\partial\nu} = 0 \text{ on } \partial\Omega_{\mathrm{N}}\}, \qquad (5.115)$$

$$\forall\psi \in D(B), \quad B\psi = -\Delta\psi + k'(P(t))\psi \in L^2(\Omega), \qquad (5.116)$$

where $D(B)$ is endowed with the norm of $H^2(\Omega)$. Because of the assumption made in (1.57) that $\partial\Omega_{\mathrm{D}}$ is both open and closed with respect to $\partial\Omega$ and the $C^{1,1}$-regularity of $\partial\Omega$, the regularity results on elliptic equations imply that B is an isomorphism from $D(B)$ onto $L^2(\Omega)$:

$$\exists C_B > 0 \text{ such that } \frac{1}{C_B}\|\psi\|_{H^2} \le \|B\psi\|_{L^2} \le C_B\|\psi\|_{H^2} \quad \forall\psi \in D(B).$$

The same regularity results imply that the solution $A(t)$ of the variational problem (5.109) satisfies $A(t) \in H^2(\Omega)$, and hence is a strong solution of (5.105), so that $A(t) \in D(B)$, and

$$B\,A(t) = -k''(P(t))\,V(t)^2 \in L^1(\Omega) \subset L^2(\Omega).$$

Multiplication by $\psi \in D(B)$ and integration over Ω gives

$$\langle B\,A(t), \psi\rangle_{L^2(\Omega)} = -\langle k''(P(t))\,V(t)^2, \psi\rangle_{L^2(\Omega)} \quad \forall\psi \in D(B). \qquad (5.117)$$

$A(t)$ and ψ belong to $D(B) \subset H^2(\Omega)$, so one can integrate by part twice in the left-hand side of (5.117) using the Green formula, which gives

$$\langle A(t), B\,\psi\rangle_{L^2(\Omega)} = -\langle k''(P(t))\,V(t)^2, \psi\rangle_{L^2(\Omega)} \quad \forall\psi \in D(B).$$

Since B is an isomorphism from $D(B)$ onto $L^2(\Omega)$, one has

$$\begin{aligned}
|A(t)|_{L^2(\Omega)} &= \sup_{\psi\in D(B),\,|B\psi|_{L^2(\Omega)}=1} \langle A(t), B\,\psi\rangle_{L^2(\Omega)} \\
&= \sup_{\psi\in D(B),\,|B\psi|_{L^2(\Omega)}=1} \langle k''(P(t))\,V(t)^2, \psi\rangle_{L^2(\Omega)} \\
&\le \|k''\|_{L^\infty(\Omega)}|V|^2_{L^2(\Omega)} \sup_{\psi\in D(B),\,|B\psi|_{L^2(\Omega)}=1} \|\psi\|_{L^\infty(\Omega)} \\
&\le C_B C_H \|k''\|_{L^\infty(\Omega)}|V|^2_{L^2(\Omega)},
\end{aligned}$$

where C_H denotes the embedding constant of $H^2(\Omega)$ into $L^\infty(\Omega)$ (here we use the hypothesis in (1.57) that the dimension m of Ω is smaller than 3). This proves (5.111) and (5.112) with

$$C_R = C_B C_H, \qquad C_\Theta = C_B C_H C_P (C_P^2 + C_N^2)^{1/2}$$

and the proposition is proved. ■

To this point we have proved that (1.61) is a FC/LD problem when (5.114) is satisfied. We check now that the remaining hypothesis required to apply Theorem 5.1.15 on the *convergence of LMT-regularized solutions* are satisfied.

Lemma 5.2.5 *(Hypothesis and notations of Sect. 1.5). The mapping φ is injective over C, and has a Gâteaux derivative $\varphi'(x) \in \mathcal{L}(E, F)$ at every $x \in C$, which is also injective. Hence x is linearly identifiable over C, and all points x of C are qualified.*

Moreover, the range of $\varphi'(x)^* \in \mathcal{L}(F, E)$ is given by

$$Rg\,\varphi'(x)^* = \big\{(f, g) \in E : f \in D(B) \text{ and } g = \tau_N f\big\}, \tag{5.118}$$

where $D(B)$ is defined in (5.115), and τ_N denotes the trace on $\partial\Omega_N$.

Proof. Let $x_i = (f_i, g_i) \in C$, $i = 0, 1$, be given, and define $u_i = \varphi(x_i)$, $i = 0, 1$. Then $u_0 = u_1$ implies $\langle f_1 - f_0, w\rangle_{L^2(\Omega)} + \langle g_1 - g_0, w\rangle_{L^2(\partial\Omega_N)} = 0$ for all $w \in Y$, and hence $f_1 - f_0 = 0$ (take $w = \phi \in \mathcal{D}(\Omega)$) and $g_1 - g_0 = 0$ (take $w \in Y$ and use the density in $L^2(\partial\Omega_N)$ of the traces of functions of Y), which proves the injectivity of φ.

It is then easy to check that the Gâteaux differential $\delta u = \varphi'(x)(\delta f, \delta g) \in F = L^2(\Omega)$ of φ at $x = (f, g) \in C$ in the direction $\delta x = (\delta f, \delta g) \in E$ is the solution of

$$\begin{cases} \text{find } \delta u \in Y \text{ such that} \\ \langle \nabla \delta u, \nabla w\rangle_{L^2(\Omega)} + \langle k'(u)\delta u, w\rangle_{L^2(\Omega)} = \langle \delta f, w\rangle_{L^2(\Omega)} + \langle \delta g, w\rangle_{L^2(\partial\Omega_N)} \\ \text{ for all } w \in Y, \end{cases} \tag{5.119}$$

where u is the solution of (5.100). The proof of the injectivity for $\varphi'(x)$ is the same as for $\varphi(x)$.

Concerning the qualification of $x \in C$, the injectivity of $\varphi(x)$ ensures that the solution set X contains at most one element, so that $T(X, x) = \{0\}$, and the injectivity of $\varphi'(x)$ ensure that $\operatorname{Ker} \varphi'(x) = \{0\}$, so that (5.49) is trivially satisfied, and x is qualified.

We determine now $\varphi'(x)^*$ to find its range. Let $\delta v \in F = L^2(\Omega)$ and $\delta x = (\delta f, \delta g) \in E = L^2(\Omega) \times L^2(\partial\Omega_N)$ be given, and define δu and δh by

$$\delta u \in Y \quad \text{solution of} \quad (5.119), \tag{5.120}$$
$$\delta h \in D(B) \quad \text{solution of} \quad B\,\delta h = \delta v, \tag{5.121}$$

where B is the operator defined in (5.116) with $P(t)$ replaced by $u = \varphi(x)$. Then

$$\begin{aligned}
\langle \varphi'(x)^*\,\delta v, \delta x\rangle_E &= \langle \delta v, \varphi'(x)\,\delta x\rangle_F \\
&= \langle \delta v, \delta u\rangle_F \\
&= \langle B\,\delta h, \delta u\rangle_F \\
&= \langle -\Delta \delta h + k'(u)\delta h, \delta u\rangle_{F=L^2(\Omega)}.
\end{aligned}$$

Here $\delta h \in D(B) \subset H^2(\Omega)$, and $\delta u \in Y \subset H^1(\Omega)$, so we can use the Green formula to integrate one time by part:

$$\begin{aligned}
\langle \varphi'(x)^*\,\delta v, \delta x\rangle_E = \; & \langle \nabla \delta h, \nabla \delta u\rangle_{L^2(\Omega)} + \langle k'(u)\delta h, \delta u\rangle_{L^2(\Omega} \\
& + \left\langle \frac{\partial \delta h}{\partial \nu}, \delta u \right\rangle_{L^2(\partial\Omega_D)} + \left\langle \frac{\partial \delta h}{\partial \nu}, \delta u \right\rangle_{L^2(\partial\Omega_N)}.
\end{aligned} \tag{5.122}$$

Because of the boundary conditions included in the spaces Y and $D(B)$, the boundary terms vanish in (5.122), which becomes, using (5.119),

$$\begin{aligned}
\langle \varphi'(x)^*\,\delta v, \delta x\rangle_E &= \langle \delta f, \delta h\rangle_{L^2(\Omega)} + \langle \delta g, \delta h\rangle_{L^2(\partial\Omega_\mathrm{N})} \\
&= \langle (\delta h, \delta k), \underbrace{(\delta f, \delta g)}_{\delta x}\rangle_E \;,
\end{aligned} \tag{5.123}$$

where we have set

$$\delta k = \tau_\mathrm{N}\,\delta h \tag{5.124}$$

The last equation in (5.123) shows that

$$\forall \delta v \in F,\; \varphi'(x)^*, \delta v = (\delta h, \delta k) \in E, \quad \text{defined by (5.121) and (5.124)}, \tag{5.125}$$

which proves (5.118), and the lemma is proved. ■

The nonlinear source estimation problem (1.61) has hence the following properties before any regularization is applied:

Theorem 5.2.6 *Let hypothesis and notations of Sect. 1.5 hold. Then*

1. *The parameter x is identifiable on C, and (1.61) is a FC problem, with curvature and deflection bounded by $1/R$ and Θ given in (5.113).*

2. *If moreover the deflection Θ satisfies (5.114), then (1.61) is a FC/LD problem, and Proposition 4.2.7 applies: when the data z is in the neighborhood ϑ of the attainable set defined in (5.44), there exists at most one solution $\hat{x}$, when a solution exists the objective function J is unimodal over C, and the projection in the data space onto the attainable set is stable – but there is no stability estimate for x in the $L^2(\Omega) \times L^2(\partial\Omega_N)$ parameter norm.*

3. *If moreover C is bounded, the FC/LD problem (1.61) has a (unique) solution for all $z \in \vartheta$.*

Proof. Points 1 follows from Lemma 5.2.5 and Proposition 5.2.4, point 2 from Proposition 4.2.7. The proof of point 3 goes as follows: let $x_k = (f_k, g_k)$ be a minimizing sequence of J, and $u_k = \varphi(x_k)$ the associated solutions of (5.100). The sequence x_k is bounded in E and hence u_k is bounded in $Y \subset H^1(\Omega)$, which embeds compactly in $L^2(\Omega)$. Hence there exists $\hat{x} \in C$, $\hat{u} \in Y \subset H^1(\Omega)$ and subsequences, still noted x_k, u_k such that

$$x_k \rightharpoonup \hat{x} \text{ in } E,\ u_k \rightharpoonup \hat{x} \text{ in } Y \text{ and almost everywhere on } \Omega,$$

where $\rightharpoonup$ denotes weak convergence. It is then possible to pass to the limit in the variational formulation (5.100) giving u_k, which shows that $\hat{u} = \varphi(\hat{x})$. Hence $\hat{x}$ is a minimizer of J over C, which ends the proof. ■

We can now apply the LMT-regularization to the FC/LD nonlinear source problem:

Theorem 5.2.7 *Let hypothesis and notations of Sect. 1.5 as well as the deflection condition (5.114) hold, and suppose the data z belong to the neighborhood ϑ defined in (5.44). Then*

1. *The regularized problems (1.63) are all Q-wellposed for n large enough, and*

$$\epsilon_n \hat{f}_n \to 0, \qquad \epsilon_n \hat{g}_n \to 0,$$
$$u_n \ \to\ \hat{z}.$$

when $\epsilon_n \to 0$ and $\delta_n \to 0$, where $\hat{z}$ =projection of z onto $\overline{\varphi(C)}$.

2. *If the unregularized problem (1.61) admits a (necessarily unique) solution* $(\hat{f}, \hat{g})$ *– for example, if* C *is bounded in* $E = L^2(\Omega) \times L^2(\partial\Omega_N)$ *– then* $\hat{z} = \varphi(\hat{f}, \hat{g})$*, and one has, when* $\epsilon_n \to 0$ *and* $\delta_n/\epsilon_n \to 0$

$$\hat{f}_n \to \hat{f}, \qquad \hat{g}_n \to \hat{g},$$
$$\|u_n - \hat{z}\|_{L^2(\Omega)} = O(\epsilon_n).$$

3. *If moreover* $(\hat{f}, \hat{g})$ *satisfies the regularity condition:*

$$(f_0, g_0) - (\hat{f}, \hat{g}) \in \{(f \in D(B), g = \tau_n f)\} + T(C, \hat{x})^-, \tag{5.126}$$

where $D(B)$ *is defined in (5.115) and* τ_N *is the "trace on* $\partial\Omega_N$*" operator, one has, when* $\epsilon_n \to 0$ *and* $\delta_n \sim \epsilon_n^2$

$$\|\hat{f}_n - \hat{f}\|_{L^2(\Omega)} = O(\epsilon_n), \qquad \|\hat{f}_n - \hat{g}\|_{L^2(\partial\Omega_N)} = O(\epsilon_n),$$
$$\|u_n - \hat{z}\|_F = O(\epsilon_n^2).$$

Proof. Theorem 5.2.6 shows that the unregularized problem (1.61) is a FC/LD problem, so we can apply Theorem 5.1.15, which together with Lemma 5.2.5 gives the desired results. ■

We finally interpret the *regularity condition* (5.126) in two specific cases depending on the location of $(\hat{f}, \hat{g})$ within C:

- If $(\hat{f}, \hat{g})$ is in the interior of C, then $T(C, \hat{x})^- = \{0\}$, and (5.126) reduces to
$$(f_0, g_0) - (\hat{f}, \hat{g}) \in \{(f \in D(B), g = \tau_n f)\}. \tag{5.127}$$
- If C is a closed ball centered at the origin, and $(\hat{f}, \hat{g})$ lies on the boundary of C, then $T(C, \hat{x})^- = \{(f, g) : (f, g,) = \lambda(\hat{f}, \hat{g}),\ \lambda > 0\}$, and (5.126) becomes
$$\exists \lambda > 0 : (f_0, g_0) - (1 + \lambda)(\hat{f}, \hat{g}) \in \{(f \in D(B), g = \tau_n f)\}. \tag{5.128}$$

If zero is used as a-priori guess, then both "regularity conditions" (5.127) and (5.128) are equivalent to

$$\begin{cases} \hat{f} \in H^2(\Omega),\ \hat{f} = 0 \text{ on } \partial\Omega_D, \text{ and } \dfrac{\partial \hat{f}}{\partial \nu} = 0 \text{ on } \partial\Omega_N, \\ \hat{g} = \tau_N \hat{f} \in H^{3/2}(\partial\Omega_N). \end{cases} \tag{5.129}$$

This justifies the terminology "regularity condition" used for (5.51) or (5.126).

Remark 5.2.8 *Problem (1.61) is also a FC/LD problem for the stronger observation space* $F = H^1(\Omega)$*). In this case, it suffices to require in (1.57) that* $m \leq 4$*, and the condition that* $\partial\Omega_{\mathrm{D}}$ *is both open and closed is not needed. This follows from Lemma 5.2.3 and its proof, (5.106), and the following estimate:*

$$\begin{aligned} \|A(t)\|_Y &\leq \|k''\|_{L^\infty(I\!R)} |V(t)|^2_{L^2(\Omega)} \\ &\leq \|k''\|_{L^\infty(I\!R)} |V(t)|^2_{L^4(\Omega)} \\ &\leq \text{const} \|k''\|_{L^\infty(I\!R)} |V(t)|^2_Y, \end{aligned}$$

where const *is the embedding constant of* Y *into* $L^4(\Omega)$ *for* $m \leq 4$*. The proof of the existence of the Gâteaux derivative* $\varphi'(x)$ *is the same, and the characterization of its adjoint follows the same arguments, with now* $B \in \mathcal{L}(Y, Y')$ *and* $D(B) = Y$*. Then*

$$Rg\, \varphi'(x)^* = \big\{(f, g) \in E : f \in Y \text{ and } g = \tau_N f\big\},$$

and (5.129) is replaced by

$$\begin{cases} \hat{f} \in H^1(\Omega), \; \hat{f} = 0 \textit{ on } \partial\Omega_{\mathrm{D}}, \\ \hat{g} = \tau_N \hat{f} \in H^{1/2}(\partial\Omega_{\mathrm{N}}). \end{cases}$$

■

5.3 State-Space Regularization

We have seen in Sect. 5.1.3 that LMT-regularization failed to produce Q-wellposed problems for small ϵ's when applied to infinite curvature problems, that is, to problems where no upper bound to the curvature of the curves P of $\varphi(C)$ could be proved. So we consider in this section a special class of infinite curvature problem, where the origin of the ill-posedness resides in a poor measurement of the state of the system. To make this precise, we suppose that a *state-space decomposition* (2.1) (2.2) of the direct mapping $\varphi : x \in C \rightsquigarrow v \in F$ exists such that:

- The *parameter-to-state* map

$$\phi : x \in C \rightsquigarrow y \in F \;\; \text{solution of } e(x, y) = 0 \qquad (5.130)$$

can be "inverted" in a sense to be made precise. Hence x would be "identifiable" if a full measurement of the state $y \in Y$ were available,

- But the only observations available are measures of $v \in F$ related to y by a linear continuous *observation operator*, which incurs a loss of information,

$$M : y \in Y \rightsquigarrow v = My \in F, \qquad M \in \mathcal{L}(Y, F). \tag{5.131}$$

So the NLS problems (5.1) considered here are of the form

$$\hat{x} \quad \text{minimizes} \quad J(x) = \frac{1}{2}\| \underbrace{M\phi(x)}_{\varphi(x)} - z\|_F^2 \quad \text{over} \quad C, \tag{5.132}$$

where the data z is supposed to be *attainable*:

$$\exists\, \hat{x} \in C, \qquad z = \hat{z} = M\hat{y} \in F, \qquad \hat{y} = \phi(\hat{x}) \in Y. \tag{5.133}$$

For such problems, the *state-space* regularization [26, 29] consists in choosing an a-priori guess $y_0 \in Y$ of the state $y = \phi(x)$, and to use this additional information to build up the *state space regularized problem*:

$$\hat{x}_\epsilon \text{ minimizes } J_\epsilon(x) = \frac{\epsilon^2}{2}\|\phi(x) - y_0\|_Y^2 + J(x) \text{ over } C, \tag{5.134}$$

to be compared with the LMT-regularized problems (5.2).

State-space regularization corresponds to a widespread approach in the engineering community, which consists in smoothing or interpolating first the data before performing inversion. We analyze the behavior of problems (5.134) for two levels of hypotheses:

- *Strong hypotheses*: we suppose that the NLS problem associated to (C, ϕ) is Q-wellposed, and that *the observation is dense*, that is, the observation operator M is injective, and so there is no irremediable information loss. We shall prove in this case that the state-space regularized problems (5.134) are, after localization, Q-wellposed and converging when $\epsilon \to 0$. We shall use for this a *geometric approach*, along the lines of Chap. 4.

- *Weak hypotheses*: we suppose only that ϕ is invertible on C, and consider the case of a *incomplete observation*, where M is allowed to be noninjective – in which case there can be a definite loss of information. Here, Q-wellposedness of the state-space regularized problems will be lost, and only weak convergence results to a "state-space minimum-norm" solution will be available for subsequences of solutions. The tools there will be *soft analysis*.

5.3.1 Dense Observation: Geometric Approach

A typical example of this situation is $M =$ canonical injection from H^1 in L^2. It corresponds to parameters x, which are known to be OLS-identifiable from measurements of y in H^1, but where the only data at hand are measurements of y in L^2. For example, the one-dimensional elliptic inverse problem of Sect. 1.4 has been shown in Sect. 4.8 to be Q-wellposed when a strong H^1 observation was available, and so state-space regularization is indicated when only L^2 observation is available. A similar situation for L^2 observation occurs with the diffusion coefficient estimation problem in a 2D elliptic equation of Sect. 1.6, whose Q-wellposedness for H^1 observation is studied in Sects. 4.9 and 5.4.

Properties of (C, ϕ) and M:

We suppose that C and ϕ satisfy

$$\begin{cases} E = \text{Banach space, } Y = \text{Hilbert space,} \\ C = \text{closed, convex subset of } E, \\ \text{there exists } 0 < \alpha_m \le \alpha_M,\ R > 0 \text{ and } \Theta < \pi/2 \ \text{ such that} \\ \forall x_0, x_1 \in C, \quad P\ :\ t \rightsquigarrow \phi((1 - x_0)t + tx_1) \text{ is in } W^{2,\infty}([0,1]; F) \\ \text{and, for a.e. } t \in [0,1], \\ \alpha_m \|x_1 - x_0\|_E \le \|V(t)\|_Y \le \alpha_M \|x_1 - x_0\|_E, \\ \|A(t)\|_Y \le (1/R)\|V(t)\|_Y^2, \\ \|A(t)\|_Y \le \Theta \|V(t)\|_Y, \\ \text{where } V(t) = P'(t),\ A(t) = P''(t). \end{cases} \tag{5.135}$$

This ensures by Theorem 4.4.1 that the NLS problem

$$\hat{x} \quad \text{minimizes} \quad \frac{1}{2}\|\phi(x) - y\|_Y^2 \quad \text{over} \quad C \tag{5.136}$$

is Q-wellposed for y in a neighborhood of $\phi(C)$ of size R in Y. The strict inequality $\Theta < \pi/2$ is not required by Theorem 4.4.1, but it will simplify (somehow...) the calculations as it will allow to use the simple deflection condition $\Theta_n \le \pi/2$ to ensure Q-wellposedness of the regularized problems.

Then we suppose that the observation operator satisfies

$$M \in \mathcal{L}(Y, F), \qquad \text{with } M \text{ injective and } F = \text{Hilbert space,} \tag{5.137}$$

so that the true state $\hat{y}$ and parameter $\hat{x}$ associated to $z = \hat{z}$ by (5.133) are uniquely defined.

Properties of Data:

We suppose that a sequence $z_n \in F$, $n = 1, 2 \ldots$ of noise corrupted measurements is available, which converges to the attainable noise free data $z = \hat{z}$:

$$z_n \in F, \qquad \|z_n - \hat{z}\|_F \leq \delta_n, \qquad \delta_n \to 0 \text{ when } n \to \infty, \tag{5.138}$$

and that an a-priori guess

$$y_0 \in Y \tag{5.139}$$

of the true state $\hat{y}$ has been chosen. This a-priori guess can be enhanced by application of the linear LMT-regularization theory of Sect. 5.1.1 to the estimation of $\hat{y}$ from z_n, $n = 1, 2 \ldots$: let $\epsilon_n \to 0$ be a sequence of regularization parameters, and define y_n by the *auxiliary problem*

$$y_n \in Y \quad \text{minimizes} \quad \frac{\epsilon_n^2}{2}\|y - y_0\|_Y^2 + \frac{1}{2}\|M(y) - z_n\|_F^2 \quad \text{over} \quad Y. \tag{5.140}$$

Suppose that $\hat{y}$ satisfies the regularity condition (5.24) of Theorem 5.1.8 (iii):

$$\hat{y} - y_0 = M^* w \quad \text{for some } w \in F, \tag{5.141}$$

and that the regularization parameters ϵ_n^2 goes to zero more slowly than the error δ_n on the data:

$$\exists \lambda > 0 \text{ such that } \forall n \in I\!N : \lambda \, \|w\|_F \, \epsilon_n^2 \geq \delta_n. \tag{5.142}$$

The error estimate (5.41) gives then

$$\|y_n - \hat{y}\|_Y \leq \epsilon_n(\lambda + 1)\|w\|_F, \tag{5.143}$$

which leads us to use y_n as a priori guess for the state-space in lieu of y_0 in (5.134). So we replace in a first step (5.134) by

$$\hat{x}_n \text{ minimizes } J_n(x) = \frac{\epsilon_n^2}{2}\|\phi(x) - y_n\|_Y^2 + \frac{1}{2}\|M\,\phi(x) - z_n\|_F^2 \text{ over } C. \tag{5.144}$$

Properties of State-Space Regularized Problem on C:

To study the Q-wellposedness of (5.144), we define

$$\left\{ \begin{array}{ll} F_n = Y \times F \text{ equipped with the norm: } \|(y,z)\|_n^2 = \epsilon_n^2\|y\|_Y^2 + \|z\|_F^2 & \\ \varphi_n : x \in C \rightsquigarrow \big(\phi(x), M\,\phi(x)\big) \in F_n, & \\ Z_n = \big(y_n, z_n\big) & \text{(noisy data)}, \\ \widehat{Z} = \big(\hat{y}, \hat{z}\big) = \varphi_n(\hat{x}) & \text{(error free attainable data)}, \end{array} \right. \tag{5.145}$$

and rewrite it as

$$\hat{x}_n \quad \text{minimizes} \quad J_n(x) = \frac{1}{2}\|\varphi_n(x) - Z_n\|_n^2 \quad \text{over} \quad C. \tag{5.146}$$

A simple calculation shows that the quantities $\alpha_{m,n}, \alpha_{M,n}, R_n, \Theta_n$ associated to C and φ_n by (5.135) are

$$\begin{aligned}
\alpha_{m,n} &= \alpha_m \, \epsilon_n > 0, &(5.147)\\
\alpha_{M,n} &= \alpha_M(\epsilon_n^2 + \|M\|^2)^{\frac{1}{2}}, &(5.148)\\
R_n &= \frac{\epsilon_n^2}{(\epsilon_n^2 + \|M\|^2)^{1/2}} R > 0, \ R_n \to 0 \ \text{ when } n \to \infty, &(5.149)\\
\Theta_n &= \frac{(\epsilon_n^2 + \|M\|^2)^{1/2}}{\epsilon_n} \Theta \to \infty \ \text{ when } n \to \infty, &(5.150)
\end{aligned}$$

which satisfy, for all n,

$$\frac{\epsilon_{n+1}^2}{\epsilon_n^2} \leq \frac{R_{n+1}}{R_n} \leq \frac{\epsilon_{n+1}}{\epsilon_n}. \tag{5.151}$$

We see from (5.147)–(5.149) that (5.146) is a linearly stable FC-problem, which is good, but (5.150) shows also that its deflection Θ_n will become larger than $\pi/2$ for n large enough, which is bad! However, Θ_n decreases to $\Theta < \pi/2$ when $\epsilon_n \to +\infty$, so one chooses ϵ_0 such that

$$(\epsilon_0^2 + \|M\|^2)^{\frac{1}{2}} \geq \left((1 - \Theta^2/(\pi/2)^2\right)^{-\frac{1}{2}}\|M\| \quad \Leftrightarrow \quad \Theta_0 \leq \pi/2, \tag{5.152}$$

which ensures that $\varphi_0(C)$ is s.q.c. in F_0, and guarantees at least that (5.146) is Q-wellposed for $n = 0$!

Localizing the State-Space Problem to $C_n \subset C$:

Then for $n = 1, 2 \ldots$, we decide to control the deflection by limiting the size of the attainable set, and replace in a final step (5.146) by

$$\hat{x}_n \quad \text{minimizes} \quad J_n(x) = \frac{1}{2}\|\varphi_n(x) - Z_n\|_n^2 \quad \text{over} \quad C_n, \tag{5.153}$$

where we have restricted the search to the subsets C_n of C defined by

$$\begin{aligned}
C_0 &= C &(5.154)\\
C_n &= \{x \in C : \|\varphi_n(x) - \widehat{Z}\|_n \leq \chi_C \, R_n\} \quad \text{for } n = 1, 2 \ldots, &(5.155)
\end{aligned}$$

where $0 < \chi_C < 1$ will be chosen later.

Of course, this *localization constraint* cannot be implemented in practice, as the center of the ball is the error-free data $\widehat{Z}$, which by definition is unknown! But we shall see that the algorithm can be so tuned that this constraint is never active during the course of the resolution.

Lemma 5.3.1 *Let ϵ_0 satisfy (5.152), $\epsilon_n, n = 1, 2, \dots$ satisfy*

$$\exists\, 0 < \mu < 1 \quad \text{such that} \quad \epsilon_n^2 > \epsilon_{n+1}^2 \geq \mu\, \epsilon_n^2, \tag{5.156}$$

and $0 < \chi_C < \mu$ be chosen such that

$$\frac{2\chi_C}{(1 - \chi_C/\mu)^{\frac{1}{2}}} \leq \frac{\pi}{2}. \tag{5.157}$$

Then for all $n = 0, 1, \dots$ one has

$$C_n \supset C_{n+1}, \tag{5.158}$$
$$C_n \text{ is closed and convex in } E, \tag{5.159}$$
$$\varphi_n(C_n) \text{ is closed and s.q.c. in } F_n, \tag{5.160}$$

where $\varphi_n(C_n)$ is equipped with the family of paths image of the segments of C_n by φ_n.

Proof. We proceed by induction. The properties are clearly satisfied for $n = 0$. We suppose they hold for some $n \in I\!N$, and we prove that they hold for $n+1$:

- $C_n \supset C_{n+1}$: Let $x \in C_{n+1}$ be given. One has

$$\|\varphi_{n+1}(x) - \widehat{Z}\|_n \leq \frac{\epsilon_n}{\epsilon_{n+1}} \|\varphi_{n+1}(x) - \widehat{Z}\|_{n+1} \leq \chi_C \frac{\epsilon_n}{\epsilon_{n+1}} R_{n+1} \leq \chi_C R_n,$$

  where we have used (5.151), so that $x \in C_n$.

- C_{n+1} is convex: Let $x_0, x_1 \in C_{n+1}$ be given, and denote by P the curve $t \in [0, 1] \rightsquigarrow \varphi_{n+1}((1-t)x_0 + t x_1) \in F_{n+1}$, by p its reparameterization as function of the arc length ν, by L_{n+1} the arc length of P in F_{n+1}, and by f the function

$$f(\nu) = \|\widehat{Z} - p(\nu)\|_{n+1}^2 \text{ for } 0 \leq \nu \leq L_{n+1}. \tag{5.161}$$

  We want to show that $[x_0, x_1] \subset C_{n+1}$, that is,

$$f(\nu) \leq (\chi_C R_{n+1})^2 \text{ for } 0 \leq \nu \leq L_{n+1}. \tag{5.162}$$

Derivating twice f gives, as in the proof of the median Lemma 6.2.6

$$\begin{aligned}
f''(\nu) &= 2\big(1-\langle\widehat{Z}-p(\nu),a(\nu)\rangle_{n+1}\big) && (5.163)\\
&\geq 2\big(1-\sup_{0\leq\nu\leq L_{n+1}}\|\widehat{Z}-p(\nu)\|_{n+1}/R_{n+1}\big) && (5.164)\\
&\geq 2\big(1-\sup_{0\leq\nu\leq L_{n+1}}\|\widehat{Z}-p(\nu)\|_{n}/R_{n+1}\big) && (5.165)\\
&\geq 2\big(1-\chi_C R_n/R_{n+1}\big) && (5.166)\\
&\geq 2\big(1-\chi_c/\mu\big). && (5.167)
\end{aligned}$$

We have used in the fourth line the fact that C_{n+1} is included in the *convex* C_n, and inequalities (5.151) and (5.156) in the last line. The convexity property gives then

$$\begin{aligned}
f(\nu)+\nu(L_{n+1}-\nu)\Big(1-\frac{\chi_c}{\mu}\Big) &\leq \frac{L_{n+1}-\nu}{L_{n+1}}f(0)+\frac{\nu}{L_{n+1}}f(L_{n+1}) \quad (5.168)\\
&\leq (\chi_C R_{n+1})^2
\end{aligned}$$

for $0\leq\nu\leq L_{n+1}$. The term $1-\chi_C/\mu$ is strictly positive because of (5.157), and (5.162) is proved.

- $\varphi(C_n)$ is s.q.c. in F_{n+1}: Choosing $\nu=L_{n+1}/2$ in (5.168) gives

$$\frac{L_{n+1}^2}{4}\Big(1-\frac{\chi_c}{\mu}\Big)\leq(\chi_C R_{n+1})^2. \qquad (5.169)$$

Hence the deflection Θ_{n+1} of $\varphi_{n+1}(C_{n+1})$ in F_{n+1} satisfies

$$\Theta_{n+1}\leq\frac{L_{n+1}}{R_{n+1}}\leq\frac{2\chi_C}{(1-\chi_C/\mu)^{\frac{1}{2}}}\leq\frac{\pi}{2}, \qquad (5.170)$$

by choice of χ_C, and $\varphi_{n+1}(C_{n+1})$ is s.q.c. in F_{n+1}.

- C_n and $\varphi_n(C_n)$ are closed: C is closed, so the linear stability property of φ_n – see (5.147) – implies that $\varphi_n(C)$ is closed. Hence $\varphi_n(C_n)$ is closed as the intersection of two closed sets, and C_n is closed as the preimage of a closed set by a continuous mapping. ■

Main Result

We can now summarize the hypothesis we have made so far and state the main result of this section:

Theorem 5.3.2 *Let the following hypotheses and notations hold:*

- *C and $\varphi = M \circ \phi$ satisfy (5.135) and (5.137)*
- *The true data $\hat{z}$ is attainable: $\hat{z} = M\hat{y}$ with $\hat{y} = \phi(\hat{x})$* (5.133)
- *The noisy data $z_n \in F$ satisfy $\|z_n - \hat{z}\|_F \leq \delta_n$ with $\delta_n \to 0$* (5.138)
- *An a-priori guess y_0 of the true state $\hat{y}$ is available such that*

  $\hat{y} - y_0 = M^* w$ *for some* $w \in F$ (5.141)

- *A sequence of regularization parameters $\epsilon_n \to 0$ is chosen such that*

$$(\epsilon_0^2 + \|M\|^2)^{\frac{1}{2}} \geq \big((1 - \Theta^2/(\pi/2)^2\big)^{-\frac{1}{2}} \|M\|, \tag{5.152}$$
$$\exists\lambda \geq 0 : \forall n,\ \lambda \|w\|_F\, \epsilon_n^2 \geq \delta_n, \tag{5.142}$$
$$\exists\mu > 0 : \forall n,\ \epsilon_n^2 > \epsilon_{n+1}^2 \geq \mu\, \epsilon_n^2, \tag{5.156}$$

- *A sequence C_n of subsets C_n of C is defined by*

$$C_0 = C, \tag{5.154}$$
$$C_n = \big\{x \in C : \|\varphi_n(x) - \widehat{Z}\|_n \leq \chi_C\, R_n\big\} \quad n = 1, 2, \ldots, \tag{5.155}$$

  where $0 < \chi_C < \mu$ is chosen such that

$$0 < 2\chi_C/(1 - \chi_C/\mu)^{\frac{1}{2}} \leq \pi/2 \tag{5.157}$$

- *The a-priori guess y_0 is close enough to $\hat{y}$ so that*

$$\chi_Z/(1 - \chi_Z)^{1/2} \leq \mu\, \chi_C \quad \text{and} \quad \chi_Z < 1 - \chi_C, \tag{5.171}$$

  where χ_Z is defined by

$$\chi_Z = \frac{\|M\|}{R} \Big(\frac{(\lambda + 1)^2 + \lambda^2}{1 - \Theta^2/(\pi/2)^2} \Big)^{1/2} \|w\|_F \tag{5.172}$$

Then, for all $n = 0, 1, 2, \ldots,$ one has

1. *C_n is closed and convex in E*

2. *The* state-space regularized problem *(5.153)*

$$\hat{x}_n \textit{ minimizes } J_n(x) = \frac{\epsilon_n^2}{2}\|\phi(x) - y_n\|_Y^2 + \frac{1}{2}\|M\,\phi(x) - z_n\|_F^2 \textit{ over } C_n, \quad (5.173)$$

where y_n *is computed from* y_0 *by the* auxiliary problem

$$y_n \in Y \textit{ minimizes } \frac{\epsilon_n^2}{2}\|y - y_0\|_Y^2 + \frac{1}{2}\|M(y) - z_n\|_F^2 \textit{ over } Y \quad (5.140)$$

is Q-wellposed in F_n*: it has a unique solution* $\hat{x}_n$*, and* J_n *has no parasitic local minimum*

3. *The localization constraint are not active at* $\hat{x}_n$

$$\|\varphi_n(\hat{x}_n) - \widehat{Z}\|_n < \chi_C\, R_n. \quad (5.174)$$

4. $\hat{x}_n \in C_{n+1}$*, so it can be used as initial guess for the* $(n+1)$*-th optimization*

5. J_n *is strictly convex on* C_n *for* $n \geq 1$

Proof. The two first points follow from Lemma 5.3.1, which shows that C_n is convex, and that (5.173) is a FC/LD problem with an enlargement neighborhood of size R_n given by (5.149) and a linearly stable problem as one can see in (5.147) – and hence a Q-wellposed problem according to Theorem 4.4.1. It remains to check that the distance of the data $Z_n = (y_n, z_n)$ to $\varphi_n(C_n)$ is strictly smaller than R_n. Inequality (5.143) and (5.142) gives

$$\begin{aligned}
\|Z_n - \widehat{Z}\|_n^2 &= \epsilon_n^2\|y_n - \hat{y}\|_Y^2 + \|z_n - \hat{z}\|_F^2 \quad (5.175)\\
&\leq \epsilon_n^4\{(\lambda+1)^2 + \lambda^2\}\|w\|_F^2\\
&\leq \Big(\frac{R_n}{R}\Big)^2(\epsilon_n^2 + \|M\|^2)\{(\lambda+1)^2 + \lambda^2\}\|w\|_F^2\\
&\leq \Big(\frac{R_n}{R}\Big)^2(\epsilon_0^2 + \|M\|^2)\{(\lambda+1)^2 + \lambda^2\}\|w\|_F^2\\
&\leq \Big(\frac{R_n}{R}\Big)^2\|M\|^2\frac{(\lambda+1)^2 + \lambda^2}{1 - \Theta/(\pi/2)^2}\|w\|_F^2\\
&= (\chi_Z R_n)^2,
\end{aligned}$$

where we have used (5.149) and (5.152) in the third and fifth lines. Hence

$$d_n(Z_n, \varphi_n(C_n)) \leq \chi_Z R_n < R_n, \quad (5.176)$$

which ends the proof of the two first points.

We turn now to points three and four. Define

$$\begin{array}{rcl} \widehat{Z}_n & = & \varphi_n(\hat{x}_n) \\ P_n & : & t \rightsquigarrow \varphi_n((1-t)\hat{x}_n + t\hat{x}) \\ L_n & = & \text{arc length of } P_n \text{ in } F_n \\ d_n(t) & = & \|Z_n - P_n(t)\|_n \\ d_n & = & \sup_{0\le t\le 1} d_n(t). \end{array}$$

By construction, d_n has a minimum at $t = 0$, so we can apply the obtuse angle Lemma 6.2.9

$$\|Z_n - \widehat{Z}_n\|_n^2 + \Big(1 - \frac{d_n}{R_n}\Big) L_n^2 \le \|Z_n - \widehat{Z}\|_n^2.$$

But $d_n(Z_n, \varphi_n(C_n)) \le \|Z_n - \widehat{Z}\|_n \le \chi_C R_n < R_n$ with $\varphi_n(C_n)$ s.q.c., and Proposition 7.2.12 shows that $t \rightsquigarrow d_n(t)$ is an s.q.c. function. Hence $d_n = \max\{d_n(0), d_n(1)\} = \|Z_n - \widehat{Z}\|_n \le \chi_Z\, R_n$, which shows that

$$L_n \le \frac{\chi_Z}{(1-\chi_Z)^{1/2}} R_n \le \mu\, \chi_C\, R_n.$$

But $\mu < 1$ gives

$$\|\widehat{Z}_n - \widehat{Z}\|_n \le L_n < \chi_C\, R_n,$$

which is (5.174), and point three is proved. To prove point four, we notice that

$$\|\widehat{Z}_n - \widehat{Z}\|_{n+1} \le \|\widehat{Z}_n - \widehat{Z}\|_n \le L_n < \chi_C\, \mu\, R_n \le \chi_C\, R_{n+1},$$

where we have used (5.151), and point four is proved.

We prove now point five. Hypothesis (5.135) on C, ϕ and the properties (5.149) and (5.147) of φ_n show that, for $x_0, x_1 \in C_n$, the second Gâteaux derivative of J_n at x_0 in the direction $x_1 - x_0$ is, with the usual meaning for V_n and A_n,

$$\begin{array}{rcl} J_n''(x_0)(x_1 - x_0)^2 & = & \|V_n(0)\|_n^2 + \langle \varphi_n(x_0) - Z_n, A_n(0)\rangle_n, \\ & \ge & \|V_n(0)\|_n^2 \Big(1 - \dfrac{\|\varphi_n(x_0) - Z_n\|_n}{R_n}\Big), \\ & \ge & \epsilon_n^2 \alpha_m^2 \Big(1 - \dfrac{\|\varphi_n(x_0) - Z_n\|_n}{R_n}\Big). \end{array}$$

But

$$\|\varphi_n(x_0) - Z_n\|_n \leq \underbrace{\|\varphi_n(x_0) - \widehat{Z}\|_n}_{\leq \chi_C R_n \text{ for } n \geq 1} + \|\widehat{Z} - Z_n\|_n \leq (\chi_C + \chi_Z) R_n < R_n.$$

Hence J''_n is positive definite over C_n, which proves the last point. ■

5.3.2 Incomplete Observation: Soft Analysis

We suppose in this section that C, ϕ verify only

$$\begin{cases} E,\, Y = \text{reflexive Banach spaces,} \\ C = \text{closed, convex, bounded subset of } E, \\ \phi \,:\, C \rightsquigarrow Y \text{ is weakly sequentially closed, that is,} \\ x_n \rightharpoonup x \text{ in } E \text{ with } x_n \in C, \text{ and } \phi(x_n) \rightharpoonup \hat{\phi} \text{ in } Y \text{ imply} \\ x \in C \text{ and } \hat{\phi} = \phi(x), \end{cases} \tag{5.177}$$

$$\begin{cases} \phi \text{ is continuously invertible at } \hat{x} \in C, \text{ that is,} \\ \text{if } \phi(x_n) \to \phi(\hat{x}) \text{ in } Y \text{ then } x_n \to \hat{x} \text{ in } E, \end{cases} \tag{5.178}$$

where the symbols $\rightharpoonup$ and $\to$ denote, respectively, weak and strong convergence, and that the observation operator M satisfies

$$M \in \mathcal{L}(Y, F), \qquad \text{with } F = \text{normed linear space.} \tag{5.179}$$

In applications, M may be an imbedding, a restriction, a point evaluation, or a boundary observation.

We suppose that the true data $z = \hat{z}$ is *attainable*

$$\exists\, x^* \in C \quad : \quad \hat{z} = \varphi(x^*) \;\Leftrightarrow\; \hat{z} = My^* \in F \text{ with } y^* = \phi(x^*) \in Y, \tag{5.180}$$

but now x^* and y^* can be multiply defined as M, and possibly ϕ, is not necessarily injective. We also denote by $z_n \in F$ a sequence of noisy data

$$z_n \in F, \qquad \|z_n - \hat{z}\|_F \leq \delta_n, \qquad \delta_n \to 0 \text{ when } n \to \infty, \tag{5.138}$$

which converges to $\hat{z}$, and by

$$y_0 \in F \tag{5.139}$$

an a-priori guess for a "true" state y^*. The main difference with the previous section is that, due to the underdetermination for y^*, it is not possible to deduce from y_0 an upgraded a-priori guess y_n, which converges to one specific "true" state y^*! There are, however, often natural choices for y_0: for example, if $\hat{z}$ represent pointwise data in a finite dimensional space F and Y is a function space, then y_0 can be an interpolation in Y of the pointwise data. If both Y and F are function spaces with Y strictly embedded in F and $\hat{z} \in F$ but $\hat{z} \notin Y$, then y_0 would arise from $\hat{z}$ by a smoothing process, for example, by solving one auxiliary problem (5.140).

Hence the *state-space regularized problems* are here (compare with (5.173))

$$\hat{x}_n \text{ minimizes } J_n(x) = \frac{\epsilon_n^2}{2}\|\phi(x) - y_0\|_Y^2 + \frac{1}{2}\|M\,\phi(x) - z_n\|_F^2 \text{ over } C, \quad (5.181)$$

where the regularization parameters ϵ_n are chosen such that

$$\epsilon_n > 0, \qquad \epsilon_n \to 0, \qquad \delta_n/\epsilon_n \text{ bounded.} \quad (5.182)$$

Theorem 5.3.3 *Hypothesis and notations (5.177), (5.179), (5.180), (5.138), (5.139), and (5.182). Let $\hat{x}_n$ be a solution of (5.181) for $n = 1, 2, \ldots$ Then there exists a weakly convergent subsequence of $\{\hat{x}_n\}$, and every weak limit $\hat{x}$ of such a sequence $\{\hat{x}_{n_k}\}$ satisfies*

1. *$\hat{x}$ is a solution to the unregularized problem (5.132)*

$$\varphi(\hat{x}) = M\,\phi(\hat{x}) = \hat{z}.$$

2. *When $k \to \infty$,*

$$\phi(x_{n_k}) \rightharpoonup \phi(\hat{x}) \text{ weakly in } Y,$$

$$\|M\,\phi(\hat{x}_{n_k}) - \hat{z}\|_F = O(\epsilon_{n_k}).$$

3. *If $\delta_n/\epsilon_n \to 0$,*

$$\phi(x_{n_k}) \to \phi(\hat{x}) \quad \text{strongly in } Y,$$

and $\hat{x}$ is a state-space y_0-minimum-norm *solution:*

$$\|\phi(\hat{x}) - y_0\|_Y \le \min\Big\{\|\phi(x) - y_0\|_Y : x \in C \text{ and } M\,\phi(x) = \hat{z}\Big\}.$$

4. *If in addition ϕ satisfies the invertibility hypothesis (5.178) on C, then*

$$\hat{x}_{n_k} \to \hat{x} \quad \text{strongly in} \quad E.$$

Proof. We follow the proof of [26]. Because of (5.180), there exists $x^* \in C$ such that $M\,\phi(x^*) = \hat{z}$, so we have for all n

$$\frac{\epsilon_n^2}{2}\|\phi(\hat{x}_n) - y_0\|_Y^2 + \frac{1}{2}\|M\,\phi(\hat{x}_n) - z_n\|_F^2 \le \frac{\epsilon_n^2}{2}\|\phi(x^*) - y_0\|_Y^2 + \frac{1}{2}\|\hat{z} - z_n\|_F^2 \quad (5.183)$$
$$\le \frac{\epsilon_n^2}{2}\|\phi(x^*) - y_0\|_Y^2 + \frac{\delta_n^2}{2}.$$

Boundedness of δ_n/ϵ_n implies that $\phi(x_n)$ is bounded in Y. Since C is bounded as well, there exists a weakly convergent subsequence of x_n, again denoted by x_n, and $(\hat{x}, \hat{\phi}) \in E \times Y$ such that $(x_n, \phi(x_n)) \rightharpoonup (\hat{x}, \phi(\hat{x}))$ weakly in $E \times Y$. Then (5.177) implies that $\hat{x} \in C$ and $\hat{\phi} = \phi(\hat{x})$. Since $M\,\phi(x_n) \to \hat{z}$ in F, it follows also that $M\,\phi(\hat{x}) = \hat{z}$. This proves point 1 and the first assertion of point 2.

Since $\|z_n - \hat{z}\|_F \le \delta_n$ and $\delta_{n/}\epsilon_n$ is bounded, it follows that

$$\|M\,\phi(x_n) - \hat{z}\|_F \le \epsilon_n\|\phi(x^*) - y_0\|_Y + 2\delta_n = O(\epsilon_n),$$

which ends the proof of Part 2. Next we assume that $\delta_n/\epsilon_n \to 0$. Choosing $x^* = \hat{x}$ in (5.183) gives

$$\|\phi(\hat{x}_n) - y_0\|_Y^2 \le \|\phi(\hat{x}) - y_0\|_Y^2 + \frac{\delta_n^2}{\epsilon_n^2},$$

and consequently

$$\overline{\lim}\,\|\phi(\hat{x}_n) - y_0\|_Y \le \|\phi(\hat{x}) - y_0\|_Y \le \underline{\lim}\,\|\phi(\hat{x}_n) - y_0\|_Y.$$

This implies that $\phi(\hat{x}_n) \to \varphi(\hat{x})$ strongly in Y. Then (5.183) gives, for an arbitrary element x^* satisfying $M\,\phi(x^*) = \hat{z}$,

$$\|\phi(\hat{x}) - y_0\|_Y^2 = \lim_{n\to+\infty} \|\phi(\hat{x}_n) - y_0\|_Y^2, \quad (5.184)$$
$$\le \lim_{n\to+\infty} \left\{\|\phi(x^*) - y_0\|_Y^2 + \frac{\delta_n^2}{\epsilon_n^2}\right\}, \quad (5.185)$$
$$= \|\phi(x^*) - y_0\|_Y^2, \quad (5.186)$$

and Part 3 is proved. Finally Part 4 follows immediately from the hypothesis (5.178). ■

Because of the much weaker hypothesis involved on C, ϕ, and M, Theorem 5.3.3 is more widely applicable than Theorem 5.3.2: it applies of course to the same examples (one-dimensional elliptic inverse problem of Sect. 1.4, diffusion coefficient estimation problem in a 2D elliptic equation of Sect. 1.6) for point or boundary observation, but also for problems where Q-wellposedness for a state-space observation cannot be proved. Examples can be found in the original paper [26].

5.4 Adapted Regularization for Example 4: 2D Parameter Estimation with H^1 Observation

As we have seen in Sect. 1.3.4, *adapted regularization* tries to supply only the information on the parameter that cannot be retrieved from the data. It is expected to create less bias in the regularized estimate, and is the most desirable regularization. But its implementation is possible only after the missing information has been identified, which is a difficult task. So we illustrate this approach on the 2D elliptic parameter estimation problem described in Sect. 1.6, where this happens to be possible. Q-well-posedness of this problem was studied in Sect. 4.9, where linear stability and finite curvature could be obtained only at the price of reduction to finite dimension. We show now that adapted regularization permits to restore at least linear stability for the *infinite dimensional parameter set* C defined in (4.106). The material is adapted from the two references [30, 21].

Let notations and hypothesis (1.65) and (4.105) through (4.107) hold, so that Propositions 4.9.2 (a is linearly identifiable) and 4.9.3 (deflection condition) hold. Let $x_0, x_1 \in C$ be given, and for all $t \in [0,1]$ denote by $\eta(t)$ the associated velocity given by (4.114), which we rewrite as

$$\int_\Omega (a_1 - a_0)\nabla u_{a(t)} \cdot \nabla v = -\int_\Omega a(t)\nabla \eta(t) \cdot \nabla v, \quad \text{for all } v \in Y, \tag{5.187}$$

where $a(t) = (1-t)a_0 + ta_1$.

To study the linear stability property (4.39), one has first to equip the parameter space E with a norm for which this stability has some chance to hold. As can be seen in (4.111) for u and (5.187) for η, the diffusion coefficient a appears always as a coefficient of ∇u in the equations. Though $\nabla u \neq 0$ a.e. in Ω, it can vanish at stagnation points (see the proof of Proposition 4.9.2), and so there is little hope to obtain stability for a norm involving uniquely a.

So we shall rather consider stability for a gradient weighted "distance" in the parameter space:

$$d_{\text{grad}}(a_0, a_1) = \int_0^1 |(a_1 - a_0)\nabla u_{a(t)}|_{\mathbb{L}^2(\Omega)}\, \mathrm{d}t. \tag{5.188}$$

This quantity satisfies the two first axioms of a distance, but we do not know whether it satisfies the third one (triangular inequality). This is why we have written the word *distance* between quotes.

5.4.1 Which Part of a is Constrained by the Data?

To implement adapted regularization, we have to understand which part of $d_{\text{grad}}(a_0, a_1)$ is controlled by the arc length $\int_0^1 |\nabla\eta(t)|_{\mathbb{L}^2}\, dt$ in the data space of the curve image by φ of the $[a_0, a_1]$ segment. Once this will be done, it will be possible to introduce the *missing information* through the addition of an ad-hoc regularization term.

Equation (5.187) for η suggests to define an equivalence relation $\sim$ of vector fields in $\mathbb{L}^2(\Omega)$ by

$$\vec{q} \sim \vec{q}' \;\text{ if }\; (\vec{q}, \nabla v) = (\vec{q}', \nabla v) \;\text{ for all }\; v \in Y, \tag{5.189}$$

where $(\cdot,\cdot)$ denotes the scalar product in $\mathbb{L}^2(\Omega)$, and to decompose accordingly $\mathbb{L}^2(\Omega)$ into the sum of two orthogonal subspaces

$$\mathbb{L}^2(\Omega) = G \oplus G^{\perp},$$

where

$$\begin{cases} G = \mathbb{L}^2(\Omega)/_{\sim} & \text{the quotient space} \\ G^{\perp} & \text{the orthogonal complement,} \end{cases} \tag{5.190}$$

with orthogonal projections P and $P^{\perp}$.

Equation (5.187) becomes then

$$(a_1 - a_0)\nabla u_{a(t)} \sim -a(t)\nabla\eta(t) \quad \forall t \in [0, 1],$$

or equivalently

$$P\big((a_1 - a_0)\nabla u_{a(t)}\big) = P\big(-a(t)\nabla\eta(t)\big),$$

and hence

$$|P\big((a_1 - a_0)\nabla u_{a(t)}\big)|_{\mathbb{L}^2(\Omega)} \le a_M |\nabla \eta(t)|_{\mathbb{L}^2(\Omega)}. \tag{5.191}$$

Hence we see that $|\nabla\eta(t)|_{\mathbb{L}^2(\Omega)}$ constrains only the norm of the component in G of $(a_1 - a_0)\nabla u_{a(t)}$. There is no direct information on the norm of its component in $G^\perp$, it is exactly this information that has to be supplied by the regularization term.

Remark 5.4.1 *Before providing the missing information through a regularization term, one should investigate whether, by chance, the component in $G^\perp$ is not controlled by its component in G through the constraints that define the set C in (4.106). This would mean the existence of some $M > 0$ such that*

$$\begin{cases} \forall a_0, a_1 \in C \quad \textit{one has} \\ |P^\perp\big((a_1 - a_0)\nabla u_{a(t)}\big)|_{\mathbb{L}^2(\Omega)} \le M \; |P\big((a_1 - a_0)\nabla u_{a(t)}\big)|_{\mathbb{L}^2(\Omega)}. \end{cases} \tag{5.192}$$

A partial counter example to this property has been given in [30] for a simple diffusion problem on the unit square with a diffusion coefficient $a = 1$, no source or sink boundaries inside the domain, and flow lines parallel to the x_1-axis. It was shown there that for any perturbation h orthogonal to the flow line – and hence function of x_2 only – and satisfying

$$h \in \mathcal{C}^{0,1}([0,1]), \; h(0) = h(1) = 0, \; \int_0^1 h = 0, \tag{5.193}$$

one has, for some constant c,

$$\begin{cases} \frac{2}{3}|h|_{L^2([0,1])} \le |P^\perp(h\nabla u_a)|_{\mathbb{L}^2(\Omega)} \\ \qquad\qquad\qquad\quad |P(h\nabla u_a)|_{\mathbb{L}^2(\Omega)} \le 2c|h|_{H^{-1/2}([0,1])}. \end{cases}$$

But given $\epsilon > 0$, one can always construct a sequence h_n, $n \in \mathbb{N}$ of perturbations satisfying (5.193) and such that

$$|h_n|_{L^2([0,1])} = \epsilon > 0, \qquad |h_n|_{H^{-1/2}([0,1])} \to 0, \tag{5.194}$$

so that (5.192) cannot be satisfied for large n when $a_1 - a_0$ is replaced by h_n. However, for this construction to be a complete counter example to (5.192) would require the existence for each n of $a_{0,n}, a_{1,n} \in C$ and $t_n \in [0,1]$ such that

$$h_n = a_{1,n} - a_{0,n}, \qquad a = 1 = (1 - t_n)a_{0,n} + t_n a_{1,n}. \tag{5.195}$$

This in turn would require that h_n satisfies the additional constraints

$$\|h_n\|_{C^0[0,1]} \le a_M - a_m, \qquad Lip(h_n) \le 2b_M \quad \forall n \in \mathbb{N}, \tag{5.196}$$

where $Lip(h_n)$ denotes the Lipschitz constant of h_n.

Let g be a function on the real axis defined by

$$\begin{cases} g \in C^{0,1}(\mathbb{R}) \text{ with } g \text{ periodic of period one,} \\ g(0) = g(1) = 0, \qquad \int_0^1 g = 0, \\ g \text{ not identically zero,} \end{cases} \tag{5.197}$$

and define a sequence of perturbations h_n by

$$h_n(x_2) = g(nx_2) \quad \forall x_2 \in \mathbb{R}. \tag{5.198}$$

This sequence satisfies (5.193) and (5.194) with ϵ defined by $\epsilon^2 = \int_0^1 g^2$. It satisfies also the left part of the additional constraints (5.196), provided g is chosen such that $\|g\|_{C^0[0,1]} \le a_M - a_m$, which is always possible. However, the Lipschitz constant of h_n is $Lip(h_n) = nLip(g)$, so that the right part of (5.196) will necessarily be violated for n large enough. Hence a sequence h_n defined by (5.197) and (5.198) does not contradict (5.192), as it cannot satisfy (5.196) and hence cannot be of the form (5.195).

It is not known whether there exists a sequence h_n that satisfies (5.193), (5.194), and (5.195), and hence would contradicts (5.192). On the other side, there is for the time no proof that (5.192) holds true, and property (5.192), which would imply the linear stability for the d_{grad} "distance" on the infinite dimensional set C, remains undecided. ■

5.4.2 How to Control the Unconstrained Part?

To see which regularization term can be used to bring the missing information on $P^\perp\big((a_1 - a_0)\nabla u_{a(t)}\big)$, we use the *grad-rot decomposition* of $\mathbb{L}^2(\Omega) = G \oplus G^\perp$. We introduce for this the space

$$W = \{\psi \in H^1(\Omega) \colon \psi|_{\partial\Omega_N} = 0\},$$

(where the condition $\int_\Omega \psi = 0$ is added to the definition of W in the case where the Neuman boundary $\partial\Omega_N$ is empty), and recall the definition of rotational

$$\forall \varphi \in H^1(\Omega), \qquad \vec{\text{rot}}\ \varphi = \begin{pmatrix} \frac{\partial \varphi}{\partial x_2} \\ -\frac{\partial \varphi}{\partial x_1} \end{pmatrix},$$

$$\forall \vec{\psi} \in H^1(\Omega) \times H^1(\Omega), \qquad \text{rot}\ \vec{\psi} = \frac{\partial \psi_2}{\partial x_1} - \frac{\partial \psi_1}{\partial x_2}.$$

Lemma 5.4.2 *Let (1.65) and (4.105) through (4.107) and (5.189) and (5.190) hold. Then*

$$G = \{\nabla \varphi : \varphi \in Y\}, \qquad G^\perp = \{\vec{rot}\ \psi : \psi \in W\}, \tag{5.199}$$

and, for every $\vec{q} \in \mathbb{L}^2(\Omega)$, one has

$$P\vec{q} = \nabla \varphi, \qquad P^\perp \vec{q} = \vec{rot}\ \psi,$$

where $\varphi \in V$ and $\psi \in W$ are given by

$$\begin{cases} (\nabla \varphi, \nabla v) & = (\vec{q}, \nabla v) \quad \text{for all } v \in Y, \\ (\vec{rot}\ \psi, \vec{rot}\ w) & = (\vec{q}, \vec{rot}\ w) \quad \text{for all } w \in W. \end{cases} \tag{5.200}$$

Proof. Except for the atypical boundary conditions, this decomposition is rather standard. A outline of the proof, which is adapted from [39], Chap. 1, can be found in [30]. ■

Let then $\nabla \phi, \vec{\text{rot}}\ \psi$ be the grad-rot decomposition of $(a_1 - a_0)\nabla u_{a(t)}$ given by Lemma 5.4.2. We evaluate $|P^\perp\big((a_1 - a_0)\nabla u_{a(t)}\big)|_{\mathbb{L}^2(\Omega)} = |\vec{\text{rot}}\psi|_{\mathbb{L}^2(\Omega)}$:

$$\begin{aligned} |\vec{\text{rot}}\psi|_{\mathbb{L}^2(\Omega)} &= \sup_{\vec{s} \in \mathbb{L}^2(\Omega), |\vec{s}|_{\mathbb{L}^2(\Omega)}=1} (\vec{\text{rot}}\psi, \vec{s}) \\ &= \sup_{v \in Y, w \in W, |\nabla v|^2_{\mathbb{L}^2(\Omega)} + |\vec{\text{rot}}w|^2_{\mathbb{L}^2(\Omega)}=1} (\vec{\text{rot}}\psi, \nabla v + \vec{\text{rot}}w) \\ &= \sup_{w \in W, |\nabla w|_{\mathbb{L}^2(\Omega)}=1} (\vec{\text{rot}}\psi, \vec{\text{rot}}w) \\ |\vec{\text{rot}}\psi|_{\mathbb{L}^2(\Omega)} &= \sup_{w \in W, |\nabla w|_{\mathbb{L}^2(\Omega)}=1} \int_\Omega (a_1 - a_0)\nabla u_{a(t)} \cdot \vec{\text{rot}}w, \end{aligned} \tag{5.201}$$

where we have used the grad-rot decomposition $\nabla v + \vec{\text{rot}}\ w$ of $\vec{s}$, the orthogonality of G and $G^\perp$, the fact that $|\vec{\text{rot}}\ w| = |\nabla v|$, and the Definition (5.200) of ψ. Because of the smoothness hypotheses (4.105) and (4.106) on the parameters, the gradient – and hence the rotational – of u are in $\mathbb{L}^\infty(\Omega)$, and we can use the Green's formula in the right-hand side:

$$\int_\Omega (a_1 - a_0)\nabla u_{a(t)} \cdot \vec{\text{rot}}\, w = \int_\Omega \vec{\text{rot}}(a_1 - a_0) \cdot \nabla u_{a(t)}\, w - \int_{\partial\Omega} (a_1 - a_0) \frac{\partial u_{a(t)}}{\partial \tau}\, w. \tag{5.202}$$

where $\partial/\partial\tau$ denotes the tangential derivative along $\partial\Omega$. Since $u_a = 0$ on $\partial\Omega_{\text{D}}$ and $u_a = \text{const}$ on each $\partial\Omega_i$, we have $\partial u_a/\partial\tau = 0$ on $\partial\Omega_{\text{D}}$ and $\partial\Omega_i, i = 1, \cdots, N$, and $w = 0$ on $\partial\Omega_{\text{N}}$, so that the boundary term vanishes in (5.202). Combining (5.201) and (5.202) one obtains

$$|P^\perp((a_1 - a_0)\nabla u_{a(t)})|_{\mathbb{L}^2} = \sup_{w\in W,\, |\nabla w|_{\mathbb{L}^2(\Omega)}=1} \int_\Omega \vec{\text{rot}}(a_1 - a_0) \cdot \nabla u_{a(t)}\, w,$$

which, by Poincarés inequality in W,

$$|w|_{L^2(\Omega)} \le C_W |\nabla w|_{\mathbb{L}^2(\Omega)} \quad \forall w \in W \tag{5.203}$$

shows that

$$|P^\perp((a_1 - a_0)\nabla u_{a(t)})|_{\mathbb{L}^2(\Omega)} \le C_W |\vec{\text{rot}}(a_1 - a_0) \cdot \nabla u_{a(t)}|_{L^2(\Omega)}. \tag{5.204}$$

This shows that the part $P^\perp((a_1 - a_0)\nabla u_{a(t)})$ of parameter perturbation, which is not controlled by the data $\nabla u_{a(t)}$, is controlled by $\vec{\text{rot}}(a_1 - a_0) \cdot \nabla u_{a(t)}$, that is, by the *variation of the diffusion coefficient orthogonal to the flow lines*. This observation is in phase with the intuition that the knowledge of the pressure field $\nabla u_{a(t)}$ gives little information on the diffusion parameter a orthogonal to flow lines.

5.4.3 The Adapted-Regularized Problem

In absence of more specific a-priori information, one can then decide to search for diffusion coefficients which, loosely speaking, are "smooth orthogonal to the flow lines." This can be done by replacing the forward map φ defined in (4.112) by the *adapted-regularized* forward map:

$$\varphi_\epsilon : \quad a \leadsto (\nabla u_{a(t)}\,,\, \epsilon\, \vec{\text{rot}}\, a \cdot \nabla u_a) \in \mathbb{L}^2(\Omega) \times L^2(\Omega), \tag{5.205}$$

where $\epsilon > 0$ is the regularization parameter. The corresponding *adapted-regularized* problem is then (compare to (4.113)

$$\hat{a}_\epsilon \quad \text{minimizes} \quad \frac{1}{2}|\nabla u_a - z|^2_{\mathbb{L}^2} + \frac{\epsilon^2}{2}|\vec{\mathrm{rot}}a \cdot \nabla u_a|^2_{L^2} \quad \text{over } C. \tag{5.206}$$

5.4.4 Infinite Dimensional Linear Stability and Deflection Estimates

As expected, problem (5.206) is linearly stable:

Proposition 5.4.3 *Let (1.65) and (4.105) through (4.107) and (5.188) hold. Then the adapted-regularized forward map (5.205) satisfies the* linear stability *estimate (compare with (4.39)*

$$\begin{cases} \forall a_0, a_1 \in C \ \textit{one has} \\ \alpha_m(\epsilon)\ d_{\mathrm{grad}}(a_0, a_1) \leq \int_0^1 \|V_\epsilon(t)\|_{\mathbb{L}^2 \times L^2}\, \mathrm{d}t, \end{cases} \tag{5.207}$$

where d_{grad} is defined in (5.188), and where

$$\alpha_m(\epsilon)^2 = \mathrm{Min}\Big\{\frac{\epsilon^2}{2C_W^2}, \frac{1}{(1+2C_W^2)a_M^2}\Big\}, \tag{5.208}$$

with C_W equal to the Poincaré constant for the space W defined in (5.203).

Proof. Let $a_0, a_1 \in C$. For $t \in [0,1]$, derivation with respect to t of φ_ϵ evaluated at $a(t) = (1-t)a_0 + ta_1$ shows that the velocity $V_\epsilon(t)$ of the adapted-regularized problem is

$$V_\epsilon(t) = \Big(\nabla\eta(t),\ \epsilon\big(\vec{\mathrm{rot}}\ a(t) \cdot \nabla\eta(t) + \vec{\mathrm{rot}}\ (a_1 - a_0) \cdot \nabla u_{a(t)}\big)\Big), \tag{5.209}$$

where $\eta(t)$ is given by (5.187).

On the other hand, we obtain from the orthogonal decomposition $G \oplus G^\perp$ of $\mathbb{L}^2(\Omega)$ that

$$|(a_1 - a_0)\nabla u_{a(t)}|^2_{\mathbb{L}^2} = |P((a_1 - a_0)\nabla u_{a(t)})|^2_{\mathbb{L}^2} + \|P^\perp((a_1 - a_0)\nabla u_{a(t)})|^2_{\mathbb{L}^2},$$

and, using (5.191) and (5.204),

$$|(a_1 - a_0)\nabla u_{a(t)}|_{\mathbb{L}^2} \leq |(a_M\nabla\eta(t), C_W\ \vec{\mathrm{rot}}(a_1 - a_0) \cdot \nabla u_{a(t)})|_{\mathbb{L}^2 \times L^2}. \tag{5.210}$$

We define now, for each $t \in [0,1]$, a linear mapping G_t from $\mathbb{L}^2 \times L^2$ into itself by

$$\begin{cases} G_t(q,v) = (a_M\, q\ ,\ C_W(v/\epsilon - \vec{\text{rot}}\ a(t)\cdot q)), \\ \text{which satisfies} \\ \alpha_m(\epsilon)\|G_t(q,v)\|_{\mathbb{L}^2\times L^2} \le \|(q,v)\|_{\mathbb{L}^2\times L^2}, \end{cases} \tag{5.211}$$

so that

$$G(V_\epsilon(t)) = (a_M \nabla\eta(t),\ C_W \vec{\text{rot}}(a_1 - a_0)\cdot \nabla u_{a(t)}). \tag{5.212}$$

Combining then (5.210) with (5.212) and (5.211) gives

$$\alpha_m(\epsilon)|(a_1 - a_0)\nabla u_{a(t)}|_{\mathbb{L}^2} \le \|V_\epsilon(t)\|_{\mathbb{L}^2\times L^2}, \tag{5.213}$$

which gives (5.207) after integration between 0 and t. ■

So we see that adapted regularization ensures the stability of the diffusion coefficient a in the infinite dimensional set C for the L^2 norm weighted by $|\nabla u_a|$, which is what one could reasonably expect at best. As we have seen in the proof of Proposition 4.9.2, $|\nabla u_a(x)|$ can only vanish on a set of zero measure. In this case, the weighted stability constrains a almost everywhere on Ω, but with a strength that decreases when one approaches one (necessarily isolated) stationary point where $|\nabla u_a(x)| = 0$.

Remark 5.4.4 *Unweighted L^2 stability of a can only be obtained on subsets of Ω where a uniform lower bound γ to $|\nabla u_a|$ exists. In the very particular case where there is only one source or sink boundary $\partial\Omega_1$ with a flow rate $Q_1 \neq 0$, and where the remaining boundary is equipped with a Dirichlet condition (in short $\partial\Omega_N = \emptyset$), such a $\gamma \ge 0$ exists over the entire domain Ω, so that*

$$\gamma|a_1 - a_0|_{L^2(\Omega)} \le d_{grad}(a_0, a_1)$$

and the unweighted $L^2(\Omega)$-stability holds for a in the infinite dimensional set C. ■

The next step towards OLS-identifiability is to check whether the deflection condition $\Theta \le \pi/2$ can be satisfied by the adapted regularized problem. The velocity V_ϵ for this problem has been already given in (5.209), and the acceleration is

$$A_\epsilon(t) = \Big(\nabla\zeta(t),\ \epsilon\big(\vec{\text{rot}}\ a(t)\cdot\nabla\zeta(t) + 2\vec{\text{rot}}\ (a_1 - a_0)\cdot\nabla\eta(t)\big)\Big), \tag{5.214}$$

where $\eta(t)$ is given by (5.187).

Proposition 5.4.5 *Let (1.65) and (4.105) through (4.107) hold. The* deflection condition $\Theta \leq \pi/2$ *is satisfied for the adapted-regularized problem (5.206) as soon as*

$$\big((a_M - a_m)^2 + \epsilon^2 b_M^2 (a_M + a_m)^2\big)^{1/2} \leq \frac{\pi}{4}\, a_m, \tag{5.215}$$

where a_m, a_M, and b_M are the constants that define the admissible parameter set C in (4.106).

Proof. We use (5.214) to evaluate the norm of $A_\epsilon(t)$ to see whether it can satisfy the deflection sufficient condition (4.25)

$$\begin{aligned} a_m^2 \|A_\epsilon(t)\|_{F_\epsilon}^2 \;\leq\;& a_m^2 |\nabla\zeta(t)|_{\mathbb{L}^2}^2 + a_m^2 \epsilon^2 \big(\|\nabla a(t)\|_{\mathbb{L}^\infty} |\nabla\zeta(t)|_{\mathbb{L}^2} \\ & +2\|\nabla(a_1 - a_0)\|_{\mathbb{L}^\infty} |\nabla\eta(t)|_{\mathbb{L}^2}\big)^2 \\ \leq\;& \Big(4\|a_1 - a_0\|_{C^0}^2 + \epsilon^2 \big(2b_M \|a_1 - a_0\|_{C^0} \\ & +2a_m \|\nabla(a_1 - a_0)\|_{\mathbb{L}^\infty}\big)^2\Big) |\nabla\eta(t)|_{\mathbb{L}^2}^2, \end{aligned} \tag{5.216}$$

where we have used the majoration (4.120) of $|\nabla\zeta(t)|_{\mathbb{L}^2}$ by $2\|a_1 - a_0\|_{C^0} |\nabla\eta(t)|_{\mathbb{L}^2}$.

Using the Definition (4.106) of C one obtains

$$a_m \|A_\epsilon(t)\|_{F_\epsilon} \leq 2\big((a_M - a_m)^2 + \epsilon^2 b_M^2 (a_M + a_m)^2\big)^{1/2} |\nabla\eta(t)|_{\mathbb{L}^2}. \tag{5.217}$$

But (5.209) shows that $|\nabla\eta(t)|_{\mathbb{L}^2} \leq \|V_\epsilon\|_{F_\epsilon}$, so that (4.25) follows from (5.215) and (5.217), which ends the proof. ■

This is as far as one can currently go for the problem (5.206) on the infinite dimensional set C.

5.4.5 Finite Curvature Estimate

It is an open problem to know whether a finite curvature estimate (4.13) exists for the infinite dimensional admissible parameter set C. So one has to regularize one more time the problem by reduction to finite dimension:

Proposition 5.4.6 *Let (1.65) and (4.105) through (4.107) hold, and let $\mathbf{C}$ be the finite dimensional parameter set defined in (4.121). Then the adapted-regularized problem (5.206) has a finite curvature over $\mathbf{C}$, given by*

$$\frac{1}{R} = \frac{2}{a_m \alpha_m(\epsilon)}\, M(\epsilon, \mathbf{C}), \tag{5.218}$$

where $\alpha_m(\epsilon)$ given in (5.208) is independent of the dimension of $\boldsymbol{C}$, and where $M(\epsilon, \boldsymbol{C})$ is defined by

$$M(\epsilon, \boldsymbol{C}) = \sup_{a_0, a_1 \in \boldsymbol{C}\,,\ t \in [0,1]} M_{a_0,a_1,t}, \tag{5.219}$$

where

$$M_{a_0,a_1,t} = \frac{\|a_1 - a_0\|_{L^\infty}}{|(a_1 - a_0)\nabla u_{a(t)}|_{I\!L^2}} \Big(1 + \epsilon^2 (b_M + a_m \frac{\|\nabla(a_1 - a_0)\|_{L^\infty}}{\|a_1 - a_0\|_{L^\infty}})^2\Big)^{1/2}.$$

When $\boldsymbol{C}$ is defined through a finite element approximation with mesh size $h > 0$, the curvature $1/R$ blows up to infinity like $1/h^2$.

Proof. From (5.216) and $|\nabla\eta(t)|_{I\!L^2} \le \|V_\epsilon\|_{F_\epsilon}$ one obtains, with the notation $\|\cdot\|_{L^\infty}$ instead of $\|\cdot\|_{C^0}$,

$$\begin{aligned} a_m \|A_\epsilon(t)\|_{F_\epsilon} \le\ & 2\Big(\|a_1 - a_0\|^2_{L^\infty} \\ & + \epsilon^2 \big(b_M \|a_1 - a_0\|_{L^\infty} + a_m \|\nabla(a_1 - a_0)\|_{I\!L^\infty}\big)^2\Big)^{1/2} \|V_\epsilon(t)\|_{F_\epsilon}. \end{aligned}$$

Combining with (5.213) gives

$$\|A_\epsilon(t)\|_{F_\epsilon} \le \frac{2}{a_m \alpha_m(\epsilon)} M_{a_0,a_1,t}\ \|V_\epsilon(t)\|^2_{F_\epsilon},$$

which proves (5.218) using the finite curvature estimate (4.13).

When $\boldsymbol{C}$ is defined via finite elements with mesh size h, the ratios $\|\cdot\|_{L^\infty}/\|\cdot\|_{L^2}$ and $\|\nabla\cdot\|_{L^\infty}/\|\cdot\|_{L^\infty}$ blow up like $1/h$ when $h \to 0$, which proves the last part of the theorem. ■

5.4.6 OLS-Identifiability for the Adapted Regularized Problem

One can now combine Propositions 5.4.3, 5.4.5, and 5.4.6 with Theorem 4.4.1 to obtain *finite dimensional OLS-identifiability* for the adapted regularized problem. We shall need the diameter D of the attainable set for the adapted-regularization term

$$D = \sup_{a \in \boldsymbol{C}} |(\vec{\mathrm{rot}}\ a \cdot \nabla u_a)|_{L^2} \le b_M \sqrt{2} \sum_{i=1}^{N} C_i |Q_i|, \tag{5.220}$$

where C_i is the continuity constant of the linear mapping $v \in Y \rightsquigarrow v|_{\partial\Omega_i} \in I\!R$.

Theorem 5.4.7 *Let (1.65) and (4.105) through (4.107) hold, and suppose that*

1. *The deflection condition (5.215) is satisfied*
2. *The admissible parameter set* $\boldsymbol{C}$ *is finite dimensional as in (4.121)*

Let $1/R < +\infty$ *be the curvature of the adapted-regularized problem given by (5.218),* D *the diameter introduced in (5.220), and suppose that* ϵ *is such that*

$$0 < \epsilon < \bar{\epsilon}\ ,\ \text{where } \epsilon \text{ is defined by } \bar{\epsilon}D = R. \tag{5.221}$$

Then a *is OLS-identifiable via the adapted-regularized problem (5.206) on the finite dimensional parameter set* $\boldsymbol{C}$ *for the* d_{grad} *"distance" in* E*:*

1. Existence, uniqueness, and stability:

$$\text{for any } z_0, z_1 \text{ in } \vartheta = \{z \in \mathbb{L}^2(\Omega) \colon d_{\mathbb{L}^2}(z, \varphi(\boldsymbol{C})) < R - \epsilon\, D\} \tag{5.222}$$

such that

$$\|z_0 - z_1\|_{\mathbb{L}^2} + \max_{j=0,1} d_{\mathbb{L}^2}(z_j, \varphi(\boldsymbol{C})) + \epsilon D \le d < R, \tag{5.223}$$

the least-squares problem (5.206) stated on $\boldsymbol{C}$ *admits unique solutions* $\hat{a}_j$, $j = 0, 1$, *which satisfy*

$$\alpha_m(\epsilon)\, d_{\text{grad}}(\hat{a}_1 - \hat{a}_0) \le (1 - \frac{d}{R})^{-1} \|z_1 - z_0\|_{\mathbb{L}^2}, \tag{5.224}$$

where $\alpha_m(\epsilon)$ *is given by (5.208) and* d_{grad} *by (5.188).*

2. Optimizability:

   $\forall z \in \vartheta$, *the least squares problem (5.206) has no parasitic local minimum over* $\boldsymbol{C}$.

Proof. Propositions 5.4.3, 5.4.5, and 5.4.6 ensure that the hypotheses of Theorem 4.4.1 are satisfied for the adapted regularized problem (5.206). Hence existence, uniqueness, stability, and optimizability will hold for this problem as soon as the distance of the regularized data $z_\epsilon = (z, 0)$ to the regularized attainable set is strictly less than R:

$$d_{F_\epsilon}\big((z,0), \varphi_\epsilon(\boldsymbol{C})\big) < R. \tag{5.225}$$

Using the Definition 5.220 of D we obtain

$$\begin{aligned} d_{F_\epsilon}\big((z,0),\varphi_\epsilon(\boldsymbol{C})\big) &= \inf_{a\in\boldsymbol{C}} \big(\|z-\varphi(a)\|^2_{\mathbb{L}^2} + \epsilon^2 |\vec{\mathrm{rot}}a \cdot \nabla u_a|_{L^2}\big)^{1/2} \\ &\le \inf_{a\in\boldsymbol{C}} \big(\|z-\varphi(a)\|^2_{\mathbb{L}^2} + \epsilon^2 D^2\big)^{1/2} \\ &\le \big(\inf_{a\in\boldsymbol{C}} \|z-\varphi(a)\|^2_{\mathbb{L}^2} + \epsilon^2 D^2\big)^{1/2} \\ &\le \inf_{a\in\boldsymbol{C}} \|z-\varphi(a)\|_{\mathbb{L}^2} + \epsilon D \\ &\le d_{\mathbb{L}^2}\big(z,\varphi(\boldsymbol{C})\big) + \epsilon D. \end{aligned}$$

Hence condition (5.225) is satisfied as soon as $d_{\mathbb{L}^2}\big(z,\varphi(\boldsymbol{C})\big) \le R - \epsilon D$, and the announced existence, uniqueness, stability, and optimizability properties are simply a rewriting of (4.42)–(4.44) from Theorem 4.4.1. ■

Part II

A Generalization of Convex Sets

Chapters 6 and 7 give a comprehensive presentation of the theory of quasi-convex (q.c.) and strictly quasi-convex (s.q.c.) sets, which retain the desirable properties of convex sets listed at the beginning of Chap. 4.

Chapter 8 develops *deflection*-sufficient conditions for the strict quasi-convexity of a set, which are then applied to the case of the attainable set of a nonlinear least squares problem.

The presentation of s.q.c sets given there is different – and hopefully more natural – than that in the original material [20, 19, 28]. These chapters can be read independently of the rest of the book if one is interested in the projection on nonconvex sets in Hilbert spaces.

Chapter 6

Quasi-Convex Sets

In this chapter, we define a new class of subsets of a Hilbert space, called the quasi-convex sets to which properties (i) (uniqueness), (iii) (stability) and (iv) (existence as soon as the set is closed) of Proposition 4.1.1 can be generalized, provided they are required to hold only on some neighborhood. Technically, the whole chapter will consist in adapting the classical proofs for convex sets to the case where the segments are replaced by paths with finite curvature.

We postpone to Chap. 7 the generalization of property (ii) on the absence of parasitic stationary points.

There exists already many results concerning the projection on nonconvex sets. For example, given $\eta > 0$, a point X is called an *η-projection* of some point $z \in F$ on D if it satisfies

$$X \in D, \ \|X - z\| \leq d(z, D) + \eta. \tag{6.1}$$

So if one defines the so-called Edelstein set of D by

$$\begin{aligned} \mathcal{E}(D) \;=\; \Big\{ z \in F \mid \forall \varepsilon > 0, \ \exists \eta > 0 \quad \text{such that} \\ X_0, X_1 = \eta - \text{projections of } z \implies \|X_0 - X_1\| \leq \varepsilon \Big\}, \end{aligned} \tag{6.2}$$

then clearly any $z \in \mathcal{E}(D)$ has the property that any sequence that minimizes the distance to z over D is a Cauchy sequence. This implies that, when D is closed, all points of the Edelstein set $\mathcal{E}(D)$ have a unique projection on D. The interesting result is (cf, e.g., Aubin 1979) that this set fills out almost

G. Chavent, *Nonlinear Least Squares for Inverse Problems: Theoretical Foundations and Step-by-Step Guide for Applications*, Scientific Computation,
DOI 10.1007/978-90-481-2785-6_6,

completely F, in the sense that it is a dense countable intersection of open sets. This result alone does not allow to generalize Proposition 4.1.1 as desired for at least two reasons: the Edelstein result does not guarantee that the "bad points," which have no or more than one projection, will stay outside the neighborhood of D: the set $\mathcal{E}(D)$ will, in general, contain no neighborhood of D.

The second reason is that there is no guarantee, for a point $z \in \mathcal{E}(D)$, that the distance to z has no parasitic stationary points, which is one of the properties of convex sets we want to generalize.

So our generalizations of convex sets, the quasi-convex sets of this chapter and the strictly quasi-convex sets of Chap. 7, will be chosen in such a way that their Edelstein set $\mathcal{E}(D)$ contains a neighborhood ϑ of D. This will ensure an easy generalization of the properties (i) and (iv) of Proposition 4.1.1 (uniqueness and existence of the projection). Then (ii) and (iii) (stationary points and stability) will be generalized by other arguments.

In the applications to inverse problems we have in mind, D is the attainable set

$$D = \{\varphi(x) | x \in C\}, \tag{6.3}$$

image of the usually closed and convex admissible parameter set C by the nonlinear mapping φ to be inverted. So it will be easy, in the applications, to draw curves on D using the above parametric representation of D, for example, the images by φ of the segments of C. We shall call such a curve a "*path*," which, under certain conditions to be made precise at the end of Chap. 8, will have a finite curvature.

We take advantage of this at the abstract level, and suppose throughout the present chapter and the next one that the set D of F is equipped with a family $\mathcal{P}$ of finite curvature paths connecting any two points of D. These paths will play the role of the segments for D in our proofs when D happens to be convex. The *family of paths* $\mathcal{P}$ associated with the set $D \subset F$ will be defined in an axiomatic way, without reference to C and φ. So the results on quasi-convex and strictly quasi-convex sets can be used for other applications than for the analysis of nonlinear least squares problems.

The organization of the chapter is as follows:

- In Sect. 6.1, we define precisely the axioms that a family $\mathcal{P}$ of paths equipping D has to satisfy, as well as the geometric attributes that are naturally associated with a path of $\mathcal{P}$ (arc length, curvature, etc...).

- In Sect. 6.2, we define the property, for a set D equipped with a family of paths $\mathcal{P}$ – in short a set $(D,\mathcal{P})$ – to be quasi-convex. Then we re-do, for such a set, all classical proofs for the projection of a point on a convex set, with the necessary adaptations. This leads to the conclusion that the properties (i), (iii) and (iv) of Proposition 4.1.1 can be generalized to neighborhoods of quasi-convex sets. However, parasitic stationary points may still exist, and (ii) does not generalize to quasi-convex sets.

6.1 Equipping the Set D with Paths

The first step of our construction consists in choosing, in the possibly nonconvex set D, a *family* $\mathcal{P}$ *of paths* p, which will play for D the role the segments play for a convex set.

In a convex set, the segments are "pieces of straight line," and so have a zero curvature. We shall relax this property by requiring only that a path p of D is a "piece of a curve with finite curvature":

Definition 6.1.1 *Given* $\ell > 0$, *a curve* $p : [0,\ell] \to F$ *is a* path *of* D *if and only if*

$$p \in W^{2,\infty}([0,\ell];F) \tag{6.4}$$

$$p(\nu) \in D \quad \forall \nu \in [0,\ell] \tag{6.5}$$

$$\|p'(\nu)\|_F = 1 \text{ for a.e. } \nu \in [0,\ell] \tag{6.6}$$

In (6.4), $W^{2,\infty}([0,\ell];F)$ is the space of function from $[0,\ell]$ into F, whose two first distributional derivatives happen to be $L^\infty([0,\ell];F)$ functions.
Condition (6.5) ensures that the path p stays in the set D. Then by definition of the *arc length along* p, one has

$$\text{arc length from } p(0) \text{ to } p(\nu) = \int_0^\nu \|p'(\nu)\|_F \, d\nu, \tag{6.7}$$

which in the view of (6.6) can be rewritten simply as

$$\text{arc length from } p(0) \text{ to } p(\nu) = \nu. \tag{6.8}$$

Conversely, if ν is defined as the arc length along p, then necessarily (6.6) will hold (the proof will be given in Proposition 8.2.1 below). So the hypothesis 6.6 simply means that we have chosen to parameterize our paths p by their arc length ν, which brings considerable simplification in the evaluation of the geometric quantities associated with the path p:

Definition 6.1.2 *Let a path p be given as in Definition 6.1.1. Then* $\nu \in [0, \ell]$ *is the* arc length along p, *and*

$$\textit{the (arc) length of } p \textit{ is } \; L(p) \stackrel{\text{def}}{=} \ell, \tag{6.9}$$

$$\textit{the velocity } \; v(\nu) \stackrel{\text{def}}{=} p'(\nu) \textit{ is the unit tangent vector to } p \textit{ at } p(\nu), \tag{6.10}$$

$$\textit{the acceleration } \; a(\nu) \stackrel{\text{def}}{=} p''(\nu) \textit{ is the main normal to } p \textit{ at } p(\nu), \tag{6.11}$$

$$\textit{the radius of curvature of } p \textit{ at } \nu \textit{ is } \; \rho(\nu) \stackrel{\text{def}}{=} \|a(\nu)\|_F^{-1}, \tag{6.12}$$

where $\rho(\nu)$ *can possibly be infinite.*

So we see that, with the above choice of parametrization, the condition (6.4) on p can be rewritten as

$$1/\rho \in L^\infty([0, \ell]),$$

which is equivalent to say that p is a path with finite curvature.

Now that we have relaxed the "zero-curvature condition" for the paths of D, we want to equip D with a collection of paths, which retains the other properties of the segments of a convex set, namely the following:

- There is always one segment connecting any two distinct points of a convex
- any subsegment of a segment is a segment

So we are led to the following axiomatic definition of a family of paths for D:

Definition 6.1.3 *(Family of paths). A set of curves* $\mathcal{P}$ *is a family of paths on* D *if and only if*

$$\mathcal{P} \quad \underline{\textit{is made of paths on } D}: \tag{6.13}$$

$$\textit{all curves } p \textit{ of } P \textit{ satisfy Definition 6.1.1};$$

$$\mathcal{P} \quad \underline{\textit{is complete on } D}: \tag{6.14}$$

$$\forall X, Y \in D, \quad X \neq Y,$$

$$\exists p \in \mathcal{P} \textit{ such that } p(0) = X \textit{ and } p(L(p)) = Y;$$

$$\mathcal{P} \quad \underline{\textit{is stable with respect to restriction}}: \tag{6.15}$$

$$\forall p \in \mathcal{P}, \; \forall \nu', \nu'' \in [0, L(p)], \; \nu' < \nu'', \textit{the path}$$

$$\tilde{p} : \nu \in [0, \nu'' - \nu'] \to p(\nu' + \nu) \textit{ belongs to } \mathcal{P}.$$

From now on, we shall always consider that the set $D \subset F$ is equipped with a family of paths $\mathcal{P}$, which we shall write as $(D, \mathcal{P})$. The notions of quasi-convex and strictly quasi-convex sets will be developed for such couples $(D, \mathcal{P})$, and hence depend on the choice made for the family of paths $\mathcal{P}$, which equips D. We discuss now possible choices for the family of paths $\mathcal{P}$:

- When D is convex, one can always take $\mathcal{P} = \{\text{segments of } D\}$, and for this choice, all the results on quasi-convex sets (this chapter) and strictly quasi-convex sets (next chapter) will reduce to their classical convex counterparts.

- When D is nonconvex, a first natural idea, which is the direct generalization of one possible definition of the segments in a convex, is to define $\mathcal{P}$ as the collection of all minimum-length paths connecting any two points of D. The difficulty with this choice will be to check that the regularity properties of Definition 6.1.1 (in particular finite curvature!) are satisfied by the minimum length paths. This choice, when it is possible, is undoubtedly the most intrinsic one; the radii of curvature of the paths give in that case a direct information on those of the "manifold " D (we use quotes because we shall carefully avoid any rigorous developments involving D as a differential manifold), and so one can expect that the size$\times$curvature conditions to be derived later in Chap. 8 will then give the most precise characterization of strict quasi-convexity. However, the choice of the geodesics for $\mathcal{P}$ is not necessarily the best, even when D is convex (e.g., in the case where one searches for the best Lipschitz constant for the projection on a convex set, and where one wants to take into account the curvature of the boundary of D). The use of a geodesic or adapted collection of paths $\mathcal{P}$ is a completely open problem, which we do not consider in this book.

- In applications to nonlinear inverse problems, we will take advantage of D being the attainable set $\varphi(C)$ of the mapping φ to be inverted on C (see (6.3)), and choose for $\mathcal{P}$ the image by φ of all segments of C:

$$\mathcal{P} = \{\varphi([x, y]), x, y \in C\}. \tag{6.16}$$

It will also be sometimes convenient to suppose the existence in $\mathcal{P}$ of a *generating family of paths* $\mathcal{P}_G$, which mimics the family of segments connecting two points of the boundary of a convex set:

Definition 6.1.4 *(Generating family of paths). A subset $\mathcal{P}_G$ of $\mathcal{P}$ is said to be a generating family of paths for $\mathcal{P}$ if and only if*

$$\mathcal{P} = \cup_{p \in \mathcal{P}_G} \{p' | p' \text{ is a subpath of } p\}. \tag{6.17}$$

In the applications to nonlinear inverse problems (Chaps. 4 and 5, where D is defined by (6.3) and $\mathcal{P}$ by (6.16)), a *generating family of path* is given by

$$\mathcal{P}_G = \{\varphi([x, y]), x, y \in \partial C\}. \tag{6.18}$$

We conclude this section by indicating how the choice of a collection of paths $\mathcal{P}$ on D leads naturally to the definition of an "arc length distance" on D:

Definition 6.1.5 *Let $(D, \mathcal{P})$ be given. Then for any $X, Y \in D$, we call arc length distance in D of X and Y the quantity*

$$\delta(X, Y) = \sup_{\substack{p \in \mathcal{P} \\ p : X \to Y}} L(p), \tag{6.19}$$

with the convention that

$$\delta(X, Y) = 0 \quad \text{if there is no path from } X \text{ to } Y \tag{6.20}$$

Because of (6.14), (6.20) can arise only at points X, Y such that $Y = X$. But there may exists points X and paths p with nonzero length going from X to X, so that one would have at such points

$$\delta(X, X) > 0. \tag{6.21}$$

Such paths will not occur if $(D, \mathcal{P})$ is quasi-convex (Proposition 6.2.5 below), and a-fortiori if it is in the smaller class of strictly quasi-convex (s.q.c.) sets of Chap. 7. Notice that one always has

$$\|X - Y\| \le \delta(X, Y), \quad \forall X, Y \in D \tag{6.22}$$

and that δ does not necessarily satisfy the axiom of a distance! The reason for the sup in Definition 6.1.5 is that it will enable us to write in a simple way (see Proposition 6.2.11) the stability results for the projection on quasi-convex sets and s.q.c. sets, where the distance between the two projections will be measured by the length $L(p)$ of *any* path p of $\mathcal{P}$ connecting the two projections.

6.2 Definition and Main Properties of q.c. Sets

The word "quasi-convex" has been used with various meaning, especially when applied to functions. We use it here to qualify sets that share some of the properties of convex sets.

So let us now consider a set D of F equipped with a collection $\mathcal{P}$ of paths according to Definition 6.1.3.

Given a point $z \in F$, and a path $p \in \mathcal{P}$, we study the properties of the "distance to z along p" defined by

$$d_{z,p}(\nu) = \|z - p(\nu)\|_F, \qquad \forall \nu \in [0, L(p)] \tag{6.23}$$

(when D is convex and p is a segment, $d_{z,p}^2$ is a strictly convex function).

To that purpose, we introduce a quantity $k(z,p;\nu) \in I\!R$, which will indicate where the projection H of z on the main normal to p at $M = p(\nu)$ is located with respect to the center of curvature C of p (see Fig. 6.1). If we define

$$k(z,p;\nu) = \frac{MH}{MC},$$

where MH and MC denote the algebraic measures on the oriented normal, the condition

$$k(z,p;\nu) < 1 \tag{6.24}$$

corresponds to the "good" situation where the point H is as follows:

- Either on the "convex side" of p ($k(z,p;\nu) \leq 0$)
- Or on the "concave side" of p, but at a distance smaller than the radius of curvature ($0 \leq k(z,p;\nu) < 1$).

So one can expect that the situation deteriorates when k approaches 1, as this corresponds to H approaching C, and quasi-convex sets will be obtained by requiring that k stays uniformly away from 1. We give first a simple expression for $k(z,p;\nu)$ in terms of z and p. With the notation of Fig. 6.1,

$\rho(\nu) = MC =$ radius of curvature of p at M,
$\gamma =$ angle between $z - p(\nu)$ and $\vec{MC}$,

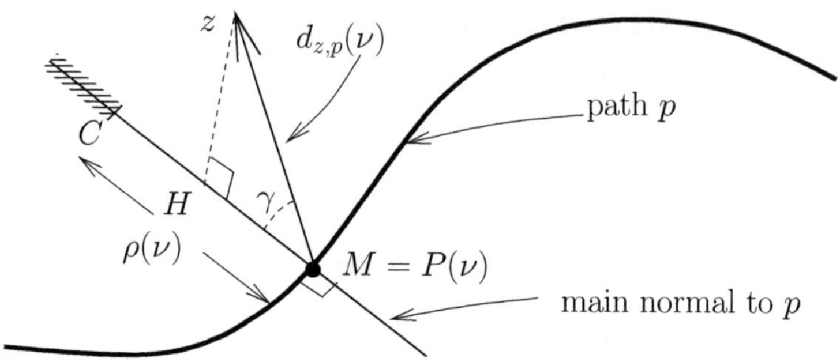

Figure 6.1: Notations for formula (6.25)

$k(z, p; \nu)$ can be rewritten as

$$k(z, p; \nu) = \frac{d_{z,p}(\nu)}{\rho(\nu)} \cos \gamma, \tag{6.25}$$

that is, using the fact that $a(\nu) = p''(\nu)$ is a normal vector with length $1/\rho(\nu)$ and pointing to the center of curvature C (see Definition 6.1.3):

$$k(z, p; \nu) = \langle z - p(\nu), a(\nu) \rangle. \tag{6.26}$$

First derivation of $d^2_{z,p}$ with respect to ν gives

$$\frac{\mathrm{d}}{\mathrm{d}\nu}\,(d^2_{z,p})(\nu) = -2\langle z - p(\nu), v(\nu) \rangle,$$

and second derivation gives, as $\|v(\nu)\| = 1$,

$$\frac{\mathrm{d}^2}{\mathrm{d}\nu^2}\,(d^2_{z,p})(\nu) = 2\Big(1 - k(z, p; \nu)\Big). \tag{6.27}$$

Then, given z and p, we shall consider the worst case when ν varies in $[0, L(P)]$, which in view of (6.24) is obtained by setting

$$k(z, p) = \sup_{\nu \in [0, L(p)]} k(z, p; \nu). \tag{6.28}$$

Finally, given z and $\eta > 0$, we will consider the worst case for all paths $p \in \mathcal{P}$, which connect two η-projections of z on D, that is, which are in the subset

$$\mathcal{P}(z, \eta) = \{p \in \mathcal{P} |\ \|p(j) - z\|_F \leq d(z, D) + \eta,\ j = 0, L(p)\} \tag{6.29}$$

of $\mathcal{P}$. This leads to define

$$k(z,\eta) = \sup_{p\in\mathcal{P}(z,\eta)} k(z,p). \tag{6.30}$$

The function $\eta \to k(z,\eta)$ is nondecreasing, and so it has a right limit at $\eta = 0$, which we shall denote by $k(z,0)$:

$$k(z,0) = \lim_{\eta\to 0_+} k(z,\eta). \tag{6.31}$$

By construction, $k(z,0)$ satisfies

$$k(z,0) \leq k(z,\eta), \quad \forall \eta > 0. \tag{6.32}$$

Definition 6.2.1 *(Quasi-convex set). Let $D \subset F$ be equipped with a family of path $\mathcal{P}$. The set $(D,\mathcal{P})$ is quasi-convex if and only if there exists a neighborhood ϑ of D in F and a lower semi-continuous (l.s.c.) function $\eta_{\max} : \vartheta \to]0,+\infty]$, such that*

$$\begin{cases} \forall z \in \vartheta,\ \forall \eta, \text{ such that } 0 < \eta < \eta_{\max}(z), \text{ one has} \\ k(z,\eta) < 1 \end{cases} \tag{6.33}$$

or equivalently

$$\begin{cases} \forall z \in \vartheta,\ \forall \eta, \text{ such that } 0 < \eta < \eta_{\max}(z), \text{ one has} \\ d^2_{z,p} \text{ is uniformly } \ \alpha - \text{convex over } \mathcal{P}(z,\eta). \end{cases} \tag{6.34}$$

We shall call ϑ a regular (q.c.) neighborhood *of $(D,\mathcal{P})$.*

The equivalence between the conditions (6.33) and (6.34) results from

$$\begin{aligned} \frac{\mathrm{d}^2}{\mathrm{d}\nu^2}\,(d^2_{z,p})(\nu) &= 2\Big(1 - k(z,p;\nu)\Big), && (6.35)\\ &\geq 2(1-k(z,p)), && (6.36)\\ &\geq 2(1-k(z,\eta)), \qquad \forall p \in \mathcal{P}(z,\eta). && (6.37) \end{aligned}$$

We shall see soon that, as its name suggests, the neighborhood ϑ is the one on which the "projection on D" is well-behaved. But before getting into this, we check first that Definition 6.2.1 ensures the existence of a largest open regular neighborhood ϑ:

Proposition 6.2.2 *Let $(D,\mathcal{P})$ be quasi-convex. Then there exists a* largest *open regular neighborhood ϑ of D, and a* largest l.s.c. function $\eta_{\max} : \vartheta \to]0,+\infty]$ *satisfying the definition 6.2.1 of quasi-convex sets.*

Proof. Let us denote by $\vartheta_i, \eta_{\max,i}, i \in I$ all open neighborhoods and l.s.c. functions satisfying Definition 6.2.1. Then

$$\vartheta = \cup_{i\in I}\vartheta_i \tag{6.38}$$

is an open neighborhood of D. If we denote by $\tilde{\eta}_{\max,i}$ the extension of $\eta_{\max,i}$ to ϑ by zero outside ϑ_i, then $\tilde{\eta}_{\max,i}$ is l.s.c. as $\eta_{\max,i}$ is l.s.c. on ϑ_i and ϑ_i is open. Define then

$$\forall z \in \vartheta : \eta_{\max}(z) = \sup_{i\in I} \tilde{\eta}_{\max,i}(z), \tag{6.39}$$

which is l.s.c. as a supremum of a family of l.s.c. functions. Hence $\vartheta, \eta_{\max}$ will satisfy the definition of quasi-convexity as soon as they satisfy (6.33), which we prove now. Let $z \in \vartheta$ and $0 < \eta < \eta_{\max}(z)$ be given, and set

$$\alpha = (\eta_{\max}(z) - \eta)/2 > 0.$$

From the Definition (6.39) of $\eta_{\max}(z)$, there exists $i_0 \in I$ such that

$$\tilde{\eta}_{\max,i_0}(z) \geq \eta_{\max}(z) - \alpha > \eta > 0.$$

This proves, as $\tilde{\eta}_{\max,i_0}(z) > 0$, that $z \in \vartheta_{i_0}$. But ϑ_{i_0} and $\eta_{\max,i_0}$ satisfy (6.33) by hypothesis, so that

$$k(z,\eta) < 1,$$

which proves that ϑ and $\eta_{\max}$ satisfy also (6.33). ■

We give now two very simple examples:

Example 6.2.3 *If D is convex and $\mathcal{P}$ made of the family of all segments of D, then $(D,\mathcal{P})$ is quasi-convex, with*

$$\begin{cases} \vartheta = F \\ \eta_{\max}(z) = +\infty \quad \forall z \in F. \end{cases}$$

■

Example 6.2.4 *Let D be an "arc of circle" of F of radius R and arc length L (Fig. 6.2), equipped with the sole family of path $\mathcal{P}$ available, made of all sub-arcs of circles included in D.*

Then D is quasi-convex as soon as

$$L < 2\pi R,$$

and the largest associated q.c. regular neighborhood ϑ is shown in Fig. 6.2, together with a graphical illustration of the way $\eta_{\max}(z)$ is determined. ■

We investigate now the properties of quasi-convex sets. We begin with the

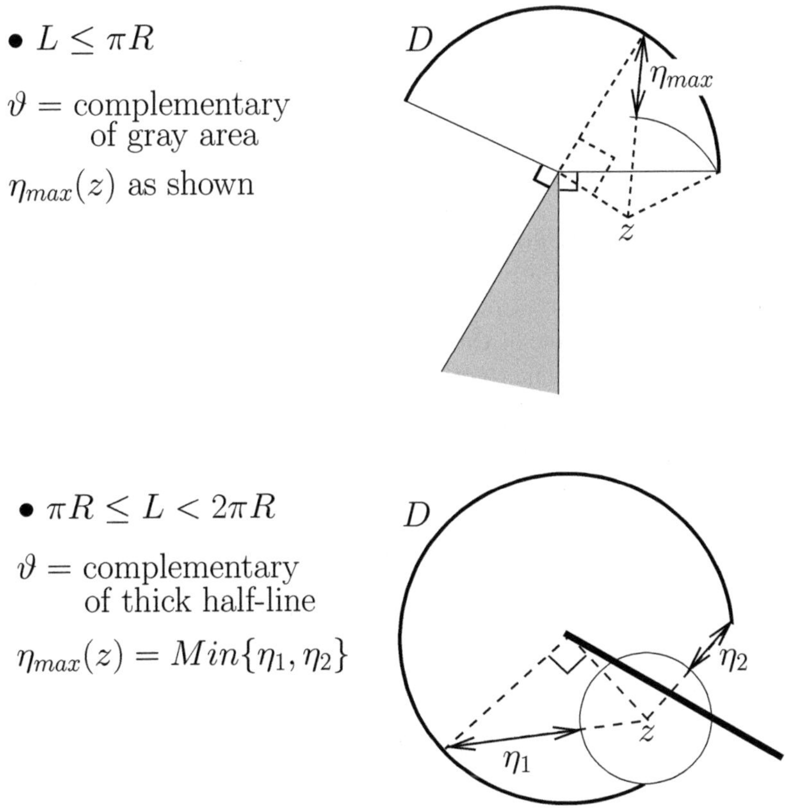

Figure 6.2: Quasi-convex arcs of circle

Proposition 6.2.5 *Let $(D, \mathcal{P})$ be quasi-convex, and $\vartheta, \eta_{\max}$ be associated regular neighborhood and function. Then the paths p satisfy*

$$\nu \to p(\nu) \text{ is injective} \quad \forall p \in \mathcal{P},$$

or equivalently, in term of arc length distance

$$\delta(X, X) = 0 \quad \forall X \in D.$$

Proof. If $\nu \to p(\nu)$ were not injective, there would exist $\nu', \nu'' \in [0, L(p)], \nu' < \nu''$, such that $p(\nu') = p(\nu'')$. Let us call this point of D as X

$$X = p(\nu) = p(\nu') \in D \quad \nu' < \nu'',$$

and $\tilde{p}$ the path

$$\tilde{p} : \nu \in [0, \nu'' - \nu'] \rightsquigarrow p(\nu' + \nu),$$

which is in $\mathcal{P}$ because of property (6.15) of the collection of paths $\mathcal{P}$. The path $\tilde{p}$ has both ends at X, and so $\tilde{p} \in \mathcal{P}(X; \eta)$ for any $0 < \eta < \eta_{\max}(X)$. This implies, using the Definition 6.2.1 of quasi-convex sets, that the function $\tilde{d}(\nu)^2 = \|\tilde{p}(\nu) - X\|^2$ is strictly convex, which is contradictory to the fact that $\tilde{d}(\nu) \geq 0$ and $\tilde{d}(0) = \tilde{d}(\nu'' - \nu') = 0$. This ends the proof of the injectivity of $\nu \to p(\nu)$, which in view of the Definition 6.1.5 of the arc length distance $\delta(X, Y)$ is clearly equivalent to $\delta(X, X) = 0$. ∎

We give now a lemma that generalizes the median lemma to triangles with one curvilinear side:

Lemma 6.2.6 *(median lemma). Let D be equipped with a collection of paths $\mathcal{P}$, and let $z \in F$ and $p \in \mathcal{P}$ be given. Then*

$$d_{1/2}^2 + (1 - k)\frac{L(p)^2}{4} \leq \frac{1}{2}d_0^2 + \frac{1}{2}d_1^2, \tag{6.40}$$

where (see Fig. 6.3)

$$d_t = \|z - p(tL(p))\|_F \quad \forall t \in [0, 1], \tag{6.41}$$

$$k = k(z, p) \text{ defined by } (6.28) \tag{6.42}$$

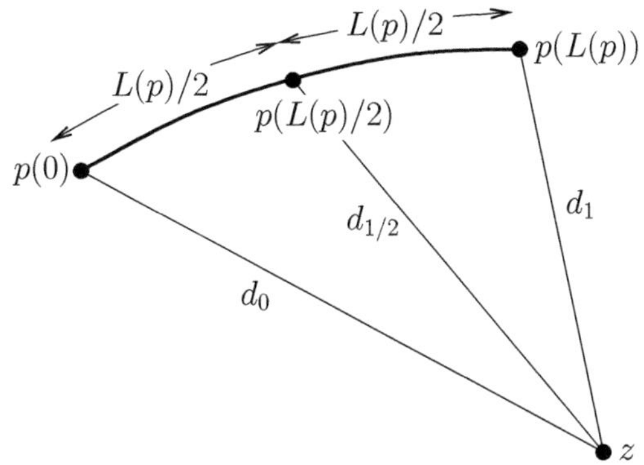

Figure 6.3: Notations for the median lemma

Proof. Define, for any $\nu \in [0, L(p)]$,

$$f(\nu) = \|z - p(\nu)\|_F^2 = (d_{\nu/L(p)})^2,$$

which satisfies

$$f''(\nu) \geq 2(1 - k(z,p)).$$

This proves that the function $\nu \to g(\nu) \stackrel{\text{def}}{=} f(\nu) + \nu(L(p) - \nu)(1 - k(z,p))$ is convex over $[0, L(p)]$. This implies that

$$g\Big(\frac{L(p)}{2}\Big) \leq \frac{1}{2}g(0) + \frac{1}{2}g(L(p)),$$

which is (6.40). ■

Of course, under the sole hypothesis that $z \in F$ and $p \in \mathcal{P}$, it can very well happen that $k \geq 1$, in which case the formula (6.40) is not very useful! So we shall use the median lemma in situations where one can ensure that $k < 1$, as, for example, in the

Corollary 6.2.7 *Let $(D, \mathcal{P})$ be quasi-convex, $\vartheta, \eta_{\max}$ be an associated regular neighborhood and function. Then for any $z \in \vartheta$ and any $\varepsilon > 0$, there exists $\eta = \eta(z, \varepsilon) \in]0, \eta_{\max}(z)[$ such that, for any η-projections X_0 and X_1 of z on D, one has*

$$\|X_0 - X_1\|_F \leq \delta(X_0, X_1) \leq \varepsilon. \tag{6.43}$$

This implies in particular that ϑ is included in the Edelstein set $\mathcal{E}(D)$ of D.

Proof. Let $z \in \vartheta$ and $\varepsilon > 0$ be given. For $\eta \in]0, \eta_{\max}(z)]$, let X_0 and X_1 be two η-projections of z on D. Then

either $X_0 = X_1$, in which case one has using Proposition 6.2.5

$$\|X_0 - X_1\|_F = \delta(X_0, X_1) = 0 < \varepsilon \text{ for any } \eta \in]0, \eta_{\max}(z)[. \qquad (6.44)$$

or $X_0 \neq X_1$, in which case there exists necessarily a path $p \in \mathcal{P}$ connecting X_0 to X_1. As X_0 and X_1 are η-projections of z, the path p is in $\mathcal{P}(z, \eta)$, so that $k(z, p) \leq k(z, \eta)$, with $k(z, \eta) < 1$ because of the quasi-convexity of $(D, \mathcal{P})$. Application of the median lemma 1 to z and p shows that (6.40) holds with

$$\begin{array}{lll} d_{1/2} & \geq d(z, D), & \\ 1 > k & = k(z, \eta) & \geq k(z, p), \\ d_0 & = \|X_0 - z\|_F & \leq d(z, D) + \eta, \\ d_1 & = \|X_1 - z\|_F & \leq d(z, D) + \eta, \end{array}$$

which gives

$$d(z, D)^2 + (1 - k(z, \eta))\frac{L(p)^2}{4} \leq (d(z, D) + \eta)^2 = d(z, D)^2 + 2\eta d(z, D) + \eta^2,$$

that is,

$$L(p)^2 \leq 4\eta \frac{2d(z, D) + \eta}{1 - k(z, \eta)}. \qquad (6.45)$$

When $\eta \to 0+$, $k(z, \eta) \to k(z, 0) < 1$ and the right-hand side of (6.45) goes to zero.

Hence there exists $\eta(z, \varepsilon) \in]0, \eta_{\max}(z)[$ such that

$$L(p) \leq \varepsilon \text{ as soon as } \eta = \eta(z, \varepsilon),$$

which implies

$$\|X_0 - X_1\|_F \leq L(p) \leq \varepsilon \text{ as soon as } \eta = \eta(z, \varepsilon).$$

Taking the supremum for all paths p, connecting X_0 to X_1 gives

$$\|X_0 - X_1\|_F \leq \delta(X_0, X_1) \leq \varepsilon \text{ as soon } \eta = \eta(z, \varepsilon). \qquad (6.46)$$

Comparing (6.44) and (6.46) shows that (6.43) holds in all cases as soon as η is taken equal to the $\eta(z,\varepsilon)$ determined in the $X_0 \neq X_1$ case, which ends the proof of the Corollary 6.2.7. ■

Notice that in Corollary 6.2.7, the estimation (6.43) on the proximity of two η-projections X_0 and X_1 of z is obtained not only for $\|X_0 - X_1\|_F$ (which corresponds exactly to saying that $\vartheta \subset \mathcal{E}(D)$), but also for the arc length distance $\delta(X_0, X_1)$: this stronger result will be very important in the applications we have in mind to nonlinear least-squares inversion where $D = \varphi(C)$, as in this case $\delta(X,Y)$ can be made equivalent in a natural way to the distance in the parameter set C, whereas $\|X - Y\|$ cannot.

We prove now that some of the properties of the *projection on* convex sets recalled in Proposition 4.1.1 generalize to *quasi-convex sets.* We begin with the

Proposition 6.2.8 *Let $(D,\mathcal{P})$ be quasi-convex, and $\vartheta, \eta_{\max}$ be a pair of associated regular neighborhood and function. Then properties (i) and (iv) of Proposition 4.1.1 generalize as follows:*

(i) Uniqueness: *for any $z \in \vartheta$, there exists at most one projection $\widehat{X}$ of z on D*

(ii) Existence: *if $z \in \vartheta$, any minimizing sequence $X_n \in D$ of the "distance to z" function over D is a Cauchy sequence in for both $\|X - Y\|_F$ and $\delta(X,Y)$. Hence X_n converges in F to the (unique) projection $\widehat{X}$ of z on the closure $\overline{D}$ of D.*

If D is closed, then $\widehat{X} \in D$, and $\delta(X_n, \widehat{X}) \to 0$ when $n \to 0$.

Proof. We prove first (i). Let $z \in \vartheta$ be such that it admits two projections $\widehat{X}_0$ and $\widehat{X}_1$ on D. As $\widehat{X}_0$ and $\widehat{X}_1$ are η-projections of z for any $\eta \in]0, \eta_{\max}(z)[$, we see from Corollary 6.2.7 that

$$\|\widehat{X}_0 - \widehat{X}_1\|_F \leq \delta(\widehat{X}_0, \widehat{X}_1) \leq \varepsilon \quad \text{for any } \varepsilon > 0,$$

which proves that $\widehat{X}_0 = \widehat{X}_1$. Hence z has at most one projection on D.

We prove now (ii). Let $z \in \vartheta$ and $\varepsilon > 0$ be given, and let $\eta(z,\varepsilon)$ be the associated value of η defined in Corollary 6.2.7. Let $\{X_n \in D, n \in I\!N\}$ be a minimizing sequence of the "distance to z function" over D, which satisfies by definition

$$\|X_n - z\|_F \to d(z,D) = \inf_{X \in D} \|X - z\|_F. \tag{6.47}$$

Then there exists $N(z,\varepsilon) \in I\!\!N$ such that

$$\forall n \geq N(z,\varepsilon), \qquad \|X_n - z\|_F \leq d(z,D) + \eta(z,\varepsilon), \tag{6.48}$$

which proves that all $X_n, n \geq N(z,\varepsilon)$ are $\eta(z,\varepsilon)$-projections of z on D. Then Corollary 6.2.7 implies that

$$\forall p,q \geq N(z,\varepsilon), \qquad \|X_p - X_q\|_F \leq \delta(X_p,X_q) \leq \varepsilon,$$

which shows that $\{X_n\}$ is a Cauchy sequence for $\|X-Y\|_F$ and $\delta(X,Y)$.

As F is complete (Hilbert space), the Cauchy sequence $\{X_n\}$ has a limit $\widehat{X} \in \overline{D}$, and using (6.47),

$$\|\widehat{X} - z\| = d(z,D) = \inf_{X \in D} \|X - z\|_F,$$

which shows that $\widehat{X}$ is the (necessarily unique as we have seen earlier) projection of z on $\overline{D}$.
When $\varphi(C)$ is closed, it remains to prove that X_n converges to $\widehat{X}$ also in the stronger arc length distance $\delta(X,Y)$.
This results once again from the Corollary 6.2.7:
$\widehat{X}$ is an η-projection for any $\eta \in]0,\eta_{\max}(z)[$, and X_n is an $\eta(z,\varepsilon)$-projection for all $n \geq N(z,\varepsilon)$, as we have seen in (6.48), so one has

$$\forall n \geq N(z,\varepsilon), \qquad \|X_n - \widehat{X}\|_F \leq \delta(X_n,\widehat{X}) \leq \varepsilon,$$

which proves that $\delta(X_n,\widehat{X}) \to 0$. ■

Notice that, if we choose for $(D,\mathcal{P})$ a convex set D equipped with the family $\mathcal{P}$ of its segments, then $\vartheta = F$ and $\|X-Y\|_F = \delta(X,Y)$, and the Proposition 6.2.8 reduces exactly to the corresponding results of Proposition 4.1.1!

We turn now to the generalization of the stability property (iii) of Proposition 4.1.1 to quasi-convex sets. We begin with two lemma:

Lemma 6.2.9 *(obtuse angle lemma). Let D be equipped with a collection of paths $\mathcal{P}$, and let $z \in F$ and $p \in \mathcal{P}$ be given.*
If

$$t \to d_t \stackrel{\text{def}}{=} \|z - p(tL(p))\|_F \quad \text{has a local minimum at } t = 0,$$

then

$$d_0^2 + (1 - k(z,p))L(p)^2 \leq d_1^2, \tag{6.49}$$

(where $k(z,p)$ is not necessarily smaller than one).

Proof. Define, as in the proof of Lemma 6.2.6, for any $\nu \in [0, L(p)]$:

$$f(\nu) = \|z - p(\nu)\|_F^2 = (d_{\nu/L(p)})^2.$$

Then $f(\nu)$ has also a local minimum at $\nu = 0$, so that

$$f'(0) \geq 0.$$

Derivating twice $f(\nu)$ gives, as in (6.36),

$$f''(\nu) \geq 2(1 - k(z,p)) \quad \forall \nu \in [0, L(p)].$$

Hence, the Taylor expansion

$$\begin{cases} f(L(p)) = f(0) + f'(0).L(p) + \frac{1}{2} f''(\nu_0).L(p)^2, \\ \text{where } \nu_0 \in [0, L(p)] \end{cases}$$

becomes

$$f(L(p)) \geq f(0) + (1 - k(z,p))L(p)^2,$$

which is (6.49). ■

We have illustrated in Fig. 6.4 the geometric interpretation of this lemma: formula (6.49) is the analogous for the curvilinear triangle $(z, p(0), p(L(p)))$ of the property that, in a triangle, the sum of squared length of edges adjacent to an obtuse angle is smaller than the squared length of the opposite edge. Of course, this analogy holds only in situations where z and p are such that $k(z,p) < 1$!

Lemma 6.2.10 *(Continuity lemma). Let D be equipped with a collection of paths $\mathcal{P}$, and let $z_0, z_1, \in F$ be two points of F admitting projections $X_0 \neq X_1$ on D.*

Then, for any path $p \in \mathcal{P}$ from X_0 to X_1, one has

$$(1 - k)L(p) \leq \|z_0 - z_1\|_F, \tag{6.50}$$

where

$$k = \frac{k(z_0,p) + k(z_1,p)}{2}. \tag{6.51}$$

Proof. Let z_0, z_1, X_0, X_1, and p be given as in the lemma. We define a function (see Fig. 6.5)

$$t \in [0,1] \rightsquigarrow d_t = \|(1-t)z_0 + tz_1 - p(t\ell)\|_F,$$

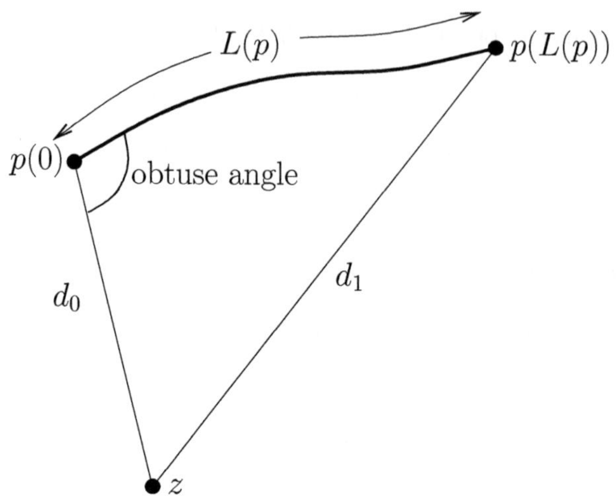

Figure 6.4: Illustration of the obtuse angle lemma

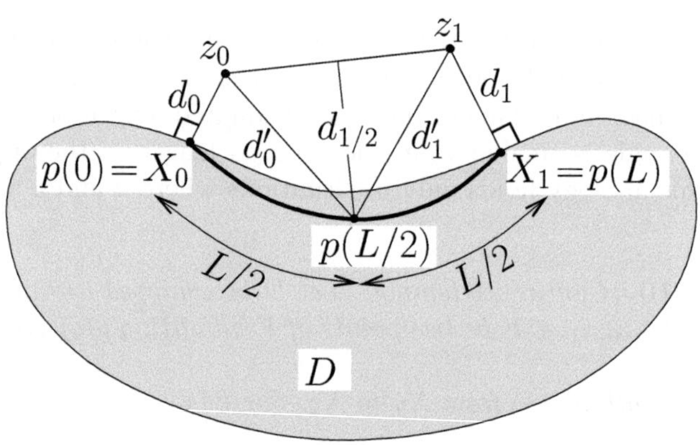

Figure 6.5: Notations for the continuity lemma

where we have used the notation

$$\ell = L(p)$$

for the length of the path p.

The second derivative of d_t^2 with respect to t is then (with the usual notation $v = p'$ and $a = p''$)

$$(d_t^2)'' = 2\|z_1 - z_0 - \ell v(t\ell)\|_F^2 - 2\langle (1-t)z_0 + tz_1 - p(t\ell), \ell^2 a(t\ell)\rangle_F,$$

which can be rewritten as

$$\begin{aligned}(d_t^2)'' &= 2\|z_1 - z_0\|_F^2 - 4\ell\langle z_1 - z_0, v(t\ell)\rangle_F + 2\ell^2\|v(t\ell)\|_F^2 \\ &-2(1-t)\ell^2\langle z_0 - p(t\ell), a(t\ell)\rangle_F \\ &-2t\ell^2\langle z_1 - p(t\ell), a(t\ell)\rangle_F.\end{aligned}$$

Using the Cauchy–Schwarz inequality, the fact that $\|v(t\ell)\|_F = 1$, and the inequality

$$\langle z_j - p(t\ell), a(t\ell)\rangle_F = k(z_j, p; t\ell) \le k(z_j, p) \quad j = 0, 1,$$

one can minorate $(d_t^2)''$ as follows:

$$(d_t^2)'' \ge 2\ell^2\Big\{1 - 2\frac{\|z_0 - z_1\|_F}{\ell} + \frac{\|z_0 - z_1\|_F^2}{\ell^2} - (1-t)k_0 - tk_1\Big\},$$

where

$$k_j = k(z_j, p), \quad j = 0, 1.$$

This implies the convexity of the function

$$t \rightsquigarrow d_t^2 + t(1-t)\ell^2\Big\{1 - \frac{2}{3}(k_0 + k_1) + \frac{1}{3}((1-t)k_1 + tk_0) - 2\frac{\|z_0 - z_1\|_F}{\ell} + \frac{\|z_0 - z_1\|_F^2}{\ell^2}\Big\}$$

over the $[0, 1]$ interval. Hence

$$d_{\frac{1}{2}}^2 + \frac{\ell^2}{4}\Big\{1 - \frac{1}{2}(k_0 + k_1) - 2\frac{\|z_0 - z_1\|_F}{\ell} + \frac{\|z_0 - z_1\|_F^2}{\ell^2}\Big\} \le \frac{1}{2}d_0^2 + \frac{1}{2}d_1^2. \quad (6.52)$$

We now use the hypothesis that X_0 is a projection of z_0 on D. This implies that the "distance to z_0" function has a local minimum on p at $\nu = 0$. Hence the curvilinear triangle $(z_0, X_0, p(\ell/2))$ has an "*obtuse angle*" at X_0. Application of the Lemma 6.2.9 gives then, with the notation of Fig. 6.5,

$$d_0^2 + (1 - k_0)\frac{\ell^2}{4} \le d_0'^2 .$$

In a similar way, the fact that X_1 is a projection of z_1 on D implies that

$$d_1^2 + (1 - k_1)\frac{\ell^2}{4} \le d_1'^2.$$

Combining the two gives

$$\frac{1}{2}d_0^2 + \frac{1}{2}d_1^2 \le \frac{1}{2}d_0'^2 + \frac{1}{2}d_1'^2 - \frac{\ell^2}{4}\Big\{1 - \frac{1}{2}(k_0 + k_1)\Big\} ,$$

which, combined with (6.52), shows that

$$\frac{\ell^2}{2}\Big\{1 - \frac{1}{2}(k_0 + k_1)\Big\} - \frac{\ell}{2}\|z_0 - z_1\|_F \le \frac{1}{2}d_0'^2 + \frac{1}{2}d_1'^2 - d_{\frac{1}{2}}^2 - \frac{1}{4}\|z_0 - z_1\|^2 . \quad (6.53)$$

But F is a Hilbert space, and from the median theorem in the triangle $(z_0, p(\ell/2), z_1)$, we see that the right-hand side of (6.53) is equal to zero. Dividing then (6.53) by $\ell/2$, we obtain

$$\Big\{1 - \frac{1}{2}(k_0 + k_1)\Big\}\ell \le \|z_0 - z_1\|_F,$$

which is the announced result. ■

We can now prove the

Proposition 6.2.11 *Let $(D, \mathcal{P})$ be quasi-convex, and $\vartheta, \eta_{\max}$ be a pair of associated regular neighborhood and function. Then property (iii) of Proposition 4.1.1 on the stability of the projection generalizes as follows:*
If

$$z_0, z_1, \in \vartheta \text{ admit projections } X_0, X_1 \text{ on } D, \quad (6.54)$$

and are close enough so that

$$\begin{cases} \exists d \ge 0 \text{ such that} \\ \|z_0 - z_1\|_F + \max_{j=0,1} d(z_j, D) \\ \qquad \le d < \min_{j=0,1}\{d(z_j, D) + \eta_{\max}(z_j)\}, \end{cases} \quad (6.55)$$

then, for any d chosen as above, one has

$$\|X_0 - X_1\|_F \le \delta(X_0, X_1) \le (1-k)^{-1}\|z_0 - z_1\|_F, \quad (6.56)$$

where $k < 1$ is defined by

$$k = (k(z_0, \eta_0) + k(z_1, \eta_1))/2, \quad (6.57)$$

with $\eta_j, j = 0, 1$ defined by

$$0 \le \eta_j = d - d(z_j, D) < \eta_{\max}(z_j). \quad (6.58)$$

Proof. We check first that the hypothesis (6.55) is satisfied as soon as $\|z_0 - z_1\|$ is small enough: if $z_0 \in \vartheta$ and $z_1 \to z_0$, then

$$\|z_0 - z_1\|_F + \max_{j=0,1} d(z_j, D) \to d(z_0, D)$$

because of the continuity of the norm and the "distance to D" function, and

$$\lim_{\varepsilon \to 0} \inf_{\|z_0 - z_1\| \le \varepsilon} \min_{j=0,1} \{d(z_j, D) + \eta_{\max}(z_j)\} \to d(z_0, D) + \eta_{\max}(z_0) > d(z_0 p)$$

because of the lower semi continuity of the function $f : z \rightsquigarrow d(z, D) + \eta_{\max}(z)$, which implies that $z \rightsquigarrow \min\{f(z), f(z_0)\}$ is l.s.c. at $z = z_0$. This ensures the existence of a d satisfying (6.55).

We check now that the functions η_0, η_1 defined by (6.58) satisfy inequalities announced there, namely

$$0 \le \eta_j < \eta_{\max}(z_j), \qquad j = 0, 1. \tag{6.59}$$

As d satisfies (6.55), we see that

$$\|z_0 - z_1\|_F + d(z_j, D) \le d < d(z_j, D) + \eta_{\max}(z_j) \quad j = 0, 1,$$

which proves (6.59) by subtracting $d(z_j, D)$ from all three terms and using the definition

$$\eta_j = d - d(z_j, D), \qquad j = 0, 1$$

of η_j. Majoration (6.59) implies, by definition of quasi-convex sets, that $k(z_j, \eta_j) < 1$ for $j = 0, 1$, so that k defined by (6.57) satisfies also $k < 1$.

We check now that X_0 is an η_1-projection of z_1 on D:

$$\begin{aligned} \|X_0 - z_1\|_F &\le \|X_0 - z_0\|_F + \|z_0 - z_1\|_F \\ &= d(z_0, D) + \|z_0 - z_1\|_F \\ &\le d = d(z_1, D) + \eta_1 \ . \end{aligned}$$

A similar proof shows that X_1 is an η_0-projection of z_0 on D. Hence

$$X_0, X_1 \text{ are } \eta_j - \text{projections of } z_j \text{ on } D, \quad j = 0, 1. \tag{6.60}$$

We conclude with the proof of the stability result (6.56):

- If $X_0 = X_1$, then the Proposition 6.2.5 ensures that $\delta(X_0, X_1) = 0$, so that (6.56) holds trivially.

- If $X_0 \neq X_1$, then there exists at least one path $p \in \mathcal{P}$ going from X_0 to X_1, so that we can apply the continuity Lemma 6.2.10 to z_0, z_1, and p:

$$(1 - k(p))L(p) \leq \|z_0 - z_1\|_F, \tag{6.61}$$

where

$$k(p) = (k(z_0, p) + k(z_1, p))/2.$$

But from (6.60) we see that $p \in \mathcal{P}(z_j, \eta_j)$ for $j = 0, 1$ so that, by definition of $k(z_j, \eta_j)$, one has

$$k(z_j, p) \leq k(z_j, \eta_j),$$

which proves that

$$k(p) \leq k < 1,$$

with k defined by (6.57). Hence (6.61) can be rewritten

$$\begin{cases} L(p) \leq (1-k)^{-1}\|z_0 - z_1\|_F \\ \text{for any path } p \in \mathcal{P} \text{ from } X_0 \text{ to } X_1, \end{cases}$$

which proves (6.56) by taking the supremum over all paths p from X_0 to X_1. This ends the proof of Proposition 6.2.11. ■

Corollary 6.2.12 *Let $(D, \mathcal{P})$ be quasi-convex and ϑ be one associated regular neighborhood. Then*

(i) *The projection on D is (whenever it exists) locally uniformly Lipschitz continuous on ϑ*

(ii) *The injection of $(D, \| \, \|_F)$ in (D, δ) is continuous, that is,*

$$\left. \begin{array}{l} X_n, X \in D \\ \|X_n - X\|_F \to 0 \end{array} \right\} \Rightarrow \delta(X_n, X) \to 0. \tag{6.62}$$

Proof. We prove first (i). Let $z_0 \in \vartheta$ be given, and suppose that $z_1 \to z_0$. For z_1 close enough to z_0, hypothesis (6.55) will hold, so that (6.56) will hold (provided z_0 and z_1 admit projections of course), with the following upper bound for k:

$$k \leq \frac{k_0 + 1}{2} < 1,$$

which is independent from z_1 and hence proves the continuity of the projection.

Then, (ii) results immediately from the application of Proposition 6.2.11 to the case $z_0 = X, z_1 = X_n$, in which case the projection of z_j on D obviously exists! ■

If we summarize now the situation, we see that quasi-convex sets provide a satisfactory generalization of the following properties of convex sets listed in Proposition 4.1.1: uniqueness and existence (Proposition 6.2.8), and stability (Proposition 6.2.11). However, the absence of parasitic stationary points (property ii) of Proposition 4.1.1) is not guaranteed for quasi-convex sets, as it appears clearly on Fig. 6.2: for the two quasi-convex arcs of circles D, and for the chosen $z \in \vartheta$, the "distance to z" function admits a global minimum at one end of the arc, and one parasitic local minimum at the other end. In both cases, one sees in Fig. 6.2 that the value of the parasitic local minimum is larger than (top) or equal (bottom) to $d(z, D) + \eta_{\max}(z)$. This is in fact general:

Proposition 6.2.13 *Let $(D, \mathcal{P})$ be quasi-convex, and ϑ be an associated regular neighborhood. Then if, for a given $z \in \vartheta$, the function*

$$d_z^2(X) = \|z - X\|^2$$

admits two distinct stationary points over D, then necessarily one of them gives to d_z a value larger than or equal to $d(z, D) + \eta_{\max}(z)$.

Proof. Let $z \in \vartheta$ be given such that the "distance to z" function admits two distinct stationary points at $X_0, X_1 \in D$, with values strictly smaller than $d(z, D) + \eta_{\max}(z)$:

$$\|X_j - z\|_F < d(z, D) + \eta_{\max}(z), \quad j = 0, 1. \tag{6.63}$$

As X_0 and X_1, are distinct, there exists $p \in \mathcal{P}$ going from X_0 to X_1, because of (6.63) is in $\mathcal{P}(z, \eta)$ for some $\eta \in]0, \eta_{\max}(z)[$. Hence from Definition 6.2.1, we see that $\nu \rightsquigarrow d_{z,p}^2(\nu) = \|p(\nu) - z\|_F^2$ is strictly convex. But, as X_0 and X_1 are stationary points of d_z^2 over D, necessarily the function $d_{z,p}^2$, which is its trace on p, has two distinct stationary points at $\nu = 0$ and $\nu = L(p)$, which is impossible for a strictly convex (Definition 6.2.1). Hence the assumption (6.63) is false, and the proposition is proved. ■

So we introduce in Chap. 7 for a stronger notion, which will eliminate the parasitic stationary points.

Chapter 7

Strictly Quasi-Convex Sets

The quasi-convex sets introduced in Chap. 6 do a good job in generalizing the properties of convex sets with respect to uniqueness, stability, and existence of the projection. But they miss their point on the subject of parasitic stationary points. So we shall start in this chapter from a complementary point of view, and introduce in Sect. 7.1 another family of sets, called the *strictly quasi-convex sets* (s.q.c. sets in short) which, almost by definition, will ensure the absence of parasitic stationary points. Note that the name "s.q.c." has provisorily to be taken as a whole, as it will not be clear at all from the definition that s.q.c. sets are quasi-convex!

In Sect. 7.2, we shall characterize s.q.c. sets $(D, \mathcal{P})$ as those for which $R_{\mathrm{G}}(D) > 0$, where R_{G} is a new geometric attribute, called the *global radius of curvature.* As a byproduct of this characterization, we shall prove that s.q.c. sets necessarily satisfy $R(D) \geq R_{\mathrm{G}}(D) > 0$, where R is the ordinary radius of curvature, and that s.q.c. sets are quasi-convex in the sense of Chap. 6. The end of Sect. 7.2 will be devoted to summarizing the nice generalization provided by s.q.c. sets concerning uniqueness, absence of parasitic stationary points, stability, and existence of the projection. Finally, we shall discuss in Sect. 7.3 the computation of the global radius of curvature $R_{\mathrm{G}}(D)$ of a set D equipped with a family of paths $\mathcal{P}$, both from an analytical and a numerical point of view.

G. Chavent, *Nonlinear Least Squares for Inverse Problems: Theoretical Foundations and Step-by-Step Guide for Applications*, Scientific Computation,
DOI 10.1007/978-90-481-2785-6_7, © Springer Science+Business Media B.V. 2009

7.1 Definition and Main Properties of s.q.c. Sets

Let $D \subset F$ be given, and equipped with a family of paths $\mathcal{P}$ in the sense of the Definition 6.1.3. For any $z \in F$ and $p \in \mathcal{P}$, we denote as usual by

$$d_{z,p}(\nu) = \|z - p(\nu)\|_F \tag{7.1}$$

the "distance to z along p" function.

When D is convex and $\mathcal{P}$ is made of the segments of D, $d_{z,p}^2$ is a quadratic and strictly convex function, which is hence 2-convex and has a unique stationary point.

According to the Definition 6.2.1, $(D, \mathcal{P})$ is quasi-convex as soon as $d_{z,p}^2$ retains from the convex case the property that it is uniformly α-convex over $\mathcal{P}(z, \eta)$ as soon as $z \in \vartheta$ and $0 < \eta < \eta_{\max}(z)$.
This property is very local, as paths p of $\mathcal{P}(z, \eta)$ connect two η-projections of z, which become close when $\eta \to 0$ (cf Corollary 6.2.7).

So we shall require, to define strictly quasi-convex sets, that $(D, \mathcal{P})$ inherits from the convex case a different, more global property, namely that $d_{z,p}^2$ has a unique stationary point as soon as $z \in \vartheta$ and p connects any two points of D through an η-projection of z, $0 \leq \eta < \eta_{\max}(z)$.

It will be convenient to give a name to functions with a unique stationary point:

Definition 7.1.1 *A function $f \in \mathcal{C}^1([0, L])$ is called* strictly quasi-convex *(s.q.c. in short) if and only if the inequation*

$$f'(\nu)\lambda \geq 0 \quad \forall \lambda \in I\!R, \quad \nu + \lambda \in [0, L] \tag{7.2}$$

has a unique solution ν in $[0, L]$.

A function $f \in \mathcal{C}^0([0, L])$ is usually said to be quasi-convex when it satisfies

$$\begin{cases} \forall \nu_0,\ \nu_1 \in [0, L],\ \nu_0 < \nu_1, \\ \forall \nu \in]\nu_0, \nu_1[:\ f(\nu) \leq \mathrm{Max}\{f(\nu_0), f(\nu_1)\}. \end{cases} \tag{7.3}$$

A quasi-convex function has one or more global minima, and no parasitic stationary points. But it can have parasitic stationary points (in the zones where f is constant), as well as inflexion points. Sometimes, the name "strictly quasi-convex" is used for functions $f \in \mathcal{C}^0([0, L])$, which satisfy (7.3) with a strict inequality. Such functions eliminate parasitic local minima (strict or not), but not, when $f \in \mathcal{C}^1([0,L])$, parasitic stationary points where f' vanish.

Elimination of all parasitic stationary point – and hence of all parasitic local minima – is essential for our purpose, as it guaranties that local optimization algorithms used to compute the projection on D will not stop at such a stationary point. So we shall always use s.q.c. function in the sense of Definition 7.1.1.

We define now our new family of sets:

Definition 7.1.2 *(Strictly quasi-convex sets). A set $(D,\mathcal{P})$ is strictly quasi-convex (s.q.c. in short) if and only if there exists a neighborhood ϑ of D and a function $\eta_{\max} : \vartheta \to]0,+\infty]$ such that*

$$\begin{cases} \forall z \in \vartheta,\ \forall \eta,\ 0 < \eta < \eta_{\max}(z) \quad \text{one has} \\ d(z,p) \le d(z,D) + \eta \Rightarrow d^2_{z,p} \text{ is s.q.c. along } p, \end{cases} \tag{7.4}$$

and ($\eta_{\max}$ is continued by zero outside of ϑ):

$$\lim_{\epsilon \to 0} \inf_{d(z,D) \le \epsilon} \left\{ d(z,D) + \eta_{\max}(z) \right\} \; > \; 0. \tag{7.5}$$

The neighborhood ϑ is called an (s.q.c.) regular neighborhood *of D.*

Condition (7.4) of course is meant to handle the problem of parasitic stationary points, but condition (7.5) is difficult to interpret directly. As we shall see in the proof of Theorem 7.2.5, it will have the effect to ensure that the regular neighborhood ϑ associated to s.q.c. sets contains necessarily an enlargement neighborhood of D, which is a desirable property (see (4.4) in Chap. 4). Notice that (7.5) is automatically satisfied as soon as F is a finite dimensional, D bounded, and $\eta_{\max}$ an l.s.c. function.

We check first that, as for quasi-convex sets, there exists a largest neighborhood ϑ and function $\eta_{\max}$:

Proposition 7.1.3 *Let $(D,\mathcal{P})$ be a s.q.c. set. Then there exists a* largest regular open neighborhood *ϑ of D, and a* largest function $\eta_{\max} : \vartheta \to]0,+\infty]$ *satisfying the Definition 7.1.2 of s.q.c. sets.*

Proof. Let $\vartheta_i, \eta_{\max,i}, i \in I$ denote the collection of all open neighborhoods and functions satisfying Definition 7.1.2, and let ϑ and $\eta_{\max}$ be defined by

(6.38) and (6.39). Then $\eta_{\max}$ satisfies obviously (7.5). We check now that ϑ and η satisfy (7.4); let us choose $z \in \vartheta$ and $p \in \mathcal{P}$ such that

$$d(z,p) < d(z,D) + \eta_{\max}(z).$$

From the formula (6.39) defining $\eta_{\max}$, we see that there exists $\hat{\imath} \in I$ such that

$$d(z,p) < d(z,D) + \eta_{\max\,\hat{\imath}}(z). \tag{7.6}$$

As $d(z,p) \geq d(z,D)$, (7.6) implies that $\eta_{\max\,\hat{\imath}}(z) > 0$, which shows that necessarily $z \in \vartheta_{\hat{\imath}}$. Then the strict quasi-convexity of $d^2_{z,p}$ over p results from the fact that (7.4) holds, by hypothesis, for all $\vartheta = \vartheta_i$ and $\eta_{\max} = \eta_{\max\,i}$, $i \in I$, and hence in particular for $\hat{\imath}$. ■

The definition chosen for s.q.c. sets makes it very easy to handle the problem of parasitic stationary points:

Proposition 7.1.4 *Let $(D, \mathcal{P})$ be s.q.c., and ϑ be an associated regular neighborhood. Then property (ii) of Proposition 4.1.1 on stationary points generalizes as follows:*

If $z \in \vartheta$ admits a projection $\hat{X}$ on D, the "distance to z" function has no parasitic stationary point *on D (its unique stationary point is $\hat{X}$). In particular, $\hat{X}$ is necessarily unique.*

Proof. Let $z \in \vartheta$ be given such that it admits a projection X on D, and suppose that the "distance to z" function possesses, beside the global minimum on D at X, a stationary point on D at a point $Y \neq X$. Let $p \in \mathcal{P}$ be a path connecting X to Y. As $X \in p$ and X is the projection of z onto D, one has

$$d(z,p) \leq \|z - X\| = d(z,D) < d(z,D) + \eta_{\max}(z).$$

Using the Definition 7.1.2 of s.q.c. sets, this implies that the "distance to z" function $d^2_{z,p}$ is s.q.c. along the path p, which contradicts the fact that this function has a global minimum at $\nu = 0$ (because X is a global minimum on D) and a stationary point at $\nu = L(p)$ (because of the hypothesis we have made that Y is a stationary point of d^2_z on D). ■

Of course, convex sets, which were already quasi-convex sets as we have seen in Chap. 6 earlier, are also s.q.c. sets with a regular neighborhood $\vartheta = F$ and a function $\eta_{\max}(z) = +\infty$. Arcs of circle of radius R and length L,

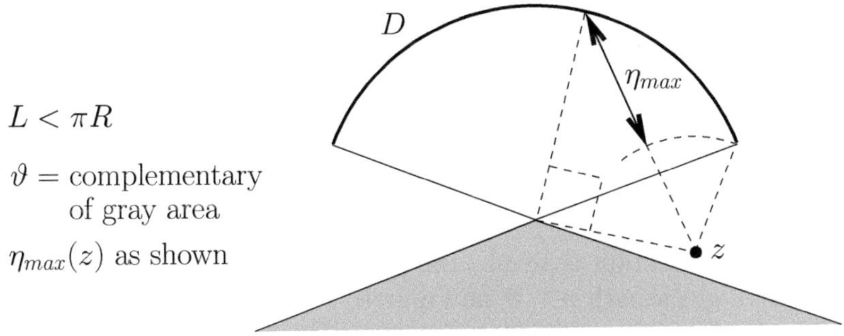

Figure 7.1: s.q.c. arc of circle

however, are s.q.c. only if $L < \pi R$, with a largest regular neighborhood ϑ shown in Fig. 7.1. Comparison with Fig. 6.2 shows the diminution in size of the neighborhood ϑ associated to the same arc of circle of length $L < \pi R$ when it is considered as s.q.c. instead of quasi-convex.

As we see in Fig. 7.1, the largest regular neighborhood ϑ associated to the arc of circle D by the Definition 7.1.2 of s.q.c. sets catches exactly all points z of the plan admitting a unique projection on D with no parasitic stationary points on D of the distance to z. So the notion of s.q.c. provides a sharp description of the sets D to which the result of Proposition 4.1.1 can be generalized, when D is an arc of circle. In fact, as we shall see in the next section, this remains true for all D made of one path p.

The definition of s.q.c. set is quite technical. We have been able to use it directly and to determine the associated largest regular neighborhood ϑ in the above examples only because they were extremely simple (D was a convex set or an arc of circle). Using the definition itself to recognize s.q.c. sets in the applications to nonlinear least-squares problems (see Chap. 1) seems quite impractical, if not unfeasible. So we develop in the next section a characterization of s.q.c. sets, which will be more easy to use, and will provide the size R_G of the *largest open regular enlargement neighborhood*

$$\vartheta = \{z \in F | d(z, D) < R_G\}$$

associated to the s.q.c. set (the reason for calling R_G the size of the neighborhood will become clear in the next section).

For the applications to nonlinear least-squares, the determination of such a regular enlargement neighborhood is very important, as R_G gives an upper bound to the size of *the noise level* (measurement and model errors) on data for which the least-squares problem remains Q-wellposed.

7.2 Characterization by the Global Radius of Curvature

The idea in this section is to use the Definition 7.1.1 of s.q.c. function to localize, for a given path $p \in \mathcal{P}$ and a given $\nu \in [0, L(p)]$, the points $z \in F$ such that $d^2_{z,p}$ has one stationary point at ν. This leads to

Definition 7.2.1 *(Affine normal cone)*
Given a path $p \in \mathcal{P}$ and $\nu \in [0, L(p)]$, we define the affine normal cone *$N(\nu)$ by*

$$N(\nu) = \Big\{ z \in F \big| \langle z - p(\nu), \lambda v(\nu)\rangle_F \; \leq \; 0 \quad \forall \lambda \in I\!R, \;\; \nu + \lambda \in [0, L] \Big\},$$

where as usual $v(\nu) = p'(\nu)$ is a unit tangent vector to p at $p(\nu)$.

Hence $z \in N(\nu)$ means exactly that $d^2_{z,p}$ has a stationary point at ν. So by construction, when $\nu \neq \nu'$, the intersection $N(\nu) \cap N(\nu')$ is made of points z such that $d^2_{z,p}$ has two distinct stationary points at ν and ν', so that $d^2_{z,p}$ is not s.q.c.

Figure 7.2 shows these affine normal cones $N(\nu)$ and $N(\nu')$ at two different points ν and ν' of a given path p of $I\!R^2$, and their intersection (bold point, bold line or darker grey area) in three different cases:

(a) The two points $p(\nu)$ and $p(\nu')$ are interior points of the path. Then the intersection of $N(\nu)$ and $N(\nu')$ is a point. If we call z this point, one sees clearly that the $d^2_{z,p}$ function possesses, besides a global minimum at ν, a parasitic stationary point at ν'.

(b) One of the two points (here $p(\nu)$) is an endpoint of p. Then the intersection of $N(\nu)$ and $N(\nu')$ is a half-line. If we call z the projection of $p(\nu)$ on this half line, one sees once again that the $d^2_{z,p}$ possesses, as earlier, a parasitic stationary point at ν'.

(c) The two points are the endpoints of p. Then the intersection of $N(\nu)$ and $N(\nu')$ is the bold dashed area. If we call z the projection of $p(\nu)$ on the intersection, once again $d^2_{z,p}$ has a parasitic stationary point at the endpoint $\nu' = L$.

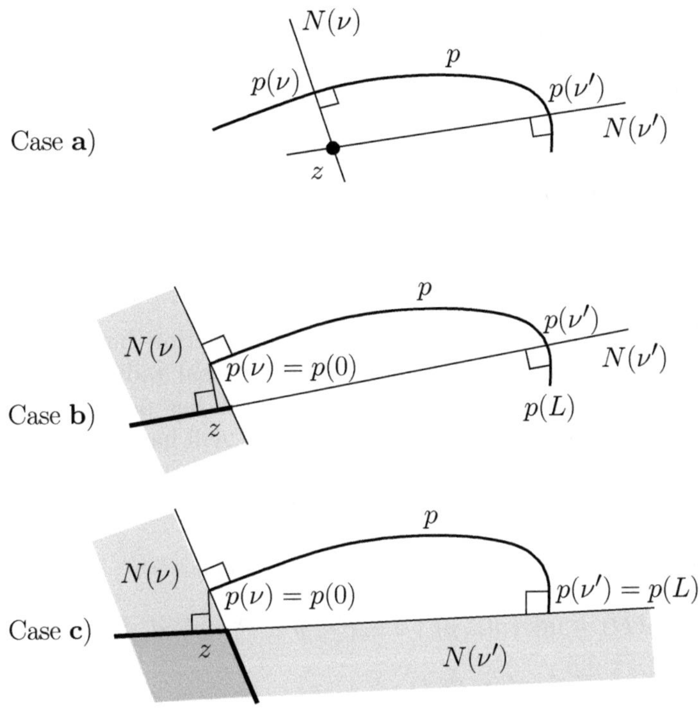

Figure 7.2: Intersection of affine normal cones at two points of a given paths p

So we see that the distance of $p(\nu)$ to $N(\nu) \cap N(\nu')$ plays a particular role, and we give it a name:

Definition 7.2.2 *(Global radius of curvature)*
Let $p \in \mathcal{P}$ be given. Then, for any $\nu, \nu' \in [0, L(p)]$ with $\nu \neq \nu'$, we define the global radius of curvature of p at ν seen from ν' by

$$\rho_G(\nu, \nu') = d\big(p(\nu), N(\nu) \cap N(\nu')\big) \in [0, +\infty], \tag{7.7}$$

with the natural convention that $\rho_G(\nu, \nu') = +\infty$ if $N(\nu)$ and $N(\nu')$ do not intersect.

Consideration of the worst case for $\nu \in [0, L(p)]$ then for all p of $\mathcal{P}$ leads to the

Definition 7.2.3 *(Global radius of curvature of a path and a set).*

Let $(D, \mathcal{P})$ be given. Then we define the global radius of curvature of a path $p \in \mathcal{P}$ by

$$R_G(p) \stackrel{\text{def}}{=} \inf_{\nu,\nu' \in [0,L(p)]\ ,\ \nu \neq \nu'} \rho_G(\nu, \nu'), \tag{7.8}$$

and that of the set $(D, \mathcal{P})$ by

$$R_G(D) \stackrel{\text{def}}{=} \inf_{p \in \mathcal{P}} R_G(p). \tag{7.9}$$

As we have seen in Fig. 7.2, the $d^2_{z,p}$ function has parasitic stationary points as soon as $d(z,p)$ is equal to the global radius of curvature at the projection $p(\nu)$ of z on p, seen from some other point $p(\nu')$ of p. It seems reasonable to make the conjecture that $d^2_{z,p}$ will have no parasitic stationary points as soon as $d(z,p)$ is strictly smaller than the infimum of all global radii of curvature! This is confirmed by the

Proposition 7.2.4 *Let $(D, \mathcal{P})$ be given. Then*

$$R_G(D) = \inf \{d(z,p) \mid z \in F\ ,\ p \in \mathcal{P} \text{ and } d^2_{z,p} \text{ not s.q.c.}\}, \tag{7.10}$$

where $d^2_{z,p}$ is defined in (7.1).

Proof. We show first that $R_G(D) \leq \inf\{...\}$.
So let $h \in \{...\}$ be given, and $z \in F$ and $p \in \mathcal{P}$ be a couple of corresponding point and path, which hence satisfy

$$d^2_{z,p} \text{ is not s.q.c.}$$

Then Definition 7.1.1 of s.q.c. functions implies that $d^2_{z,p}$ possesses at least, beside a global minimum at some ν_0, a second stationary point at $\nu_1 \neq \nu_0$. Hence (7.2) with f replaced by $d^2_{z,p}$ holds at both ν_0 and ν_1. This can be rewritten, using the Definition 7.2.1 of affine normal cones, as

$$z \in N(\nu_0) \cap N(\nu_1),$$

which implies

$$d(p(\nu_0),\ N(\nu_0) \cap N(\nu_1)) \ \leq\ \|p(\nu_0) - z\|_F. \tag{7.11}$$

But the left-hand side of (7.11) is $\rho_G(\nu_0, \nu_1)$ from the Definition 7.2.2 of the global radius of curvature, and the right-hand side is $d(z,p)$ by definition of ν_0. Hence (7.11) becomes

$$\rho_G(\nu_0, \nu_1) \;\leq\; d(z,p), \tag{7.12}$$

and, using the Definition 7.2.3 of $R_G(D)$ and the properties of z, and p

$$R_G(D) \;\leq \rho_G(\nu_0, \nu_1) \leq d(z,p), \tag{7.13}$$

which ends the first part of the proof.

We check now that $R_G(D) \geq \inf\{...\}$. Let $\epsilon > 0$ be given. From the Definition (7.9) of $R_G(D)$, we see that there exists $p \in \mathcal{P}$ such that

$$R_G(p) \;\leq\; R_G(D) + \epsilon/2, \tag{7.14}$$

and, from the definition (7.8) of $R_G(p)$, that there exists $\nu_0, \nu_1 \in [0, L(p)]$, $\nu_0 \neq \nu_1$, such that

$$\rho_G(\nu_0, \nu_1) \;\leq\; R_G(p) + \epsilon/2. \tag{7.15}$$

Let $z \in F$ be the projection on $N(\nu_0 \cap N(\nu_1))$ of $p(\nu_0)$. By construction, $d^2_{z,p}$ has two distinct stationary points at ν_0 and ν_1, and so cannot be s.q.c.! On the other hand, the definition (7.7) of ρ_G shows that

$$\rho_G(\nu_0, \nu_1) = \|z - p(\nu_0)\|_F \geq d(z,p). \tag{7.16}$$

Combining (7.14), (7.15), and (7.16) gives

$$d(z,p) \;\leq\; R_G(D) + \epsilon. \tag{7.17}$$

Hence given $\epsilon > 0$, we have been able to find a $z \in F$ and $p \in \mathcal{P}$ such that (7.17) holds and $d^2_{z,p}$ is not s.q.c., which proves that

$$\inf\{...\} \leq R_G(D) + \epsilon.$$

This holds for any $\epsilon > 0$, which proves that $\inf\{...\} \leq R_G(D)$. ■

Proposition 7.2.4 gives immediately a *characterization of s.q.c. sets*:

Theorem 7.2.5 *A set $D \subset F$ equipped with a family of paths $\mathcal{P}$ is s.q.c. if and only if*

$$R_G(D) \;>\; 0. \tag{7.18}$$

The largest *associated* open regular enlargement neighborhood ϑ *is given by*

$$\vartheta = \Big\{z \in F | d(z,D) < R_G(D)\Big\}, \tag{7.19}$$

and the corresponding $\eta_{\max}$ *function by*

$$\forall z \in \vartheta, \ \eta_{\max}(z) = R_G(D) - d(z,D) > 0. \tag{7.20}$$

Proof. We prove first the sufficient condition. So let (7.18) hold, and define ϑ and $\eta_{\max}$ by (7.19) and (7.20). Then for any z, η, and p satisfying the hypothesis of (7.4), one has, by definition of $\eta_{\max}$,

$$h = d(z,p) \ < \ R_G(D),$$

which by Proposition 7.2.4 implies that $d^2_{z,p}$ is necessarily s.q.c. So (7.4) holds, and (7.9) holds trivially as $d(z,D) + \eta_{\max}(z) = R_G(D) > 0$. Hence $(D, \mathcal{P})$ is s.q.c.

We prove now the necessary condition. So let $(D, \mathcal{P})$ be s.q.c., and suppose that $R_G(D) = 0$. Using Proposition 7.2.4, we can find $z_n \in F$, $p_n \in \mathcal{P}$ for all $n = 1, 2, \ldots$ such that

$$d(z_n, p_n) \ = \ \frac{1}{n}, \tag{7.21}$$

$$d^2_{z_n,p_n} \quad \text{not s.q.c.} \tag{7.22}$$

But (7.21) implies that $d(z_n, D) \to 0$, and, using (7.9), the existence of $\gamma > 0$ such that

$$d(z_n, D) + \eta_{\max}(z_n) \ \geq \ \gamma \ > \ 0 \quad \forall n.$$

Hence, for n large enough, one has

$$d(z_n, D) + \eta_{\max}(z_n) \ > \ \frac{2}{n}. \tag{7.23}$$

This implies, as $d(z_n, D) \leq d(z_n, p_n) = 1/n$, that

$$\eta_{\max}(z_n) \ > \ \frac{1}{n}.$$

So (7.23) can be rewritten as

$$d(z_n, p_n) \ = \ \frac{1}{n} \leq d(z_n, D) + \eta_n,$$

where $\eta_n = \eta_{\max}(z_n) - \dfrac{1}{n}$ satisfies

$$0 \;<\; \eta_n \;<\; \eta_{\max}(z_n).$$

As $(D, \mathcal{P})$ is s.q.c., the two last inequalities imply that $d^2_{z_n,p_n}$ is s.q.c. for n large enough, which contradicts (7.22); hence necessarily $R_G(D) > 0$, and Proposition 7.2.4 implies that ϑ defined in (7.19) is the largest regular enlargement neighborhood associated with the s.q.c. set $(D, \mathcal{P})$. ■

Theorem 7.2.5 is illustrated graphically in Fig. 7.3 in the case of a set D made of one s.q.c. arc of circle p: one sees that the enlargement neighborhood of size $R_G(p)$ is the largest regular enlargement included in the largest possible regular neighborhood, which is recalled from Fig. 7.1.

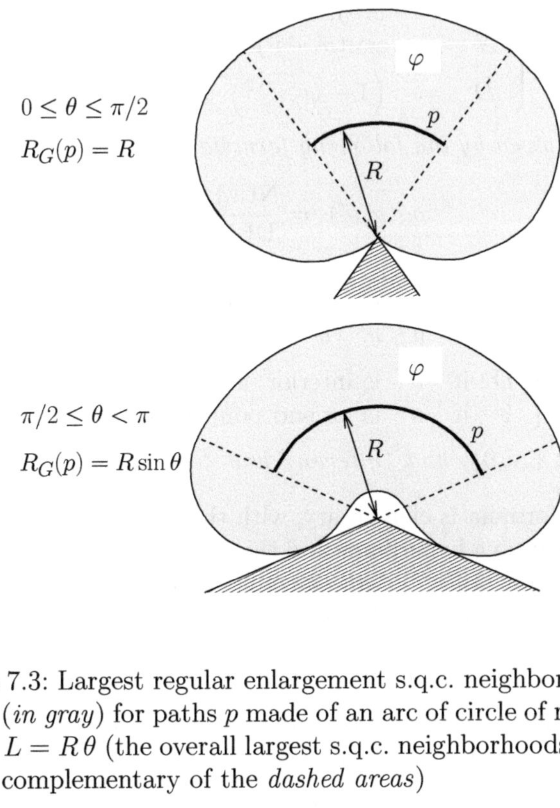

Figure 7.3: Largest regular enlargement s.q.c. neighborhood of size $R_G(p)$ (*in gray*) for paths p made of an arc of circle of radius R and length $L = R\,\theta$ (the overall largest s.q.c. neighborhoods correspond to the complementary of the *dashed areas*)

Now that we have defined, and characterized, s.q.c. sets that allow for the generalization of property (ii) of 4.1.1 (no parasitic stationary points for the projection), we would like to see whether properties (i), (iii), and (iv) (uniqueness, stability, and existence) can also be generalized to s.q.c. sets. According to the results of Chap. 6, this would be true if s.q.c. sets happened to be quasi-convex sets; to prove that this is indeed the case, we need to study further the analytical properties of the global radius of curvature $\rho_{\mathrm{G}}(\nu, \nu')$ and its relation with the usual, local radius of curvature $\rho(\nu) = \|a(\nu)\|^{-1}$ defined in Chap. 6.

Proposition 7.2.6 *Let a path $p \in \mathcal{P}$ and $\nu, \nu' \in [0, L(p)]$, $\nu \neq \nu'$ be given, and denote*

$$\left\{ \begin{array}{lllll} X & = & p(\nu), & X' & = p(\nu'), \\ v & = & v(\nu), & v' & = v(\nu'), \\ N & = & \operatorname{sgn}(\nu' - \nu)\langle X' - X, v'\rangle, & & \\ D & = & \left(1 - \langle v, v'\rangle^2\right)^{1/2}. & & \end{array} \right. \tag{7.24}$$

Then $\rho_{\mathrm{G}}(\nu, \nu')$ is given by the following formula:

$$\rho_{\mathrm{G}}(\nu, \nu') = \frac{\mathrm{NUM}}{\mathrm{DEN}}, \tag{7.25}$$

where:

$$\mathrm{NUM} = \left\{ \begin{array}{lll} |N| & \text{if} \quad \nu' & \text{is interior,} \\ N^+ & \text{if} \quad \nu' & \text{is an end point,} \end{array} \right. \tag{7.26}$$

$$\mathrm{DEN} = \left\{ \begin{array}{lll} D & \text{if} \quad \nu & \text{is interior or} \quad N\langle v, v'\rangle \geq 0, \\ 1 & \text{if} \quad \nu & \text{is an end point and } N\langle v, v'\rangle \leq 0, \end{array} \right. \tag{7.27}$$

where $N^+ = \mathrm{Max}\,\{N, 0\}$, and "interior" and "end point" are relative to p.

The proof of this formula is elementary, with the basic ingredients being the projection of a point on a hyperplane and the angle between two hyperplanes. If the reader looks now at Fig. 7.2 and imagines what is going to happen when $\nu' \to \nu$, he will not be surprised by the

Proposition 7.2.7 *Let a path $p \in \mathcal{P}$ be given. Then*

(i) For any $\nu \in [0, L(p)]$, there exists an open neighborhood $I(\nu)$ in $[0, L(p)]$ such that, for any $\nu' \in I(\nu)$ one has, with the notation (7.24),

$$\rho_{\mathrm{G}}(\nu, \nu') = N/D, \; \rho_{\mathrm{G}}(\nu', \nu) = N'/D, \tag{7.28}$$

(ii) For almost every $\nu \in [0, L(p)]$, we have

$$\rho_G(\nu, \nu') \to \rho(\nu), \ \ \rho_G(\nu', \nu) \to \rho(\nu), \tag{7.29}$$

when $\nu' \to \nu$ in $I(\nu)$, where $\rho(\nu) = \|a(\nu)\|^{-1}$ is the usual radius of curvature of p at ν.

Proof. Property (i) follows immediately from Proposition 7.2.6 if we prove that N and $\langle v, v' \rangle$ are positive when ν' is close enough to ν. But, with the notation $\mathrm{d}\nu = \nu' - \nu$, one has

$$N = \operatorname{sgn} \mathrm{d}\nu \ \langle p(\nu + p\nu) - p(\nu), \quad v(\nu + \mathrm{d}\nu) \rangle,$$

that is, using a Taylor formula

$$N \ = \ |\mathrm{d}\nu| \ \langle \ v(\nu + \theta \, \mathrm{d}\nu), \ v(\nu + \mathrm{d}\nu) \rangle$$

for some $0 \le \theta \le 1$, and

$$\langle v, v' \rangle \ = \ \langle v(\nu), \ v(\nu + \mathrm{d}\nu) \rangle.$$

As the velocity v is a continuous function over $[0, L(p)]$, we see that

$$\begin{cases} N \,/\, |\delta\nu| \ \to 1 \\ \\ \langle \nu, \nu' \rangle \to 1 \end{cases} \quad \text{when } \delta\nu \to 0, \tag{7.30}$$

which ends the proof of (i).

We compute now the limit of ρ_G when $\nu' \to \nu$. Because of (i), this amounts to searching for the limit of N/D. We have already found in (7.30) an equivalent of N in term of $\delta\nu = \nu' - \nu$, and we give now an equivalent of D in terms of $\delta\mu = \|v' - v\|$. The theorem of the median applied to v and v' implies that

$$\langle v, v' \rangle \ = \ 1 - \delta\mu^2 \,/2.$$

Hence

$$D \ = \ \delta\mu(1 - \delta\mu^2 \,/4)^{1/2},$$

and, as $\delta\mu \to 0$ when $\delta\nu \to 0$,

$$D/\delta\mu \to 1 \quad \text{when} \quad \delta\nu \to 0. \tag{7.31}$$

Using (7.30) and (7.31), we can replace the search for the limit of $\rho_G(\nu, \nu+d\nu)$ by that of the limit of $|d\nu| / \delta\mu$ when $d\nu \to 0$. As our guess for this limit is $\rho(\nu) = \|a(\nu)\|_F^{-1}$, let us choose one realization of $a(\nu)$ in $L^\infty\Big([0, L(p)]\ ;\ F\Big)$ – which will still be denoted by $a(\nu)$ – so that $a(\nu)$ is well-defined, and compare $\delta\mu / |d\nu|$ to $\|a(\nu)\|$

$$\frac{\delta\mu}{|d\nu|} = \frac{\|v' - v\|}{|d\nu|} = \Big\|\frac{1}{d\nu} \int_0^{d\nu} a(\nu + \tau) d\tau\Big\|_F.$$

The triangular inequality implies then

$$\Big|\frac{\delta\mu}{|d\nu|} - \|a(\nu)\|_F\Big| \le \Big\|\frac{1}{d\nu} \int_0^{d\nu} a(\nu + \tau) d\tau - a(\nu)\Big\|_F. \tag{7.32}$$

But the right-hand side of (7.32) tends to zero when $d\nu \to 0$ each time ν is a Lebesgue point for the chosen realization of a (see, e.g., Theorem 8.8 of [Rudin 1987]). As almost every point of $[0, L(p)]$ is a Lebesgue point, we have proven the left part of (7.29). To prove the right part, we remark that

$$\rho_G(\nu, \nu') - d\beta \le \rho_G(\nu', \nu) \le \rho_G(\nu, \nu') + d\beta,$$

where

$$0 \le d\beta = \frac{|\langle X' - X, v - v'\rangle|}{D} = \frac{|d\nu|\ |\langle v(\nu + \theta\, d\nu, v - v'\rangle|}{D}$$

for some $0 \le \theta \le 1$. The Cauchy–Schwarz inequality implies then

$$0 \le d\beta \le \frac{|d\nu|\ \delta\mu}{D},$$

which, as $D / \delta\mu \to 1$ (see (7.31)), shows that

$$\delta\beta \to 0 \quad \text{when} \quad d\nu \to 0.$$

This ends the proof of the theorem. ■

We can now compare the global radius of curvature of a path and a set to the usual ones:

Definition 7.2.8 *(Radius of curvature of a path and a set)*

Let $(D, \mathcal{P})$ be given. Then we define the (local) radius of curvature of the path $p \in \mathcal{P}$ *by*

$$R(p) \stackrel{\text{def}}{=} \inf_{\nu \in [0, L(p)]} \text{ess}\, \rho(\nu), \tag{7.33}$$

and that of the set $(D, \mathcal{P})$ *by*

$$R(D) \stackrel{\text{def}}{=} \inf_{p \in \mathcal{P}} R(p). \tag{7.34}$$

Proposition 7.2.9 *Let $(D, \mathcal{P})$ be given. Then*

(i) For any $p \in \mathcal{P}$ one has

$$+\infty \ \geq \ R(p) \ > \ 0, \tag{7.35}$$

and

$$R(p) \ \geq \ R_{\mathrm{G}}(p) \ \geq \ 0, \tag{7.36}$$

(ii) The local and global radius of curvatures of D satisfy

$$R(D) \ \geq \ R_{\mathrm{G}}(D) \ \geq \ 0. \tag{7.37}$$

Proof. Inequality (7.35) holds because $p \in W^{2,\infty}\big([0, L(p)]\big)$ (cf. Definition 6.1.1), and inequalities (7.36) and (7.37) follow from Proposition 7.2.7 (ii). ■

We can now state the second main result of this chapter:

Theorem 7.2.10 *Let $(D, \mathcal{P})$ be given. Then*

$$(D, \mathcal{P}) \quad \text{s.q.c.} \Longrightarrow (D, \mathcal{P}) \quad \text{quasi} - \text{convex}. \tag{7.38}$$

Proof. Let $(D, \mathcal{P})$ be s.q.c. Then we know from Theorem 7.2.5 that

$$R_{\mathrm{G}}(D) \ > \ 0, \tag{7.39}$$

and that

$$\begin{cases} \vartheta = \big\{ z \in F \mid d(z, D) \ < \ R_{\mathrm{G}}(D) \big\}, \\ \eta_{\max}(z) \ = \ R_{\mathrm{G}}(D) - d(z, D), \end{cases} \tag{7.40}$$

satisfy (7.4).

Then Proposition 7.2.9, together with (7.39), imply that

$$R(D) \geq R_{\mathrm{G}}(D) > 0, \tag{7.41}$$

so that the curvature of the paths of $\mathcal{P}$ is uniformly bounded.

We check now that ϑ and $\eta_{\max}$ defined by (7.40) satisfy (6.33), which will prove, as $\eta_{\max}$ is continuous and hence l.s.c., that $(D, \mathcal{P})$ is quasi-convex.

So let $z \in \vartheta$ and η, $0 < \eta < \eta_{\max}(z)$ be given. We want to prove that $k(z, \eta) < 1$, that is, $k(z, p; \nu)$ stays uniformly away from 1 over all $p \in \mathcal{P}(z, \eta)$ and $\nu \in [0, L(p)]$. So let $p \in \mathcal{P}(z, \eta)$ and $\nu \in [0, L(p)]$ be given. By definition, p satisfies

$$\|z - p(j)\|_F \leq d(z, D) + \eta \quad j = 0, L(p), \tag{7.42}$$

which implies that

$$d(z, p) \leq d(z, D) + \eta,$$

and, as ϑ and $\eta_{\max}$ satisfy (7.4)

$$d^2_{z,p} \quad \text{is s.q.c.} \tag{7.43}$$

Combining (7.42) and (7.43), we obtain

$$d_{z,p}(\nu) \leq \max_{j=0,L(p)} d_{z,p}(j) \leq d(z, D) + \eta, \quad \forall \nu \in [0, L(p)]. \tag{7.44}$$

From the definition (6.26) of $k(z, p; \nu)$, we see that

$$k(z, p; \nu) \leq |k(z, p, \nu)| \leq \|z - p(\nu)\|_F \, \|a(\nu)\|_F,$$

or, with the notation $d_{z,p}(\nu) = \|z - p(\nu)\|_F$ and $\|a(\nu)\|_F = 1/\rho(\nu)$

$$k(z, p, \nu) \leq d_{z,p}(\nu)/\rho(\nu),$$

and, using (7.44) and the definition of $R(D)$

$$k(z, p, \nu) \leq \frac{d(z, D) + \eta}{R(D)}.$$

This upper bound is clearly uniform in p, ν, and satisfies, as $\eta < \eta_{\max}$ defined in (7.40)

$$\frac{d(z, D) + \eta}{R(D)} < \frac{R_{\mathrm{G}}(D)}{R(D)} \leq 1$$

because of (7.41). Hence (6.33) is satisfied, and $(D, \mathcal{P})$ is quasi-convex. ■

We summarize in the next theorem the *nice properties of s.q.c. sets with respect to projection*, which generalize those recalled for the convex case in Proposition 4.1.1:

Theorem 7.2.11 *Let $(D, \mathcal{P})$ be s.q.c., and*

$$\vartheta = \Big\{ z \in F \mid d(z, D) < R_G(D) \Big\} \tag{7.45}$$

be the largest associated open regular enlargement neighborhood. Then:

(i) Uniqueness: *for any $z \in \vartheta$, there exists at most one projection of z on D.*

(ii) Unimodality: *if $z \in \vartheta$ admits a projection $\widehat{X}$ on D, the "distance to z" function has no parasitic stationary point on D other that $\widehat{X}$.*

(iii) Stability: *if $z_0, z_1 \in \vartheta$ admit projections $\widehat{X}_0, \widehat{X}_1$ on D and are close enough so that there exists $d \geq 0$ satisfying*

$$\|z_0 - z_1\|_F + \max_{j=0,1} d(z_j, D) \leq d < R_G(D), \tag{7.46}$$

then for any path p going from $\widehat{X}_0$ to $\widehat{X}_1$

$$\|\widehat{X}_0 - \widehat{X}_1\|_F \leq L(p) \leq (1 - d\,/R(p))^{-1}\, \|z_0 - z_1\|_F, \tag{7.47}$$

where $R(p) \geq R(D) \geq R_G(D) > 0$.

(iv) existence: *if $z \in \vartheta$, any minimizing sequence $X_n \in D$ of the "distance to z" function over D is a Cauchy sequence for both distances $\|X-Y\|_F$ and $\delta(X, Y)$. Hence X_n converges to the (unique) projection $\widehat{X}$ of z onto the closure $\overline{D}$ of D.*

If moreover D is closed, then $\widehat{X} \in D$, and $\delta(X_n, \hat{X}) \to 0$ when $n \to 0$.

Proof. Theorem 7.2.11 is simply a compilation of Propositions 6.2.5, 6.2.8, and 7.1.4 and Theorems 7.2.5 and 7.2.10, with some simplifications in the stability formula brought by the simple form of the $\eta_{\max}(z)$ function. ■

We summarize also the properties of the "squared distance to z" function $d^2_{z,p}$ along paths of $\mathcal{P}$ when $(D, \mathcal{P})$ is s.q.c.:

Proposition 7.2.12 *Let $(D, \mathcal{P})$ be s.q.c. Then, for any $z \in \vartheta$ and any $p \in \mathcal{P}$ such that*

$$d(z,p) \;<\; R_{\rm G}(D),$$

the function $d^2_{z,p}$ satisfies

(i) $d^2_{z,p}$ – and hence $d_{z,p}$ – is s.q.c. over the whole path p

(ii) For any η such that $0 < \eta < R_G(D) - d(z,D)$, $d^2_{z,p}$ is α-convex between any two η-projections of z, where

$$\alpha = 2\Big(1 - \frac{d(z,D)+\eta}{R(p)}\Big) \geq 2\Big(1 - \frac{d(z,D)+\eta}{R(D)}\Big) > 0.$$

7.3 Formula for the Global Radius of Curvature

We establish in this section the formula which will be our starting point for the evaluation of $R_{\rm G}(p)$ and $R_{\rm G}(D)$ both numerically and analytically.

We remark first that, when computing $R_{\rm G}(p)$ for some path p, one does not need to consider in the computation of $\rho_{\rm G}(\nu,\nu')$ all the cases indicated in Proposition 7.2.6, according to whether ν and/or ν' are interior or end points of the paths: it is enough to evaluate $\rho_{\rm G}(\nu,\nu')$ always as if ν and ν' were endpoints. This is made precise in the next proposition.

Proposition 7.3.1 *Let $p \in \mathcal{P}$ be given. Then*

$$R_{\rm G}(p) = \inf_{\substack{\nu,\nu' \in [0,L(p)] \\ \nu \neq \nu'}} \rho^{\rm ep}_{\rm G}(\nu,\nu'), \tag{7.48}$$

where $\rho^{\rm ep}_{\rm G}$ ("ep" stands for "end points") is given by

$$\rho^{\rm ep}_{\rm G}(\nu,\nu') \;=\; \begin{cases} N^+/D & \text{if } \langle v,v'\rangle \;\geq\; 0, \\ N^+ & \text{if } \langle v,v'\rangle \;\leq\; 0, \end{cases} \tag{7.49}$$

where $N^+ = \max\{N, 0\}$ and v, v', N, and *D are defined in (7.24).*

Moreover, the infimum in (7.48) is not changed if one eliminates from the search all couples $\nu \neq \nu'$ such that

$$\begin{cases} \nu \text{ is an interior point of } \; p, \\ N > 0 \quad \text{and} \quad \langle v,v'\rangle \;<\; 0. \end{cases} \tag{7.50}$$

Proof. From the definitions of ρ_G and ρ_G^{ep} one has

$$\rho_G(\nu,\nu') \;\geq\; \rho_G^{ep}(\nu,\nu') \quad \forall \nu \neq \nu',$$

which shows that

$$\inf_{\nu,\nu' \in [0,L(p)]\;,\;\nu\neq\nu'} \rho_G^{ep}(\nu,\nu') \;\leq\; R_G(p).$$

So (7.48) will hold as soon as

$$\begin{cases} \forall \nu,\nu' \in [0,L(p)]\;,\; \nu \neq \nu',\; \exists \bar{\nu} \in [0,L(p)] \text{ such that} \\ \rho_G^{ep}(\nu,\nu') \;\geq\; \rho_G(\bar{\nu},\nu'), \end{cases} \tag{7.51}$$

which we prove now. So let $p \in \mathcal{P}$, and ν,ν' be given such that (a similar proof holds if $\nu' < \nu$)

$$0 \leq \nu < \nu' \leq L(p).$$

Let X, X', v, v', N, and D be defined as in (7.24), and set, for any $\tau \in [0,L(p)]$,

$$\begin{aligned} N(\tau,\nu') &= \langle X' - p(\tau), v'\rangle, \\ v(\tau) &= p'(\tau). \end{aligned}$$

We treat separately the cases according to the signs of N and $\langle v,v'\rangle$:

Case 1: $N = N(\nu,\nu') \leq 0$.

The mapping $\tau \to N(\tau,\nu')$ is continuous, negative at $\tau = \nu$, and positive when τ is inferior and close enough to ν'. Hence, there exists $\bar{\nu} \in [\nu,\nu'[$ such that $N(\bar{\nu},\nu') = 0$. Hence

$$\rho_G^{ep}(\nu,\nu') \;\geq\; 0 \;=\; \rho_G(\bar{\nu},\nu'),$$

and (7.51) is satisfied.

Case 2: $N(\nu,\nu') > 0$.

Subcase 2.1: $\langle v,v'\rangle \geq 0$.

Then the formula for ρ_G and ρ_G^{ep} coincide, so that

$$\rho_G^{ep}(\nu,\nu') \;=\; N/D \;=\; \rho_G(\nu,\nu'),$$

and (7.51) holds with $\bar{\nu} = \nu$.

Subcase 2.2: $\langle v,v'\rangle\; < 0$.

Then the formula for ρ_G^{ep} is

$$\rho_G^{ep}(\nu, \nu') = N, \tag{7.52}$$

but that for ρ_G depends on the nature of ν:

Subsubcase 2.2.1: $\nu = 0$ (i.e., ν is an endpoint). Then

$$\rho_G(\nu, \nu' = N = \rho_G^{ep}(\nu, \nu'),$$

and (7.51) holds with $\bar{\nu} = \nu$.

Subsubcase 2.2.2: $\nu > 0$ (i.e., ν is an interior point). Then

$$\rho_G(\nu, \nu') = N/D > N = \rho_G^{ep}(\nu, \nu'), \tag{7.53}$$

and we have to exhibit some $\bar{\nu}$ to satisfy (7.51). So let $\bar{\nu}$ be defined as

$$\begin{cases} \bar{\nu} = \inf\Big\{\tilde{\nu} \in [0, \nu[\Big| N(\tau, \nu') > 0 \text{ and } \langle v(\tau), v' \rangle < 0 \\ \forall \tau, \quad \tilde{\nu} < \tau < \nu \Big\}. \end{cases} \tag{7.54}$$

As $N(\nu, \nu') > 0$ and $\langle v, v' \rangle < 0$, one has necessarily $\bar{\nu} < \nu$, and one checks easily that the function $\tau \rightsquigarrow N(\tau, \nu')^2$ is strictly increasing over the $[\bar{\nu}, \nu]$ interval. Hence

$$N(\bar{\nu}, \nu)^2 < N(\nu, \nu')^2 = N^2. \tag{7.55}$$

As the functions $\tau \rightsquigarrow N(\tau, \nu')$ and $\langle v(\tau), v' \rangle$ are continuous, the abscissa $\bar{\nu}$ defined by (7.54) satisfies necessarily one of the two following conditions:

- Either $N(\bar{\nu}, \nu') = 0$ and $\langle v(\bar{\nu}), v' \rangle \leq 0$. Then

$$\rho_G^{ep}(\nu, \nu') = N > 0 = \rho_G(\bar{\nu}, \nu'), \tag{7.56}$$

- Or $N(\bar{\nu}, \nu') > 0$ and $\langle v(\bar{\nu}), v' \rangle = 0$. Then $\bar{D} = (1 - \langle v(\bar{\nu}), v' \rangle^2)^{1/2} = 1$, so Proposition 7.2.6 shows that

$$\rho_G(\bar{\nu}, \nu') = N(\bar{\nu}, \nu'), \tag{7.57}$$

no matter whether $\bar{\nu}$ is an endpoint or not. Combining (7.52), (7.55), and (7.57) gives

$$\rho_G^{ep}(\nu, \nu') > \rho_G(\bar{\nu}, \nu'), \tag{7.58}$$

so that (7.51) holds with a strict inequality.

This ends the proof of formula (7.48). Then the possibility of skipping in the couples $\nu \neq \nu'$ that satisfy (7.50) follows from (7.56) and (7.58), and from the remark that, in both cases,

$$\rho_G(\bar{\nu}, \nu') = \rho_G^{ep}(\bar{\nu}, \nu'). \tag{7.59}$$

This ends the proof of the proposition. ■

A second remark is that, because $\rho_G^{ep}(\nu, \nu')$ given by (7.49) does not depend on the nature of ν and ν' (interior or endpoints), one obviously has, for any $p \in \mathcal{P}$,

$$R_G(p) = \inf_{p' \subset p} R_G(p'). \tag{7.60}$$

Notice that this result cannot be seen immediately when $R_G(p)$ is computed using $\rho_G(\nu, \nu')$ as in its definition: if p' is the sub-path of p going from ν to ν', the normal cones at ν and ν' to p' are necessarily larger than the normal cones to p, so that one has in general only

$$\rho_G(\nu, \nu'; p) \geq \rho_G(\nu, \nu'; p').$$

When a generating family $\mathcal{P}_G$ of $\mathcal{P}$ is available, as defined in Chap. 6, it is possible to "limit" the search to paths of $\mathcal{P}_G$ when computing the global radius of curvature of $(D, \mathcal{P})$:

Proposition 7.3.2 *Let $(D, \mathcal{P})$ be given, and $\mathcal{P}_G$ be a generating family of $\mathcal{P}$. Then*

$$R_G(D) = \inf_{p \in \mathcal{P}_G} R_G(p) \tag{7.61}$$

$$R(D) = \inf_{p \in \mathcal{P}_G} R(p) \tag{7.62}$$

where $R_G(p)$ can be computed by (7.48) and $R(p)$ by (7.33).

From a numerical point of view, Proposition 7.3.2 allows, at least in principle, to evaluate $R(D)$ and $R_G(D)$. For $D = \varphi(C)$, this evaluation requires to discretize the boundary ∂C of C, and to choose a discretization t_j of $[0, 1]$. Then $R(D)$ and $R_G(D)$ can be approximated by

$$R(D) \simeq \inf_{x_0, x_1 \in \partial C} \inf_{i-j=1} \rho_G^{ep}(i, j) \tag{7.63}$$

$$R_G(D) \simeq \inf_{x_0, x_1 \in \partial C} \inf_{i \neq j} \rho_G^{ep}(i, j) \tag{7.64}$$

Of course, these computations become quickly unfeasible when the dimension of the unknown parameter exceeds a few units, and a precise numerical evaluation of $R(D)$ and $R_{\mathrm{G}}(D)$ is in most cases out of reach. This is why we develop in the next chapter analytical lower bounds to $R(D)$ and $R_{\mathrm{G}}(D)$. They will provide us with sufficient conditions for a set to be s.q.c., which can be checked analytically.

Chapter 8

Deflection Conditions for the Strict Quasi-convexity of Sets

We develop in this chapter sufficient conditions for a set $(D, \mathcal{P})$ to be s.q.c. As we have seen in Chap. 7, an s.q.c. set, which is characterized by the fact that $R_G(D) > 0$, has necessarily a finite curvature, as $R(D) \geq R_G(D)$ (see Proposition 7.2.9). But the condition

$$R(D) > 0 \tag{8.1}$$

is not sufficient to ensure that $(D, \mathcal{P})$ is s.q.c.

This can be seen on the simple case where $(D, \mathcal{P})$ is an arc of circle of radius R and arc length L (Fig. 7.3). The *deflection* of this arc of circle, that is, the largest angle between two of its tangents – obviously the ones at its two ends in this case – is given by

$$\Theta_{\text{circle}} = L \,/\, R. \tag{8.2}$$

In the upper part of the figure, one sees that D is s.q.c., with the *largest regular enlargement neighborhood* ϑ of size $R_G = R$, as soon as it satisfies the *deflection condition*

$$\Theta_{\text{circle}} \leq \pi/2. \tag{8.3}$$

But the lower part of the figure indicates that it is in fact enough that Θ_{circle} satisfies the *extended deflection condition*

$$\Theta_{\text{circle}} < \pi \tag{8.4}$$

G. Chavent, *Nonlinear Least Squares for Inverse Problems: Theoretical Foundations and Step-by-Step Guide for Applications*, Scientific Computation,
DOI 10.1007/978-90-481-2785-6_8,

for the set D to remain s.q.c., at the price of a *possible reduction of the size of the regular enlargement neighborhood* ϑ to $R_G \le R$ given by

$$R_G \;=\; R \times \begin{cases} 1 & \text{if } \; 0 \le \Theta_{\text{circle}} \le \pi/2, \\ \sin\Theta & \text{if } \; \pi/2 < \Theta_{\text{circle}} < \pi. \end{cases} \tag{8.5}$$

Because of (8.2), conditions (8.3) and (8.4) are *size×curvature conditions,* as they ensure strict quasi-convexity by limiting the product of the size L of the arc of circle by its curvature $1/R$.

To see how the above results can be generalized, we define the *size* and the *deflection* associated to a path $p \in \mathcal{P}$ and to a set $(D, \mathcal{P})$ equipped with a family of paths (the global radius of curvature and the radius of curvature associated to p and D have been defined in Chap. 7, Definitions 7.2.3 and 7.2.8):

Definition 8.0.3 *(Size of paths and sets) Let $(D, \mathcal{P})$ and $p \in \mathcal{P}$ be given. The* size of p *is its arc length (Definition 6.1.2):*

$$L(p) \;=\; \ell, \quad \text{where} \;\; p : [0, \ell] \rightsquigarrow F, \tag{8.6}$$

and that of D is

$$L(D) \;=\; \sup_{p \in \mathcal{P}} \; L(p). \tag{8.7}$$

Formula (8.6) for $L(p)$ is duplicated from Definition 6.1.1, which gives the geometrical quantities associated to a path. Then (8.7) defines $L(D)$ as the size of D measured along its paths.

Definition 8.0.4 *(Deflection, see Fig. 8.1)*

Let $(D, \mathcal{P})$ and $p \in \mathcal{P}$ be given. For any ν, $\nu' \in [0, L(p)]$, the deflection of p between ν and ν' *is*

$$\theta(\nu, \nu') = \text{Arg } \cos \langle v(\nu), v(\nu') \rangle_F, \tag{8.8}$$

which satisfies

$$\begin{cases} \theta(\nu, \nu') \in [0, \pi] & \forall \nu, \nu' \in [0, L(p)], \\ \theta(\nu, \nu) \;=\; 0 & \forall \nu \in [0, L(p)], \\ \theta(\nu, \nu') = \theta(\nu', \nu) & \forall \nu, \nu' \in [0, L(p)]. \end{cases} \tag{8.9}$$

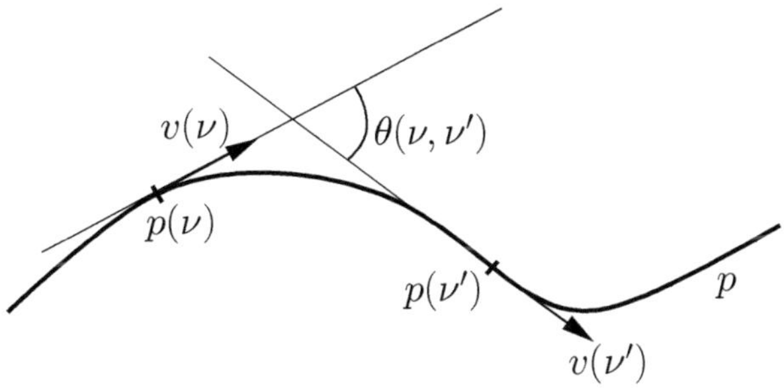

Figure 8.1: The deflection $\theta(\nu, \nu')$ between two points ν and ν' ($\nu > \nu'$) of a path p

The deflection of the path p is

$$\Theta(p) \;=\; \sup_{\nu,\nu' \in [0, L(p)]} \theta(\nu, \nu'), \tag{8.10}$$

and the deflection of $(D, \mathcal{P})$ *is*

$$\Theta(D) \;=\; \sup_{p \in \mathcal{P}} \Theta(p). \tag{8.11}$$

In Sect. 8.1, we search sufficient conditions for a $(D, \mathcal{P})$ to be s.q.c.

We study first the properties of the deflection, and prove in Proposition 8.1.2 that

$$\forall p \in \mathcal{P}, \quad \Theta(p) \leq \int_0^{L(p)} \frac{\mathrm{d}\nu}{\rho(\nu)} \leq L(p)/R(p) \leq L(D)/R(D). \tag{8.12}$$

This shows that the deflection $\Theta(D)$ of the set D is always smaller than the deflection of an arc of circle with radius $R(D)$ and arc length $L(D)$ (see (8.2)). From the point of view of deflection, arcs of circles, which "steadily turn in the same direction" are the worst sets!

In practice, it will be only possible in applications to have access to lower/upper bounds to $R(D), L(D), \Theta(D)$, and $R_G(D)$, and so we give them a name:

Definition 8.0.5 *We call* geometric attributes of the set $(D, \mathcal{P})$ *any set of numbers R (radius of curvature), L (size), Θ (deflection), R_{G} (global radius of curvature), which are lower/upper bounds to*

$$R(D) \geq 0,\ L(D) > 0,\ \Theta(D) \leq L(D)/R(D),\ R_G(D) \leq R(D). \tag{8.13}$$

and satisfy

$$R \geq 0, \qquad L > 0, \qquad \Theta \leq L/R, \qquad R_G \leq R. \tag{8.14}$$

Then we search a lower bound R_{G} to $R_{\mathrm{G}}(D)$ as a function of the lower/upper bounds R, L, and Θ. We prove in Theorem 8.1.5 that

$$\forall p \in \mathcal{P},\ R_{\mathrm{G}}(p)\ \geq\ R_{\mathrm{G}}(D)\ \geq\ R_{\mathrm{G}}\ \stackrel{\text{def}}{=}\ R_{\mathrm{G}}(R, L, \Theta), \tag{8.15}$$

where

$$R_{\mathrm{G}}(r, l, \theta) = \begin{cases} r, & 0 \leq \theta \leq \pi/2, \\ r\ \sin\theta + \big(l - r\theta\big)\cos\theta, & \pi/2 \leq \theta \leq \pi. \end{cases} \tag{8.16}$$

This shows first, using Theorem 7.2.5, that a sufficient condition for a set $(D, \mathcal{P})$ with finite curvature $1/R$ to be s.q.c. with an enlargement neighborhood ϑ of size $R_{\mathrm{G}} = R > 0$ is to satisfy the *deflection condition*

$$\Theta\ \leq\ \pi/2, \tag{8.17}$$

which generalizes condition (8.3). If one accepts the possibility of an enlargement neighborhood ϑ of reduced size $R_{\mathrm{G}} \leq R$, the same theorem shows that a finite curvature set $(D, \mathcal{P})$ is s.q.c. as soon as R, L, and $\Theta \leq L/R$ satisfy the *extended deflection condition*

$$R_{\mathrm{G}}\big(R, L, \Theta\big) > 0. \tag{8.18}$$

We show in Fig. 8.2 the set of values of the deflection Θ and the size $\times$ curvature product L/R that satisfy (8.18), and ensure that the set $(D, \mathcal{P})$ is s.q.c., with a regular enlargement neighborhood ϑ of size $0 < R_{\mathrm{G}}\big(R, L, \Theta\big) \leq R$. A simple calculation shows that (8.18) is equivalent to

$$\begin{cases} \Theta \leq \Theta_{\max} \stackrel{\text{def}}{=} L/R & \text{if } 0 \leq L/R < \pi, \\ \Theta < \Theta_{\max} \text{ such that } \Theta_{\max} - \tan\Theta_{\max} = L/R & \text{if } \pi \leq L/R. \end{cases} \tag{8.19}$$

This justifies the name given to condition (8.18): the upper limit to the *deflection* Θ is *extended* beyond $\pi/2$, up to a value $\theta_{\max}$ which depends on the estimated size$\times$curvature product L/R. This extended deflection condition reduces to the condition (8.4) obtained for an arc of circle as soon as the "worst" estimate $\Theta = L/R$ is used for the deflection.

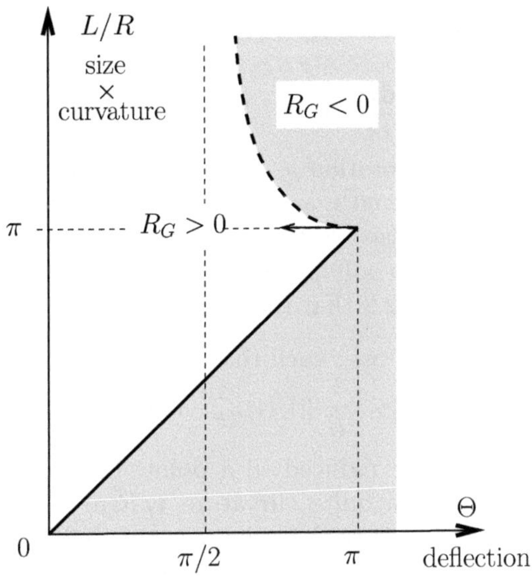

Figure 8.2: The domain of deflection Θ and size $\times$ curvature product L/R, which ensure strict quasi-convexity of a finite curvature set $(D, \mathcal{P})$ (extended deflection conditions (8.18) or (8.19))

Finally, in Sect. 8.2, we consider the case where the set D is the attainable set $\varphi(C)$ of a nonlinear least squares problem set over a convex set C of admissible parameters. It is then natural to try to equip D with the set of paths $\mathcal{P}$ made of the images by φ of the segments of C: to any $x_0, x_1 \in C$, one can always associate a curve P drawn on $D = \varphi(C)$ by

$$P : t \in [0,1] \rightsquigarrow P(t) = \varphi\big((1-t)x_0 + tx_1\big). \tag{8.20}$$

The first question is to know under which conditions the *curve* P – when it is not reduced to a point – becomes, once *reparameterized* by its arc length ν, a *path* p in the sense of Definition 6.1.1, that is, a $W^{2,\infty}$ function of ν.

A necessary condition is of course to require that

$$P \in W^{2,\infty}\big([0,1]; F\big), \tag{8.21}$$

which allows to define the *velocity* $V(t)$ and *acceleration* $A(t)$ along the curve by

$$V(t) = \frac{\mathrm{d}P}{\mathrm{d}t}(t), \qquad A(t) = \frac{\mathrm{d}^2P}{\mathrm{d}t^2}(t). \tag{8.22}$$

(We reserve the lower case notation p, v, and a to path, velocity, and acceleration with respect to arc length, as in Definition 6.1.2).
But under the sole hypothesis (8.21), the reparametrization p of P with respect to arc length satisfies only $p \in W^{1,\infty}\big([0,\ell];F\big)$. So our first task will be to show in Proposition 8.2.2 that the additional condition

$$\begin{cases} \exists R \in]0,+\infty] \quad \text{such that} \\ \|A(t)\|_F \le \dfrac{1}{R}\,\|V(t)\|_F^2 \quad \text{a.e. on }]0,1[\end{cases} \tag{8.23}$$

ensures that, when P is not reduced to a point, the reparameterization p of P by its arc length has a finite curvature $1/R(p) \le 1/R$, so that $p \in W^{2,\infty}\big([0,\ell];F\big)$, and hence is a path in the sense of Definition 6.1.1.

This will make it possible when (8.21) and (8.23) are satisfied for any $x_0, x_1 \in C$, to equip the attainable set $\varphi(C)$ with the family of paths $\mathcal{P}$ made of the reparameterizations p of the curves P, which are images of segments of C by φ and are not reduced to a point.

To apply to $(\varphi(C),\mathcal{P})$ the sufficient conditions for strict quasi-convexity of Sect. 8.1, it is necessary to estimate the geometric attributes R, L, Θ of $(\varphi(C),\mathcal{P})$. This is addressed in the last Proposition 8.2.2, where it is shown that, if there exists $\alpha_M \ge 0$ and $R > 0$ satisfying

$$\forall x_0, x_1 \in C, \quad \|V(t)\|_F \le \alpha_M \,\|x_1 - x_0\|_E \quad \forall t \in [0,1], \tag{8.24}$$

$$\forall x_0, x_1 \in C, \quad \|A(t)\|_F \le \frac{1}{R}\,\|V(t)\|_F^2 \qquad \text{a.e. on }]0,1[, \tag{8.25}$$

then R is a *lower bound to the radius of curvature* of $(\varphi(C),\mathcal{P})$:

$$\inf_{p\in\mathcal{P}} R(p) \stackrel{\text{def}}{=} R(\varphi(C)) \ge R > 0. \tag{8.26}$$

and any numbers L and Θ that satisfy

$$\forall x_0, x_1 \in C, \quad \int_0^1 \|V(t)\|_F\,\mathrm{d}t \le L \le \alpha_M\,\mathrm{diam}(C), \tag{8.27}$$

$$\begin{cases} \int_0^1 \theta(t)\,\mathrm{d}t \le \Theta \le L/R, \\ \text{where} \\ \|A(t)\|_F \le \theta(t)\|V(t)\|_F \text{ for a.e. } t \in [0,1] \text{ and all } x_0, x_1 \in C, \end{cases} \tag{8.28}$$

are upper bounds to the *arc length size* and *deflection* of $(\varphi(C), \mathcal{P})$. Hence R, L, Θ are the sought geometric attributes of $\varphi(C), \mathcal{P}$.

The sufficient condition (8.18) for strict quasi-convexity (Sect. 8.1) and the estimations (8.24) through (8.28) of R, L, Θ (Sect. 8.2) are the basis for all Q-wellposedness results of Chaps. 4 and 5. Examples of application of these conditions can be found in Sect. 5.2, where it is proved that the attainable set of the 2D elliptic nonlinear source estimation problem of Sect. 1.5 is s.q.c., and in Sects. 4.8 and 4.9, where the same result is proved for the 1D and 2D parameter estimation problems of Sects. 1.4 and 1.6.

8.1 The General Case: $D \subset F$

We consider in this section a set $D \subset F$ equipped with a family of paths $\mathcal{P}$ in the sense of Definition 6.1.3, and we search for sufficient conditions for $(D, \mathcal{P})$ to be s.q.c.

Definition 8.1.1 *Given a function* $g : [a, b] \rightsquigarrow I\!R$, *we denote by* $\mathrm{var}_{a,b} g$ *or* $\mathrm{var}_{b,a} g$ *the* total variation *(when it exists!) of* g *over the* $[a, b]$ *interval*

$$\begin{cases} \mathrm{var}_{a,b}\ g = \mathrm{var}_{b,a}\ g = \sup\Big\{ \sum_{i=1}^{N} |g(t_i) - g(t_{i-1}|,\ N \in I\!N, \\ \qquad\qquad \min\{a, b\} \le t_0 \le t_1 \le \ldots \le t_N \le \max\{a, b\}\Big\}. \end{cases} \tag{8.29}$$

When $g \in W^{1,1}([a, b])$, the total variation is given by

$$\mathrm{var}_{ab}\ g = \int_a^b |g'(t)|\,\mathrm{d}t. \tag{8.30}$$

We investigate now the regularity of the deflection $\nu \rightsquigarrow \theta(\nu, \nu')$, and its relation to arc length ν and radius of curvature $\rho(\nu)$.

Proposition 8.1.2 *(deflection estimate) Let p be a path of $\mathcal{P}$ with length $L(p)$ and smallest radius of curvature $R(p) > 0$. Then*

1. *For any $\nu' \in [0, L(p)]$, the $\nu \rightsquigarrow \theta(\nu, \nu')$ deflection function is absolutely continuous and has a bounded variation over $[0, L(p)]$. Hence $\partial\theta/\partial\nu(., \nu') \in L^1([0, L(p)])$, and the usual formula holds:*

$$\theta(\nu, \nu') = \int_{\nu'}^{\nu} \frac{\partial\theta}{\partial\nu}(t, \nu')\,\mathrm{d}t. \tag{8.31}$$

2. *Moreover,*

$$\Big|\frac{\partial \theta}{\partial \nu}(\nu,\nu')\Big| \;\leq\; \|a(\nu)\|_F \;\stackrel{\text{def}}{=}\; \frac{1}{\rho(\nu)} \quad \text{for a.e. } \;\nu \in [0, L(p)], \tag{8.32}$$

so that $\partial\theta/\partial\nu(.,\nu') \in L^\infty([0, L(p)])$.

3. *The* largest deflection $\Theta(p)$ along p *satisfies*

$$\Theta(p) \;\leq\; \|a\|_{L^1(0,L(p);F)} \;=\; \int_0^{L(p)} \frac{\partial \nu}{\rho(\nu)} \leq L(p)/R(p). \tag{8.33}$$

Proof: Let $p \in \mathcal{P}$ and $\nu' \in [0, L(p)]$ be given, and choose $0 < \overline{\theta} < \pi/2$. The function $\theta(.,.)$ is continuous over $[0, L(p)] \times [0, L(p)]$, and hence uniformly continuous. So there exists $\Delta\nu > 0$ such that

$$\nu_j \in [0, L(p)] \quad j = 1, 2 \quad \text{and} \;\; |\nu_1 - \nu_2| \leq \Delta\nu \tag{8.34}$$

implies

$$\theta(\nu_1, \nu_2) \;\leq\; \overline{\theta}. \tag{8.35}$$

A Taylor–MacLaurin development of $\cos t$ at $t = 0$ gives

$$\cos t = 1 - \frac{t^2}{2}\cos \alpha t \quad \text{for some} \;\; \alpha \in [0, 1],$$

and, as the cosine is a decreasing function over $[0, \pi]$

$$\cos t \leq \; 1 - \frac{t^2}{2}\cos t \quad \text{for} \;\; 0 \leq t \leq \pi.$$

Hence,

$$t^2 \cos t \;\leq\; 2\big(1 - \cos t\big) \quad \text{for} \;\; 0 \leq t \leq \pi.$$

Choosing $t = \theta(\nu_1, \nu_2)$, where ν_1, ν_2 satisfy (8.34) gives, as then $\cos t \geq \cos\overline{\theta} > 0$ and $\cos\theta(\nu_1, \nu_2) = \langle v_1, v_2\rangle$,

$$\theta\big(\nu_1, \nu_2\big)^2 \;\leq\; \frac{2(1 - \langle v_1, v_2\rangle)}{\cos\overline{\theta}}.$$

This can be rewritten as

$$\theta(\nu_1, \nu_2) \;\leq\; \frac{\|v_1 - v_2\|_F}{(\cos\overline{\theta})^{1/2}},$$

and, using the triangular inequality for the curvilinear triangle $v_1,\ v_2,\ v'$ on the unit sphere,

$$\big|\theta(\nu_1,\nu') - \theta(\nu_2,\nu')\big| \ \leq\ \frac{\|v_1 - v_2\|_F}{(\cos\ \overline{\theta})^{1/2}}. \tag{8.36}$$

We prove first the absolute continuity of the $\nu \rightsquigarrow \theta(\nu,\nu')$ function. Let $\epsilon > 0$ be given, and $(\alpha_i, \beta_i), i = 1, 2...N$, be disjoint segments of the interval $[0, L(p)]$ satisfying

$$\beta_i - \alpha_i \ \leq\ \Delta\nu \quad i = 1, 2...N.$$

Then we get from (8.36) that

$$\sum_{i=1}^{N} |\theta(\beta_i,\nu') - \theta(\alpha_i,\nu')| \leq \frac{1}{(\cos\overline{\theta})^{1/2}} \sum_{i=1}^{N} \|v(\beta_i) - v(\alpha_i)\|_F,$$

where, as $p \in W^{2,\infty}([0, L(p)]; F)$:

$$v(\beta_i) - v(\alpha_i) \ =\ \int_{\alpha_i}^{\beta_i} a(t)\,\mathrm{d}t$$

so that

$$\|v(\beta_i) - v(\alpha_i)\|_F \ \leq\ \int_{\alpha_i}^{\beta_i} \|a(t)\|_F\,\mathrm{d}t \ \leq\ (\beta_i - \alpha_i)\|a\|_\infty.$$

Hence,

$$\sum_{i=1}^{n} \big|\theta(\beta_i,\nu') - \theta(\alpha_i,\nu')\big| \ \leq\ \frac{\|a\|_\infty}{(\cos\overline{\theta})^{1/2}} \sum_{i=1}^{n} (\beta_i - \alpha_i), \tag{8.37}$$

which can be made smaller than ϵ by choosing the intervals (α_i, β_i) such that

$$\sum_{i=1}^{n} (\beta_i - \alpha_i) \ \leq\ \min\Big\{\Delta\nu,\ \epsilon\,\frac{(\cos\overline{\theta})^{1/2}}{\|a\|_\infty}\Big\}.$$

This proves that the function $\theta(.,\nu')$ is absolutely continuous over the $[0, L(p)]$ interval, which in turn implies that $\partial\theta/\partial\nu(.,\nu')$ is in $L^1(0, L(p))$, and that formula (8.31) holds.

We prove now that $\theta(.,\nu')$ has a bounded variation over $[0, L(p)]$. Let

$$0 \le t_0 < t_1 < .... < t_N \le L(p)$$

be given. One can always add a finite number of points to obtain a new subdivision:

$$0 \le t_0' < t_1' < ..... < t_{N'}' \le L(p),$$

with $N' \ge N$, such that

$$|t_i' - t_{i-1}'| \le \Delta\nu \qquad i = 1, 2...N'. \tag{8.38}$$

Then of course one has, because of the triangular inequality,

$$\sum_{i=1}^{N} |\theta(t_i,\nu') - \theta(t_{i-1},\nu')| \le \sum_{i=1}^{N'} |\theta(t_i',\nu') - \theta(t_{i-1}',\nu')|,$$

and because of (8.38), one obtains, as in (8.37),

$$\sum_{i=1}^{N'} |\theta(t_i',\nu') - \theta(t_{i-1}',\nu')| \le \frac{\|a\|_\infty}{(\cos\overline{\theta})^{1/2}} \sum_{i=1}^{N'} |t_i' - t_{i-1}'|.$$

The two last inequalities imply immediately that

$$\sum_{i=1}^{N} |\theta(t_i,\nu') - \theta(t_{i-1},\nu')| \le \frac{\|a\|_\infty}{(\cos\overline{\theta})^{1/2}} L(p)$$

independently of the positions and number of the points t_i, which proves that the function $\theta(.,\nu')$ has a bounded total variation.

We prove now formula (8.32). Let $\nu \in [0, L(p)]$ be a Lebesgue point for both $a \in L^\infty([0, L(p)]; F)$ and $\partial/\partial\nu(.,\nu') \in L^1([0, L(p)])$ (almost every point of $[0, L(p)]$ has this property!), and $\mathrm{d}\nu \ne 0$ such that $\nu + \mathrm{d}\nu \in [0, L(p)]$ and $|\mathrm{d}\nu| \le \Delta\nu$ defined at the beginning of the proof. Then we get from (8.36) that

$$\left|\frac{\theta(\nu + \mathrm{d}\nu,\nu') - \theta(\nu,\nu')}{\mathrm{d}\nu}\right| \le \frac{\|v(\nu + \mathrm{d}\nu) - v(\nu)\|_F \,/\, |\mathrm{d}\nu|}{(\cos\overline{\theta})^{1/2}},$$

which, by definition of the Lebesgue points, converges when $d\nu \to 0$ to

$$\left|\frac{\partial\theta}{\partial\nu}(\nu,\nu')\right| \le \frac{\|a(\nu)\|_F}{(\cos\overline{\theta})^{1/2}}.$$

But $\overline{\theta}$ can be chosen arbitrarily in the $]0, \pi/2[$ interval, which proves (8.32) as $\cos\overline{\theta}$ can be made arbitrarily close to one.

Finally, (8.33) follows immediately from (8.31) and (8.32), and Proposition 8.1.2 is proved. ■

We turn now to the estimation of a lower bound to the global radius of curvature $R_G(p)$ of a path p. Because of Proposition 7.3.1, this amounts to search for a lower bound to $\rho_G^{\mathrm{ep}}(\nu, \nu')$ independent of ν and ν', where

$$\rho_G^{\mathrm{ep}}(\nu, \nu') = \begin{cases} N^+/D & \text{if } \langle v, v' \rangle \geq 0, \\ N^+ & \text{if } \langle v, v' \rangle \leq 0, \end{cases} \tag{8.39}$$

and N and D are defined by (cf. formulas (7.24) in Proposition 7.2.6)

$$N = \mathrm{sgn}\,(\nu' - \nu)\, \langle X' - X,\ v' \rangle, \tag{8.40}$$

$$D = \left(1 - \langle v, v' \rangle^2\right)^{1/2}. \tag{8.41}$$

We give first a lower bound on the numerator N.

Lemma 8.1.3 *Let $(D, \mathcal{P})$ and $p \in \mathcal{P}$ be given. Then, for any $\nu, \nu' \in [0, L(p)]$ one has*

$$N \geq \overline{R} \sin\overline{\theta} + \left(|\nu' - \nu| - \overline{R}\,\overline{\theta}\right) \cos\overline{\theta}, \tag{8.42}$$

where

$$\overline{R} = \operatorname{inf\,ess} \quad \left\{\rho(t), \quad \min(\nu, \nu') < t < \max(\nu\nu')\right\}, \tag{8.43}$$

$$\overline{\theta} = \operatorname{sup\,ess} \quad \left\{\theta(t, \nu'), \quad \min(\nu, \nu') < t < \max(\nu\nu')\right\}, \tag{8.44}$$

Proof. We notice first that

$$\begin{aligned} \langle X - X', v' \rangle &= \int_\nu^{\nu'} \langle v(t), v' \rangle \, \mathrm{d}t \\ &= \int_\nu^{\nu'} \cos\theta(t, \nu') \, \mathrm{d}t. \end{aligned}$$

Hence, if we define

$$\nu^- = \min(\nu, \nu'), \quad \nu^+ = \max(\nu, \nu'), \tag{8.45}$$

then we obtain for N the formula

$$N = \int_{\nu^-}^{\nu^+} \cos\theta(t,\nu')\,dt,$$

and, using $\overline{\theta}$ defined by (8.44)

$$N = \int_{\nu^-}^{\nu^+} \big(\cos\theta(t,\nu') - \cos\overline{\theta}\big)\,dt \;+\; \big(\nu^+ - \nu^-\big)\cos\overline{\theta}. \tag{8.46}$$

To find a lower bound on N, we notice that, by definition of $\overline{\theta}$, one has

$$\cos\theta(t,\nu') - \cos\overline{\theta} \geq 0 \quad \text{for all} \quad t \in [\nu^-,\nu^+],$$

so that one can plug into the integral in (8.46) the inequality

$$1 \geq \rho(t)\Big|\frac{\partial\theta}{\partial\nu}\,(t,\nu')\Big| \quad \text{a.e. on} \quad]\nu^-,\nu^+[,$$

which has been proven in Proposition 8.1.2. This leads, as $\rho(t) \geq \overline{R}$ defined in (8.43), to

$$N \geq \overline{R}\int_{\nu^-}^{\nu^+} \big(\cos\theta(t,\nu') - \cos\overline{\theta}\big)\,\Big|\frac{\partial\theta}{\partial\nu}(t,\nu')\Big|dt \;+\; (\nu^+ - \nu^-)\cos\overline{\theta},$$

that is,

$$N \geq \overline{R}\int_{\nu^-}^{\nu^+} \Big|\frac{\partial}{\partial\nu}\big(\sin\theta(t,\nu') - \theta(t,\nu')\cos\overline{\theta}\big)\,\Big|dt \;+\; (\nu^+ - \nu^-)\cos\overline{\theta},$$

that is,

$$N \geq \overline{R}\,\mathrm{var}_{\nu^-,\nu^+}\Big(\sin\theta(.,\nu') - \theta(.,\nu')\cos\overline{\theta}\Big) + \big(\nu^+ - \nu^-\big)\cos\overline{\theta}. \tag{8.47}$$

But $\theta(.,\nu')$ is continuous over the $[\nu^-,\nu^+]$ interval, and so there exists $\overline{\nu} \in [\nu^-,\nu^+]$ such that

$$\theta(\overline{\nu},\nu') \;=\; \overline{\theta}.$$

The formula (8.29) defining the total variation, applied with $N = 1$, $t_0 = \overline{\nu}$, $t_1 = \nu'$, gives then, as $\theta(\nu',\nu') = 0$,

$$\mathrm{var}_{\nu^-,\nu^+}\big(\sin\theta(.,\nu') - \theta(.,\nu')\cos\overline{\theta}\,\big) \;\geq\; \big|\sin\overline{\theta} - \overline{\theta}\cos\overline{\theta}\,\big|. \tag{8.48}$$

But the function $t \rightsquigarrow \sin t - t\,\cos t$ is positive over the $[0,\pi]$ interval. Hence we can drop the absolute value in (8.48), and substitute it into (8.47), which gives the desired lower bound (8.42) on N. ■

Remark 8.1.4: *The lower bound (8.42) on N retains its maximum value only from the shape of the deflection $\theta(.,\nu')$. However, the inequality (8.47) shows that, among pieces of paths having the same size $|\nu'-\nu|$ and curvature $1/\overline{R}$, the ones whose deflection has a large total variation are more likely to have positive global radii of curvature, and hence to be s.q.c. (the $\theta \to \sin\theta - \theta\cos\overline{\theta}$ function is nondecreasing over the $[0,\overline{\theta}]$ interval, and so a large variation of $\theta(.,\nu')$ corresponds to a large variation of $\sin\theta(.,\nu') - \theta(.,\nu')\cos\overline{\theta}$, and hence is more likely to produce through (8.47) a positive lower bound to N).*

But the total variation of the deflection is difficult to estimate in applications, and so we shall retain only the less sharp estimate (8.42) based on the maximum deflection. ■

Proposition 8.1.5 *(Global radius of curvature estimate). Let p be a path of $\mathcal{P}$, and denote by $R(p) > 0$ its smallest radius of curvature, $L(p) > 0$ its length, and $0 \le \Theta(p) \le L(p)/R(p)$ its largest deflection.*

Then its smallest global radius of curvature *$R_G(p)$ satisfies*

$$R_G(p) \;\ge\; R_G\big(R(p), L(p), \Theta(p)\big), \tag{8.49}$$

where $R_G(r,l,\theta)$, defined by (8.16), is a decreasing function of $1/r$, l, and θ over the domain $0 \le \theta \le \pi$, $l - r\theta \ge 0$.

In particular, if R, L, Θ are three first geometric attributes of $(D,\mathcal{P})$ (Definition 8.0.5), a fourth one is given by

$$R_G \stackrel{\text{def}}{=} R_G(R, L, \Theta). \tag{8.50}$$

Proof. The partial derivatives of R_G are positive with respect to r, and negative with respect to l and θ over the domain of definition, which proves the announced monotonicity property.

Let now $p \in \mathcal{P}$ be given. We shall drop in the rest of the proof the argument p in $R(p)$, $R_G(p)$, $L(p)$, $\Theta(p)$, and write simply instead R, R_G, L, Θ. Let then ν, $\nu' \in [0,\ell]$ be given. Formula (8.33) shows that

$$|\nu'-\nu| \;-\; \overline{R}\,\overline{\theta} \;\ge\; 0, \tag{8.51}$$

where $\overline{R}$ and $\overline{\theta}$ are defined in (8.43) and (8.44).

We consider first the case where $0 \le \Theta \le \pi/2$. By definition of $\overline{\theta}$ and Θ one has

$$\theta(\nu,\nu') \ \le \ \overline{\theta} \ \le \ \Theta \ \le \ \pi/2, \tag{8.52}$$

and, as the cosine function is decreasing over the $[0,\pi/2]$

$$\langle v, v'\rangle \ = \ \cos\theta(\nu,\nu') \ \ge \ \cos\overline{\theta} \ \ge \ 0.$$

This implies using (8.39) that

$$\rho^{\mathrm{ep}}_{\mathrm{G}}\,(\nu,\nu') \ = \ N^+/D \ \ge \ N/D, \tag{8.53}$$

and using (8.42) and (8.51) that

$$N \ \ge \ \overline{R}\ \sin\ \overline{\theta}. \tag{8.54}$$

The sine function is increasing over $[0,\pi/2]$, and so we obtain from (8.41) that

$$D \ = \ \big(1-\langle v,v'\rangle^2\big)^{1/2} \ = \ \sin\theta(\nu,\nu') \ \le \ \sin\overline{\theta}. \tag{8.55}$$

Then (8.53), (8.54), and (8.55) imply

$$\rho^{\mathrm{ep}}_{\mathrm{G}}(\nu,\nu') \ \ge \ \overline{R}\,,$$

which implies (8.49) as $\overline{R} \ge R = R_{\mathrm{G}}(R,L,\Theta)$.

We turn now to the case where $\pi/2 < \Theta \le \pi$.
The maximum deflection $\bar{\theta}$ on p between ν and ν' satisfies $0 \le \bar{\theta} \le \Theta$. Hence two cases can happen:

- Either $0 \le \overline{\theta} \le \pi/2$, and then one finds as above that

$$\rho^{\mathrm{ep}}_{\mathrm{G}}(\nu,\nu') \ \ge \overline{R} \ \ge \ R. \tag{8.56}$$

  But $0 \le \sin\Theta \le 1$, $L/R - \Theta \ge 0$ because of (8.33), and $\cos\Theta \le 0$, and so

$$\rho^{\mathrm{ep}}_{\mathrm{G}}(\nu,\nu') \ \ge \ R\sin\Theta \ \ge \ R\sin\Theta + \big(L - R\Theta\big)\ \cos\Theta = R_{\mathrm{G}}(R,L,\Theta),$$

  which implies (8.49).

- Or $\pi/2 < \overline{\theta} \le \Theta$, and then $\langle v, v' \rangle = \cos\theta(\nu, \nu')$ can be either positive (if $0 \le \theta(\nu,\nu') \le \pi/2$) or negative (if $\pi/2 \le \theta(\nu,\nu') \le \overline{\theta}$), and so the only information we get from (8.39) is

$$\rho_{\mathrm{G}}^{\mathrm{ep}}(\nu,\nu') \ \ge N^+ \ \ge N.$$

Combined with lemma (8.31), this shows that

$$\rho_{\mathrm{G}}^{\mathrm{ep}}(\nu,\nu') \ \ge \ \overline{R}\sin\overline{\theta} + \left(|\nu'-\nu| - \overline{R}\,\overline{\theta}\right)\ \cos\overline{\theta} = R_{\mathrm{G}}(\overline{R}, |\nu'-\nu|\,, \overline{\theta}).$$

But

$$\overline{R} \ge R > 0 \qquad |\nu'-\nu| \le L \qquad \overline{\theta} \le \Theta,$$

$$0 \le \overline{\theta} \le |\nu'-\nu|/\overline{R} \qquad\qquad 0 \le \Theta \ \le L/R,$$

so the monotonicity property of R_{G} shows again that

$$\rho_{\mathrm{G}}^{\mathrm{ep}}(\nu,\nu') \ \ge R_{\mathrm{G}}(R, L, \Theta),$$

which implies (8.49). Then (8.50) follows immediately from the monotonicity properties of the $r, l, \theta \rightsquigarrow R_{\mathrm{G}}$ function. ■

We can now state the main result of this section:

Theorem 8.1.6 *Let $(D, \mathcal{P})$ be a set equipped with a family of paths, and R, L, Θ be three first geometric attributes of $(D, \mathcal{P})$ (Definition 8.0.5). Then*

- *$(D, \mathcal{P})$ is s.q.c. as soon as the fourth geometric attribute given by (8.50) satisfies $R_{\mathrm{G}} > 0$, that is,*

$$R_{\mathrm{G}} \stackrel{\text{def}}{=} R_{\mathrm{G}}(R, L, \Theta) > 0, \qquad \text{(extended deflection condition)} \quad (8.57)$$

where the function R_{G} is defined in (8.16)

- *The global radius of curvature of $(D, \mathcal{P})$ satisfies*

$$R_{\mathrm{G}}(D) \ge R_{\mathrm{G}} > 0 \qquad (8.58)$$

Proof. It follows immediately from (8.50) in Proposition 8.1.5 that $(D, \mathcal{P})$ has a strictly positive global radius of curvature. It is hence s.q.c. by virtue of Theorem 7.2.5. ■

We have already illustrated in Fig. 8.2 the domain of deflection estimates Θ and size $\times$ curvature estimates L/R, which satisfy (8.57). We show now in Fig. 8.3 that these conditions are sharp: as soon as the deflection estimate Θ becomes strictly larger than $\pi/2$, the upper bound $L/R < \Theta - \tan\Theta$ on the size $\times$ curvature estimate is the best possible. The set D of Fig. 8.3 is made of a path p, made itself of two parts: an arc of circle of radius R and deflection $\Theta \in]\pi/2, \pi[$ and, tangent at one end of the arc of circle, a segment of length $|\tan\Theta| = -\tan\Theta$. This set is not s.q.c.: let us choose for p the longest path of D, that is, $p = D$, and for z the end of p located on the segment. Then $d(z,p) = 0$, but the function $\nu \rightsquigarrow d_{z,p}(\nu)^2$ has, beside its global minimum at z (with value zero !), a parasitic stationary point, with zero derivative, at the other end of p. Such a function is not s.q.c., so that property (7.4) of the Definition 7.2.2 of s.q.c. sets cannot hold true for D!

But the best curvature, deflection, and size estimates for this set D are obviously R, Θ, and $L = R(\Theta - tan\Theta)$, which satisfy exactly (8.57) – or the second inequality of (8.19) – with the strict inequality replaced by an equality. Hence the upper bound (8.19) on L/R is the smallest, which can ensure the strict quasi-convexity of $(D, \mathcal{P})$.

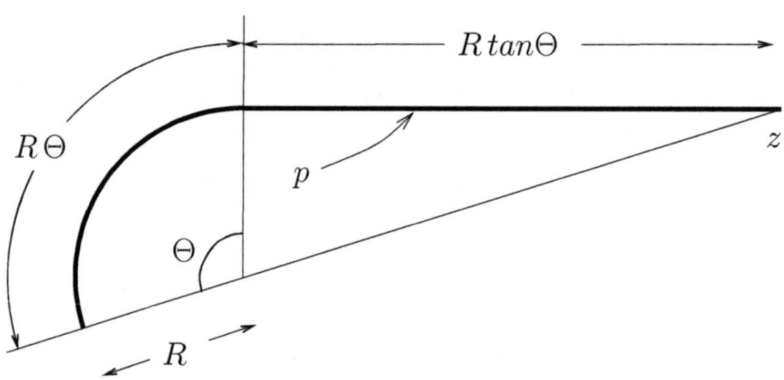

Figure 8.3: Illustration of the sharpness of the extended deflection condition (8.57)

8.2 The Case of an Attainable Set $D = \varphi\ (C)$

We consider in this section a set D of F, which is the image, by some mapping φ of a convex set C of E. In the context of inverse problem, E is the parameter space, C the set of admissible parameters, F the data space, and φ the forward map to be inverted. Strict quasi-convexity of the attainable set $D = \varphi(C)$ will be the key to Q-wellposedness of the inverse problem (see Chap. 4). So we discuss first in this section the possibility of equipping $D = \varphi(C)$ with a family of path $\mathcal{P}$, as required in the definition of s.q.c. sets, and we estimate a set of geometric attributes R, L, and Θ, as required by Theorem 8.1.6, to prove that $(\varphi(C), \mathcal{P})$ is s.q.c.

We suppose now that

$$\begin{cases} E &= \text{ Banach space,} \\ C &\subset E \quad \text{convex,} \\ \varphi : C &\rightsquigarrow F. \end{cases} \tag{8.59}$$

A natural way to equip $D = \varphi(C)$ with a family of paths is to consider the *curves* $P : [0,1] \to D$, which are image by φ of some segment of C:

$$\begin{cases} \text{to } x_0, x_1 \in C \quad \text{we associate } P : [0,1] \to D \quad \text{s.t.:} \\ \forall t \in [0,1], \quad P(t) = \varphi\big((1-t)x_0 + t x_1\big). \end{cases} \tag{8.60}$$

We suppose that the forward map φ is smooth enough so that

$$\forall x_0, x_1 \in C, \quad P \in W^{2,\infty}\big([0,1]; F\big), \tag{8.61}$$

and we denote, as indicated in (8.22), by $V(t)$ and $A(t)$ the first and second derivatives of P.

To see derivatives whether a curve P defined as above can be turned into a path p in the sense of Definition 6.1.1, we have first to reparameterize it as a function of its arc length ν, defined by

$$\forall t \in [0,1], \quad \nu(t) \;=\; \int_0^t \|V(t)\|_F \, dt,$$

which satisfies

$$0 \le \nu \le \ell \stackrel{\text{def}}{=} \nu(1),$$

and the arc length of the curve P is

$$L(P) \;=\; \ell \;=\; \nu(1) \;=\; \int_0^1 \|V(t)\|_F \, dt. \tag{8.62}$$

Definition 6.1.1 requires that paths of $\varphi(C)$ have a strictly positive length. Hence *we shall consider only in the sequel the curves P such that $L(P) > 0$* .

By construction, $t \rightsquigarrow \nu(t)$ is a nondecreasing mapping from $[0,1]$ onto $[0,\ell]$, which can be constant on some intervals of $[0,1]$. The reparameterized path $p : [0,\ell] \to D$ is hence unambiguously defined by

$$p\big(\nu(t)\big) \;=\; P(t) \quad \forall t \in [0,1]. \tag{8.63}$$

It will be convenient to denote the *velocity* and *acceleration* along p, that is, the two first derivatives of p with respect to ν, when they exist, by the *lower case* letters

$$v(\nu) \;=\; p'(\nu), \quad a(\nu) = v'(\nu) = p''(\nu) \quad \forall \nu \in [0,\ell].$$

Proposition 8.2.1 *Let (8.59) and (8.61) hold, and $x_0, x_1 \in C$ be such that the curve P defined by (8.60) has an arc length $L(P) > 0$. Then its reparameterization p by arc length defined in (8.63) satisfies*

$$p \;\in\; W^{1,\infty}\big([0,\ell];F\big), \tag{8.64}$$

$$\big\|v(\nu)\big\|_F \;=\; 1 \qquad \text{a.e. on} \quad [0,\ell], \tag{8.65}$$

and P has, at all points $t \in [0,1]$, where V is derivable and $V(t) \neq 0$, a finite curvature $1/\rho(t)$ – *that is,* a radius of curvature $\rho(t) > 0$ – *given by*

$$\frac{1}{\rho(t)} = \|a(\nu(t))\|_F = \left(\left(\frac{\|A\|_F}{\|V\|_F^2}\right)^2 - \left\langle \frac{A}{\|V\|_F^2}, \frac{V}{\|V\|_F}\right\rangle^2\right)^{\frac{1}{2}} \leq \frac{\|A\|_F}{\|V\|_F^2}. \tag{8.66}$$

Proof. The reparameterized path p is in $L^\infty\big([0,\ell];F\big)$ by construction, and hence defines a distribution on $]0,\ell[$ with values in F. We compute the derivative v of this distribution p. For any $\varphi \in \mathcal{D}\big(]0,\ell[\big)$ (the space of $\mathcal{C}^\infty\big(]0,\ell[\big)$ functions with compact support in $]0,\ell[$) one has

$$\begin{aligned}
\langle v, \varphi\rangle \;&=\; -\int_0^\ell p(\nu)\,\varphi'(\nu)\,\mathrm{d}\nu \\
&=\; -\int_0^1 p\big(\nu(t)\big)\,\varphi'\big(\nu(t)\big)\,\nu'(t)\,\mathrm{d}t \\
&=\; -\int_0^1 P(t)\,\frac{\mathrm{d}}{\mathrm{d}t}\varphi\big(\nu(t)\big)\,\mathrm{d}t
\end{aligned}$$

But $\nu(.)$ belongs to $W^{1,\infty}(]0,1[)$ and so does $\varphi\big(\nu(.)\big)$, with moreover zero values at $t=0$ and $t=1$. So we can integrate by part, as $P \in W^{2,\infty}\big(]0,1[;F\big)$,

$$\langle v, \varphi\rangle \;=\; \int_0^1 V(t)\ \varphi\big(\nu(t)\big)\,\mathrm{d}t. \tag{8.67}$$

To express $\langle v, \varphi\rangle$, as an integral with respect to ν, one has to replace in (8.67) $\mathrm{d}t$ by $\mathrm{d}\nu\ /\ \|V(t)\|_F$, which is possible only if $V(t) \neq 0$. So we define

$$I \;=\; \Big\{t \in]0,1[\ \mid V(t)\ \neq\ 0\Big\}.$$

As I is an open set, it is the reunion of a countable family of pair-wise disjoint intervals:

$$I = \bigcup_{i=1}^{\infty} I_i, \quad \text{where}\ \ I_i =]\alpha_i, \beta_i[\ , \quad i = 1, 2....$$

So we can rewrite (8.67) as

$$\begin{aligned}
\langle v, \varphi\rangle \;&=\; \int_I V(t)\ \varphi\big(\nu(t)\big)\,\mathrm{d}t \\
&=\; \sum_{i=1}^{\infty}\int_{\alpha_i}^{\beta_i} V(t)\ \varphi\big(\nu(t)\big)\,\mathrm{d}t \\
&=\; \sum_{i=1}^{\infty}\int_{\alpha_i}^{\beta_i} \frac{V(t)}{\|V(t)\|_F}\ \varphi\big(\nu(t)\big)\ \|V(t)\|_F\,\mathrm{d}t
\end{aligned}$$

We define now

$$J \;=\; \bigcup_{i=1}^{\infty} J_i, \qquad \text{where}\ \ J_i \;=\; \nu(I_i) \quad i = 1, 2....$$

The sets J_i $i = 1, 2...$ are (also pair-wise disjoints) open intervals of $]0, \ell[$. So we can associate to any $\nu \in J$ a number $t(\nu) \in]0,1[$, which is the reciprocal of the $t \rightsquigarrow \nu(t)$ function over the interval J_i containing ν. Hence we see that

$$\langle v, \varphi\rangle \;=\; \sum_{i=1}^{\infty}\int_{J_i} \frac{V(t(\nu))}{\|V(t(\nu))\|_F}\ \varphi(\nu)\,\mathrm{d}\nu,$$

$$\langle v, \varphi\rangle \;=\; \int_J \frac{V(t(\nu))}{\|V(t(\nu))\|_F}\ \varphi(\nu)\,\mathrm{d}\nu, \tag{8.68}$$

and similarly that

$$\begin{aligned}
\ell &= \int_0^1 \|V(t)\|_F \, \mathrm{d}t = \int_I \|V(t)\|_F \, \mathrm{d}t \\
\ell &= \sum_{i=1}^{\infty} \int_{\alpha_i}^{\beta_i} \|V(t)\|_F \, \mathrm{d}t = \sum_{i=1}^{\infty} \int_{J_i} \mathrm{d}\nu \\
\ell &= \int_J \mathrm{d}\nu = \text{meas } J
\end{aligned}$$

Hence we see that the complementary of J in $]0, \ell[$ has zero measure. If we continue by some $t_0 \in I$ the $\nu \longrightarrow t(\nu)$ function on this zero-measure set, we can rewrite (8.68) as

$$\langle v, \varphi \rangle = \int_0^\ell \frac{V(t(\nu))}{\|V(t(\nu))\|_F} \varphi(\nu) \, \mathrm{d}\nu,$$

which shows that the distribution $v = p'$ is in fact a function

$$v(\nu) = \frac{V(t(\nu))}{\|(Vt(\nu))\|_F} \quad \text{a.e. on }]0, \ell[. \tag{8.69}$$

This proves (8.64) and (8.65). Given any $\nu \in J$, we can differentiate (8.69) with respect to ν. This gives

$$a(\nu) = \frac{A}{\|V\|_F^2} - \frac{V}{\|V\|_F} \left\langle \frac{A}{\|V\|_F^2}, \frac{V}{\|V\|_F} \right\rangle,$$

and

$$\|a(\nu)\|_F^2 = \left(\frac{\|A\|_F}{\|V\|_F^2} \right)^2 - \left\langle \frac{A}{\|V\|_F^2}, \frac{V}{\|V\|_F} \right\rangle^2, \tag{8.70}$$

where right-hand sides are evaluated at $t = t(\nu)$. Hence for any $t \in I$, such that $\|V(t)\|_F > 0$, one has $\nu(t) \in J$, and (8.70) shows that

$$\|a(\nu(t))\| \leq \frac{\|A(t)\|_F}{\|V(t)\|_F^2} < +\infty,$$

which is (8.66). ■

So we see that the hypothesis that $P \in W^{2,\infty}([0,1]; F)$ is not enough to ensure that its reparametrization p as a function of arc length is $W^{2,\infty}([0,\ell]; F)$: in general, the derivative v of p can have discontinuities at points $\nu \notin J$!

Proposition 8.2.2 *Let (8.59) and (8.61) hold, and $x_0, x_1 \in C$ be such that the curve P associated by (8.60) satisfies*

$$\text{there exists } R_P > 0 \text{ s.t.: } \|A(t)\|_F \leq \frac{1}{R_P}\,\|V(t)\|_F^2 \quad \text{a.e. on } \,]0,1[. \quad (8.71)$$

Then one of the two following properties holds:

- **Either**

$$V(t) = 0 \quad \forall t \in [0,1] \quad \Longleftrightarrow \quad L(P) = 0, \qquad (8.72)$$

where $L(P)$ is the length of P defined in (8.62), and the curve P is reduced to one point of $\varphi(C)$,

- **or**

$$V(t) \neq 0 \quad \forall t \in [0,1] \quad \Longrightarrow \quad L(P) > 0, \qquad (8.73)$$

and the reparameterization p of P by its arc length, *defined by (8.63), is a* path *in the sense of Definition 6.1.1. So we can define the* radius of curvature of the curve P *by* $R(P) \stackrel{\text{def}}{=} R(p)$ *and its emphasize by* $\Theta(P) \stackrel{\text{def}}{=} \Theta(p)$, *which satisfy the following:*

(i) Curvature estimate:

$$R(P) \geq R_P > 0. \qquad (8.74)$$

(ii) Deflection estimate:

$$\Theta(P) \leq \int_0^1 \frac{\|A(t)\|_F}{\|V(t)\|_F}\,\mathrm{d}t \leq \frac{L(P)}{R_P}. \qquad (8.75)$$

Proof. Let $P \in W^{2,\infty}([0,1]; F)$ satisfying (8.71) be given.

To prove (8.72) and (8.73), we suppose that $V(t_0) \neq 0$ for some $t_0 \in [0,1]$, and we prove that $V(t) \neq 0 \ \forall t \in [0,1]$. Suppose this is not true, and define $I =]\alpha, \beta[\subset [0,1]$ as the largest interval containing t_0 such that $V(t) \neq 0 \, \forall t \in I$. Then necessarily $\alpha > 0$ and/or $\beta < 1$, say $\beta < 1$, in which case $V(\beta) = 0$. Then because of (8.71), the function $g(t) = \|V(t)\|_F$ satisfies

$$\left|\frac{\mathrm{d}g}{\mathrm{d}t}(t)\right| \leq \|A(t)\|_F \leq \frac{\|V(t)\|_F^2}{R_P} \leq \frac{g(t)^2}{R} \qquad \forall t \in I,$$

$$\left|\frac{\mathrm{d}}{\mathrm{d}t}\left(\frac{1}{g}\right)(t)\right| \leq \frac{1}{R_P} \qquad \forall t \in I. \qquad (8.76)$$

Hence,

$$\frac{1}{g(t)} \leq \frac{1}{g(t_0)} + \frac{|t-t_0|}{R_P} \leq \frac{1}{g(t_0)} + \frac{1}{R_P} \qquad \forall t \in I,$$

so that

$$g(t) \geq c \stackrel{\text{def}}{=} \left(\frac{1}{g(t_0)} + \frac{1}{R_P}\right)^{-1} > 0 \qquad \forall t \in I.$$

It follows that $g(\beta) = \|V(\beta)\|_F \geq c > 0$, which is a contradiction as we have seen that $V(\beta) = 0$. Hence $I =]0,1[$, and (8.72) and (8.73) are proved.

Let now (8.73) hold, and p be the reparameterization of P by arc length. As $V(t) \neq 0\ \forall t \in [0,1]$, formula (8.65) of Proposition 8.2.1 applies for all $t \in [0,1]$, where V is derivable, that is, almost everywhere on $[0,1]$. Hence $\|a(t)\|_F \leq 1/R_P$ for a.e. $t \in [0,1]$, so that $p \in W^{2,\infty}([0, L(P)]; F)$, and p is a path of curvature smaller than $1/R_P$, which proves (8.74).

Then Proposition 8.1.2 part (iii) gives the following estimate for the deflection of p:

$$\Theta(p) \leq \int_0^{L(P)} \|a(\nu)\|\, \mathrm{d}\nu. \tag{8.77}$$

Changing for the variable $t \in [0,1]$ in the integral gives

$$\Theta(p) \leq \int_0^1 \|a(\nu(t))\|\, \|V(t)\|\, \mathrm{d}t, \tag{8.78}$$

and, because of (8.66)

$$\|a(\nu(t))\|\, \|V(t)\| \leq \|A(t)\| / \|V(t)\|,$$

which proves the first inequality in (8.75). Then (8.71) gives

$$\int_0^1 \frac{\|A(t)\|}{\|V(t)\|}\, \mathrm{d}t = \int_0^1 \frac{\|A(t)\|}{\|V(t)\|^2} \|V(t)\|\, \mathrm{d}t \leq \int_0^1 \frac{1}{R_P} \|V(t)\|\, \mathrm{d}t = \frac{L(P)}{R_P},$$

and the last inequality in (8.75) is proved. ■

We can now equip $\varphi(C)$ with the family of curves $\mathcal{P}$ – or of paths p – defined by

$$\mathcal{P} = \Big\{P : t \in [0,1] \rightsquigarrow \varphi((1-t)x_0 + tx_1), x_0, x_1 \in C \mid L(P) > 0\Big\}. \tag{8.79}$$

The next theorem gives a sufficient condition, which guarantees that $\mathcal{P}$ is a family of path of $\varphi(C)$ in the sense of Definition 6.1.3, and provides a set of *geometric attributes* R, L, Θ of the *attainable set* $(\varphi(C), \mathcal{P})$, given the velocity $V(t)$ and acceleration $A(t)$ along the curves of $\mathcal{P}$.

Theorem 8.2.3 *Let C and φ be given such that (8.59) and (8.61) hold.*
(i) If there exists $R > 0$ such that

$$\forall x_0, x_1 \in C, \quad \|A(t)\|_F \ \leq \ \frac{1}{R}\ \|V(t)\|_F^2 \quad \text{a.e. in} \quad [0,1], \tag{8.80}$$

then the family of curves $\mathcal{P}$ defined in (8.79) is, once reparameterized by the arc length using (8.63), a family of paths of $\varphi(C)$ in the sense of Definition 6.1.3, and the attainable set $(\varphi(C), \mathcal{P})$ *has a* finite curvature:

$$R(\varphi(C)) \ \geq \ R \ > \ 0. \tag{8.81}$$

(ii) If there exists $\alpha_M \geq 0$ such that

$$\forall x_0, x_1 \in C, \|V(t)\|_F \ \leq \ \alpha_M \|x_1 - x_0\|_E \quad \text{a.e. in } [0,1], \tag{8.82}$$

then any number $L \geq 0$ that satisfies

$$\forall x_0, x_1 \in C, \quad \int_0^1 \|V(t)\|_F \, \mathrm{dt} \leq L \leq \alpha_M \, \mathrm{diam}(C) \tag{8.83}$$

is an upper bound to the (arc length) size $L(\varphi(C))$ *of* $(\varphi(C), \mathcal{P})$.

(iii) Any number $\Theta \geq 0$ that satisfies

$$\begin{cases} \int_0^1 \theta(t)\, \mathrm{dt} \leq \Theta \leq L/R \\ \text{where} \\ \|A(t)\|_F \leq \theta(t) \|V(t)\|_F \ \textit{for a.e. } t \in [0,1] \ \ \text{and all } x_0, x_1 \in C \end{cases} \tag{8.84}$$

is an upper bound to the deflection $\Theta(\varphi(C))$ *of* $(\varphi(C), \mathcal{P})$.

Proof. The announced results follow immediately from Proposition 8.2.2 by considering the worst case over all curves P of $\mathcal{P}$. ∎

Bibliography

[1] Aki, K., Richards, P.G., 1980, Quantitative seismology: Theory and methods, W.H. Freeman, New York 6

[2] Al Khoury, Ph., 2005, Algorithmes géométriques de résolution des moindres carrés non linéaires et problèmes inverses en spectroscopie des flammes, PhD Thesis, University of Paris 10, March 4 88, 135, 148

[3] Al Khoury, Ph., Chavent, G., 2006, Global line search strategies for nonlinear least squares problems based on curvature and projected curvature, Inverse Probl. Sci. Eng. 14(5), 495–509 135

[4] Alessandrini, G., 1986, An identification problem for an elliptic equation in two variables, Ann. Math. Pura Appl. 145, 265–296 191

[5] Alessandrini, G., Magnanini, R., 1994, Elliptic equations in divergence form, geometric critical points of solutions, and stekloff eigenfunctions, SIAM J. Math. Anal. 25(5), 1259–1268 205

[6] Anterion, F., Eymard, R., Karcher, B., 1989, Use of parameter gradients for reservoir history matching, In SPE Symposium on Reservoir Simulation, Society of Petroleum Engineers, Houston, Texas, SPE 18433 78

[7] Banks, H.T., Kunisch, K., 1989, Estimation techniques for distributed parameter systems, Birkhäuser, Boston 29

[8] Baumeister, J., 1987, Stable solutions of inverse problems, Vieweg, Braunschweig 17, 210

[9] Ben-Ameur, H., Kaltenbacher, B., 2002, Regularization of parameter estimation by adaptive discretization using refinement and coarsening indicators. J. Inverse Ill Posed Probl. 10(6), 561–583 116

[10] Ben-Ameur, H., Chavent, G., Jaffré, J., 2002, Refinement and coarsening indicators for adaptive parametrization: Application to the estimation of the hydraulic transmissivities. Inverse Probl. 18, 775–794 113, 116, 120, 123

[11] Ben-Ameur, H., Clément, F., Chavent G., Weis P., 2008, The multidimensional refinements indicators algorithm for optimal parameterization, J. Inverse Ill-Posed Probl. 16(2), 107–126 116, 122

[12] Bjork, A., 1990, Least squares methods, In Ciarlet, P.G., and Lions, J.L., eds, Handbook of Numerical Analysis, North-Holland, Amsterdam 17

[13] Bonnans, J.F., Gilbert, J.C., Lemaréchal, C., Sagastizbal, C.A., 2003, Numerical optimization: Theoretical and practical aspects, Springer, Universitext Series XIV, p 423 32, 127, 148

[14] Borzi, A., 2003, Multigrid methods for optimality systems, Habilitation Thesis, Institut fr Mathematik, Karl-Franzens-Universitt Graz, Austria 31

[15] Chardaire, C., Chavent G,. Jaffré J., Liu J., 1990, Multiscale representation for simultaneous estimation of relative permeabilities and capillary pressure, Paper SPE 20501, In Proceedings of the 65th SPE Annual Technical Conference and Exhibition, New Orleans, Louisiana, pp 303–312 96

[16] Chavent, G., 1979, Identification of distributed parameter systems: About the output least squares method, its implementation and identifiability, In Proceedings of the IFAC Symposium on Identification, Pergamon, pp 85–97 12

[17] Chavent, G., 1986, Identifiability of parameters in the output least square formulation, In Walter, E., ed, Structural Identifiability of Parametrics Model, chapter 6, Pergamon Press, pp 67–74 12

[18] Chavent, G., 1990, A new sufficient condition for the wellposedness of nonlinear least-squares problems arising in identification and control, In Bensoussan, A., and Lions, J.L., eds, Lecture Notes in Control and Information Sciences 144, Springer, Berlin, pp 452–463 211, 232

[19] Chavent, G., 1991, New size×curvature conditions for strict quasi-convexity of sets, SIAM J. Contr. Optim. 29(6), 1348–1372 12, 273

[20] Chavent, G., 1991, Quasi-convex sets and size×curvature condition, application to nonlinear inversion, J. Appl. Math. Optim. 24(1), 129–169 12, 273

[21] Chavent, G., 2002, Adapted regularization for the estimation of the diffusion coefficient in an elliptic equation, In Proceedings of Picof 02, Carthage, Tunisie 259

[22] Chavent, G., 2004, Curvature steps and geodesic moves for nonlinear least squares descent algorithms, Inverse Probl. Sci. Eng. 12(2), 173–191 135

[23] Chavent, G., Bissel, R., 1998, Indicator for the refinement of parametrization. In Tanaka, M., and Dulikravich, G.S., eds, Inverse Problems in Engineering Mechanics, Elsevier, Amsterdam, pp 309–314 113, 116, 120

[24] Chavent, G., Clement, F., 2001, Migration-based traveltime waveform inversion of 2-D simple structures: A synthetic example, Geophysics 66(3), 845–860 97

[25] Chavent, G., Kunisch, K., 1993, A geometric theory for the inverse problem in a one-dimensional elliptic equation from an H^1-observation, Appl. Math. Optim. 27, 231–260 192, 201

[26] Chavent, G., Kunisch, K., 1993, Regularization in state space, M2AN 27, 535–564 18, 247, 258, 259

[27] Chavent, G., Kunisch, K., 1994, Convergence of Tikhonov regularization for constrained ill-posed inverse problems, Inverse Probl. 10, 63–76 210, 211, 234

[28] Chavent, G., Kunisch, K., 1996, On weakly nonlinear inverse problems, SIAM J. Appl. Math. 56(2), 542–572 16, 166, 211, 237, 273

[29] Chavent, G., Kunisch, K., 1998, State space regularization: Geometric theory, Appl. Math. Opt. 37, 243–267 18, 247

[30] Chavent, G., Kunisch, K., 2002, The output least square identifiability of the diffusion coefficient from an H^1 observation in a 2-D elliptic equation, ESAIM: Contr. Optim. Calculus Variations 8, 423 97, 200, 202, 259, 261, 263

[31] Chavent, G., Lemonnier, P., 1974, Identification de la non linéarité d'une équation parabolique quasilinéaire, J. Appl. Math. Optim. 1(2), 121–162 237

[32] Chavent, G., Jaffré, J., Jégou, S., Liu, J., 1997, A symbolic code generator for parameter estimation. In Berz, M., Bischof, C., Corliss, G., and Griewank, A., eds, Computational Differentiation, SIAM, 129–136 38

[33] Chavent, G., Jaffré, J., Jan-Jégou, S., 1999, Estimation of relative permeabilities in three-phase flow in porous media, Inverse Probl. 15, 33–39 116

[34] Chicone, C., Gerlach, J., 1987, A note on the identifiability of distributed parameters in elliptic systems, SIAM J. Math. Anal. 18(5), 1378–1384 185

[35] Cominelli, A., Ferdinandi, F., De Montleau, P., Rossi, R., 2005, Using gradients to refine parameterization in field-case history match projects, In Proceedings of 2005 SPE Reservoir Simulation Symposium, paper SPE 93599, Houston, Texas, January 31st–February 2nd 116

[36] Delprat-Jannaud, F., Laiily, P., 1992, What information on the earth model do reflection travel times provide?, J. Geophys. Res. 97(B13), 19827–19844 92

[37] Engl, H.W., Kunisch, K., Neubauer, A., 1989, Convergence rates for Tikhonov regularization of nonlinear ill-posed problems, Inverse Probl. 5, 523–540 18, 211

[38] Engl, H.W., Hanke, M., Neubauer, A., 1996, Regularization of inverse problems, Kluwer, Dordrecht, p 321, (Mathematics and its applications, 375) ISBN 0-7923-4157-0 17, 117

[39] Girault, V., Raviart, P.A., 1979, Finite element methods for Navier-Stokes equations, Springer, Berlin 263

[40] Griewank, A., 1992, Achieving logarithmic growth of temporal and spatial complexity in reverse automatic differentiation, Optim. Meth. Software 1, 35–54 37, 38, 77

[41] Griewank, A., 2000, Evaluating derivatives: Principles and techniques of algorithmic differentiation (Frontiers in Applied Mathematics 19), Society for Industrial and Applied Mathematics, p 369, ISBN: 0898714516 38

[42] Grimstad, A.A., Mannseth T., Nævdal G., Urkedal H., 2003, Adaptive multiscale permeability estimation, Comput. Geosci. 7, 1–25 111

[43] Groetsch, C.W., 1984, The theory of Tykhonov regularization for Fredholm equations of the first kind, Research Notes in Mathematics 105, Pitman, Boston 17, 210

[44] Hayek, M., Lehmann, F., Ackerer, Ph., 2007, Adaptive multiscale parameterization for one-dimensional flow in unsaturated porous media, Adv. Water Resour. (to appear) 116, 270, 343, 352

[45] Hein, T., 2009, Regularization in Banach space – convergence rates by approximative source conditions, J. Inverse Ill-Posed Probl. 17, 27–41 211

[46] Isakov, V., 1998, Inverse problems for partial differential equations, Springer, Berlin, p 284 (Applied mathematical sciences, 127) ISBN 0-387-98256-6 11, 185, 191

[47] Ito, K., Kunisch, K., 1994, On the injectivity and linearization of the coefficient to solution mapping for elliptic boundary value problems, J. Math. Anal. Appl. 188(3), 1040–1066 11, 185, 191

[48] Jaffard, S., Meyer, Y., Ryan, R.D., 2001, Wavelets (Tools for science and technology), Society for Industrial and Applied Mathematics, p 256, ISBN 0-89871-448-6 104

[49] Kunisch, K., 1988, Inherent identifiability of parameters in elliptic differential equations, J. Math. Anal. Appl. 132, 453–472 191

[50] Lavaud, B., Kabir, N., Chavent, G., 1999, Pushing AVO inversion beyond linearized approximation, J. Seismic Explor. 8, 279–302 6, 88, 92

[51] Le Dimet, F.-X., Charpentier I., 1998, Méthodes de second ordre en assimilation de données, Equations aux Dérivées Partielles et Applications (Articles dédiés Jacques-Louis Lions), Gauthiers-Villars, pp 623–640

[52] Le Dimet, F.-X., Shutyaev, V., 2001, On Newton method in data assimilation, Russ. J. Numer. Anal. Math. Model. 15(5), 419–434 31

[53] Le Dimet, F.-X., Navon I.M., Daescu, D.N., 2002, Second order information in data assimilation, Mon. Weather Rev. 130(3), 629–648 31

[54] Levenberg, K., 1944, A method for the solution of certain nonlinear problems in least squares, Appl. Math. 11, 164–168 17, 209

[55] Lines, L.R., Treitel, S., Tutorial: A review of least-squares inversion and its application to geophysical problems, Geophys. Prospect. 39, 159–181 17, 270, 343, 352

[56] Lions, J.L., 1969, Quelques Méthodes de Résolution des Problèmes aux limites Non Linéaires, Dunod, Paris 25, 238

[57] Liu, J., 1993, A multiresolution method for distributed parameter estimation, SIAM J. Sci. Comput. 14, 389 96, 97, 104, 207

[58] Liu, J., 1994, A sensitivity analysis for least-squares ill-posed problems using the haar basis, SIAM J. Numer. Anal. 31, 1486 96, 97, 104

[59] Louis, A.K., 1989, Inverse und Schlecht Gestellte Probleme, Teubner, Stuttgart 17, 210

[60] Mannseth, T., 2003, Adaptive multiscale identification of the fluid conductivity function within prous-media flow. Conference on Applied Inverse Problems: Theoritical and computational aspects, Lake Arrowhead, May 18–23 111

[61] Marchand, E., Clément, F., Roberts, J.E., Pépin, G., 2008, Deterministic sensitivity analysis for a model for flow in porous media, Adv. Water Resour. 31, 1025–1037, http://dx.doi.org/10.1016/j.advwatres.2008.04.004 38

[62] Marquardt, D.W., 1963, An algorithm for least squares estimation of nonlinear parameters, J. Soc. Ind. Appl. Math. 11, 431–441 17, 210

[63] Morozov, V.A., 1984, Methods for solving incorrectly posed problems, Springer, New York 17, 210

[64] Næval, T., Mannseth, T., Brusdal, K., Nordtvedt, J.E., 2000, Multiscale estimation with spline wavelets, with application to two-phase porous media flow, Inverse Probl. 16, 315–332 96

[65] Neubauer, A., 1987, Finite dimensional approximation of constrained Tikhonov-regularized solutions of ill-posed linear operator equations, Math. Comput. 48, 565–583 210, 215

[66] Neubauer, A., 1988, Tikhonov reularization of ill-posed linear operator equations on closed convex sets, J. Approx. Theor. 53, 304–320 210

[67] Neubauer, A., 1989, Tikhonov regularization for nonlinear ill-posed problems: Optimal convergence rate and finite dimensional approximation, Inverse Probl. 5, 541–558 18, 211, 215

[68] Nocedal, J., Wright, S.J., 1999, Numerical optimization, Springer Series in Operation Research, New York 32, 127

[69] Økland Lien, M., 2005, Adaptive methods for permeability estimation and smart well management, Dr. Scient. Thesis in Applied Mathematics, Department of Mathematics, University of Bergen, Norway 96, 116

[70] Richter, G.R., 1981, An inverse problem for the steady state diffusion equation, SIAM J. Math. 4, 210–221 11

[71] Sanchez Palencia, E., 1983, Homogenization method for the study of composite media, Lecture Notes in Mathematics 985, 192–214, Springer, Berlin 200

[72] Schaaf, T., Mezghani, M., Chavent, G., 2002, Direct conditioning of fine-scale facies models to dynamic data by combining gradual deformation and numerical upscaling techniques, In Proceedings of the 8th European Conference on Mathematics of Oil Recovery (ECMOR VIII), Sept 3–6, Freiberg, Germany

[73] Schaaf, T., Mezghani, M., Chavent, G., 2003, In Search of an optimal parameterization: An innovative approach to reservoir data integration, paper SPE 84273, In Proceedings of the SPE Annual Technical Conference and Exhibition, Denver

[74] Sen, A., Srivastava, M., 1990, Regression analysis: Theory, methods and applications, Springer, Berlin 85

[75] Tikhonov, A.N., Arsenin, V., 1977, Solutions of ill-posed problems, Wiley, New York 17, 210

[76] Troianiello, G.M., 1987, Elliptic differential equations and obstacle problems, Plenum Press, New York 202

[77] van Laarhoven, P., Aarts, E., 1987, Simulated annealing, theory and practice, Kluwer, Dordrecht 31

[78] Vogel, C., 2002, Computational methods for inverse problems, Frontiers in Applied Mathematics series 23, SIAM 135, 270, 343, 352

[79] Zhang, J., Dupuy, A., Bissel, R., 1996, Use of optimal control technique for history matching, 2nd International Conference on Inverse Problems in Engineering: Theory and Practice, June 9–14, Le Croisic, France, Engineering Foundation ed. 78, 135

[80] Zhang, J., Jaffré, J., Chavent, G., 2009, Estimating nonlinearities in multiphase flow in porous media, INRIA Report 6892 88, 185

Index

MIX
Papier aus verantwortungsvollen Quellen
Paper from responsible sources
FSC® C105338

If you have any concerns about our products,
you can contact us on
ProductSafety@springernature.com

In case Publisher is established outside the EU,
the EU authorized representative is:
Springer Nature Customer Service Center GmbH
Europaplatz 3, 69115 Heidelberg, Germany

Printed by Libri Plureos GmbH
in Hamburg, Germany